Second Edition

Pearson Edexcel GCSE (9–1)

Mathematics

Higher

Student Book

Series editors: Dr Naomi Norman and Katherine Pate

2

Endorsed for Pearson Edexcel Qualifications

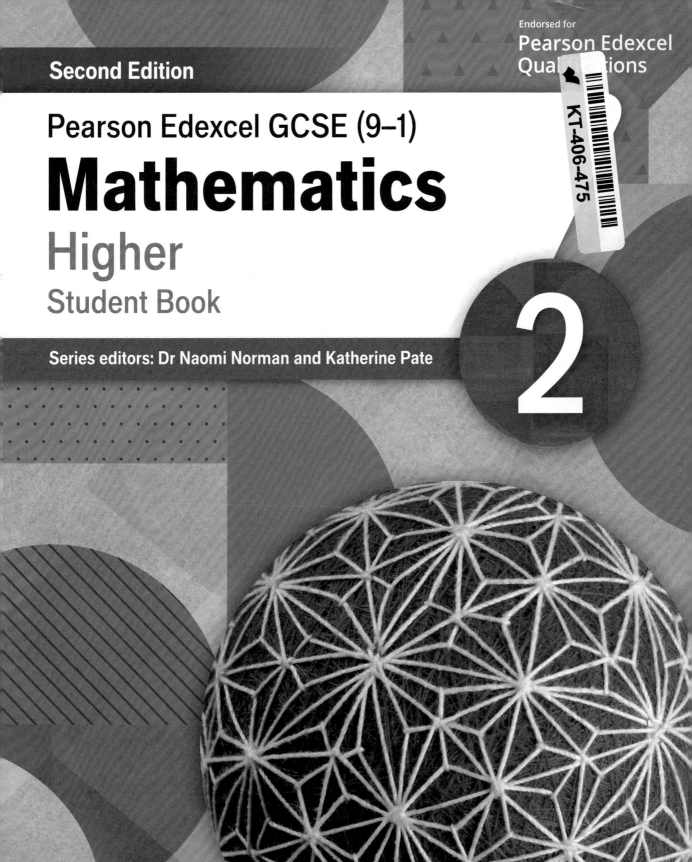

PEARSON

Published by Pearson Education Limited, 80 Strand, London, WC2R 0RL.

www.pearsonschoolsandfecolleges.co.uk

Text © Pearson Education Limited 2020
Project managed and edited by Just Content Ltd
Typeset by PDQ Digital Media Solutions Ltd
Original illustrations © Pearson Education Limited 2020
Cover photo/illustration by © David S. Rose/Shutterstock, © Julias/Shutterstock, © Attitude/Shutterstock, © Abstractor/Shutterstock, © Ozz Design/Shutterstock, © Lazartivan/Getty Images

The rights of Jack Barraclough, Chris Baston, Ian Bettison, Sharon Bolger, Ian Boote, Judith Chadwick, Ian Jacques, Catherine Murphy, Su Nicholson, Naomi Norman, Diane Oliver, Katherine Pate, Glyn Payne, Jenny Roach, Peter Sherran and Robert Ward-Penny to be identified as authors of this work have been asserted by them in accordance with the Copyright, Designs and Patents Act 1988.

First published 2020

23 22 21 20
10 9 8 7 6 5 4 3 2 1

British Library Cataloguing in Publication Data

A catalogue record for this book is available from the British Library.

ISBN 978 1 292 34639 7

Printed in Slovakia by Neografia

A note from the publisher

In order to ensure that this resource offers high-quality support for the associated Pearson qualification, it has been through a review process by the awarding body. This process confirms that this resource fully covers the teaching and learning content of the specification or part of a specification at which it is aimed. It also confirms that it demonstrates an appropriate balance between the development of subject skills, knowledge and understanding, in addition to preparation for assessment.

Endorsement does not cover any guidance on assessment activities or processes (e.g. practice questions or advice on how to answer assessment questions) included in the resource nor does it prescribe any particular approach to the teaching or delivery of a related course.

While the publishers have made every attempt to ensure that advice on the qualification and its assessment is accurate, the official specification and associated assessment guidance materials are the only authoritative source of information and should always be referred to for definitive guidance.

Pearson examiners have not contributed to any sections in this resource relevant to examination papers for which they have responsibility.

Examiners will not use endorsed resources as a source of material for any assessment set by Pearson. Endorsement of a resource does not mean that the resource is required to achieve this Pearson qualification, nor does it mean that it is the only suitable material available to support the qualification, and any resource lists produced by the awarding body shall include this and other appropriate resources.

Pearson has robust editorial processes, including answer and fact checks, to ensure the accuracy of the content in this publication, and every effort is made to ensure this publication is free of errors. We are, however, only human, and occasionally errors do occur. Pearson is not liable for any misunderstandings that arise as a result of errors in this publication, but it is our priority to ensure that the content is accurate. If you spot an error, please do contact us at resourcescorrections@pearson.com so we can make sure it is corrected.

Contents

Pearson Edexcel GCSE (9–1)
Mathematics

Second Edition

Pearson Edexcel GCSE (9–1) Mathematics Second Edition is built around a unique pedagogy that has been created by leading educational researchers and teachers in the UK. This edition has been updated to reflect six sets of live GCSE (9–1) papers, as well as feedback from thousands of teachers and students and a 2-year study into the effectiveness of the course.

The new series features a full range of print and digital resources designed to work seamlessly together so that schools can create the course that works best for their students and teachers.

*Active*Learn service

The *Active*Learn service brings together the full range of planning, teaching, learning and assessment resources.

What's in *Active*Learn for GCSE (9–1) Mathematics?

- ☑ **Front-of-class Student Books** with accompanying PowerPoints, worksheets, videos, animations and homework activities

- ☑ **254 editable and printable homework worksheets**, linked to each Master lesson

- ☑ **Online, auto-marked homework activities** with integrated videos and worked examples

- ☑ **76 assessments and online markbooks**, including end-of-unit, end-of-term, end-of-year and baseline tests

- ☑ **Interactive Scheme of Work** brings everything together, connecting your personalised scheme of work, teaching resources and assessments

- ☑ **Individual student access to videos, homework and online textbooks**

Student Books

The Student Books use a mastery approach based around a well-paced and well-sequenced curriculum. They are designed to develop mathematical fluency, while building confidence in problem-solving and reasoning.

The unique unit structure enables every student to acquire a deep and solid understanding of the subject, leaving them well-prepared for their GCSE exams, and future education or employment.

Together with the accompanying online prior knowledge sections, the Student Books cover the entire **Pearson Edexcel GCSE (9–1) Mathematics course**.

The new four-book model means that the Second Edition Student Books now contain even more meaningful practice, while still being a manageable size for use in and outside the classroom.

 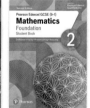

Foundation tier

 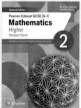

Higher tier

Pearson Edexcel GCSE (9–1)
Mathematics Second Edition
Higher Student Book

2

Building confidence

Pearson's unique unit structure has been shown to build confidence. The **front-of-class** versions of the Student Books include lots of extra features and resources for use on a whiteboard.

Master

Learn fundamental knowledge and skills over a series of lessons.

*Active*Learn **homework**

Links to online homework worksheets and exercises for every lesson.

Students can make sure they are ready for each unit by downloading the relevant **Prior knowledge check**. This can be accessed using the QR code in the Contents or via *Active*Learn.

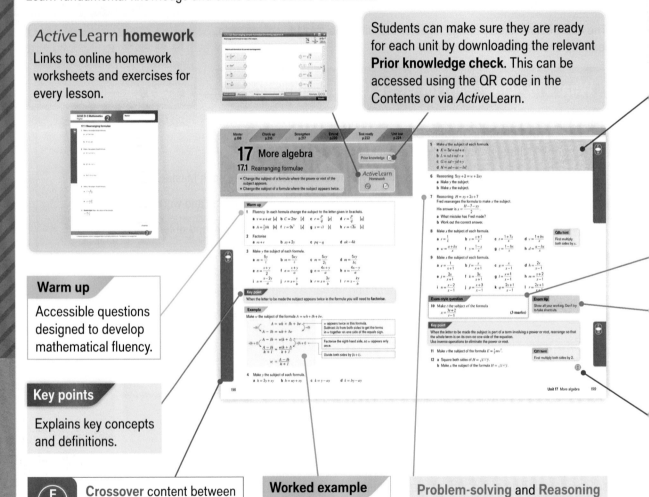

Warm up

Accessible questions designed to develop mathematical fluency.

Key points

Explains key concepts and definitions.

Crossover content between Foundation and Higher tiers is indicated in side bars.

Worked example

Step-by-step worked examples focus on the key concepts.

Problem-solving and **Reasoning** questions are clearly labelled. **Future skills** questions help prepare for life after GCSE. **Reflect** questions encourage reflection on mathematical thinking and understanding.

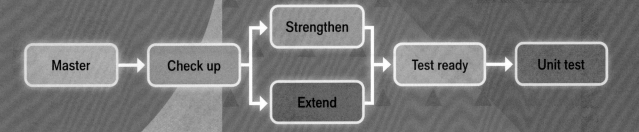

Check up

After the Master lessons, a Check up test helps students decide whether to move on to the Strengthen or Extend section.

Strengthen

Students can choose the topics they need more practice on. There are lots of hints and supporting questions to help.

Extend

Applies and develops maths from the unit in different situations.

Working towards A level

These questions take familiar ideas and extend them with styles of question that are typical of AS and A level papers. Students who enjoy tackling these questions might want to consider taking A level Mathematics.

Test ready

The **Summary of key points** is used to identify areas that need more practice and **Sample student answers** familiarise students with good exam technique.

Unit test

The exam-style Unit test helps check progress.

Mixed exercises

These sections bring topics together to help practise applying different techniques to a range of questions types, which is required in GCSE exams.

Click on any question to view it full-size, and then click 'Show' to reveal the answer.

Exam-style questions

are included throughout the books to help students prepare for GCSE exams.

Exam tips point out common errors and help with good exam technique.

Teaching and learning materials can be downloaded from the blue hotspots.

Helpful videos walk you step-by-step through answers to similar questions.

Interactive Scheme of work

The Interactive Scheme of Work makes reordering the course easy. You can view your plan for your year, term or lesson, and access all the related teaching, learning and assessment materials.

*Active*Learn Progress & Assess

The Progress & Assess service is part of the full *Active*Learn service, or can be bought as a separate subscription. It includes assessments that have been designed to ensure all students have the opportunity to show what they have learned through:

- a 2-tier assessment model
- separate calculator and non-calculator sections
- online markbooks for tracking and reporting
- mapping to indicative 9–1 grades.

Assessment Builder

Create your own classroom assessments from the bank of GCSE (9–1) Mathematics assessment questions by selecting questions on the skills and topics you have covered. Map the results of your custom assessments to indicative 9–1 grades using the custom online markbooks. Assessment Builder is available to purchase as an add-on to the *Active*Learn service or Progress & Assess subscriptions.

Purposeful Practice Books

A new kind of practice book based on cutting-edge approaches to help students make the most of practice.

With more than 4500 questions, our Pearson Edexcel GCSE (9–1) Mathematics Purposeful Practice Books are designed to be used alongside the Student Books and online resources. They:

- use minimal variation to build in small steps, consolidating knowledge and boosting confidence
- focus on strengthening problem-solving skills and strategies
- feature targeted exam practice with questions modified from real GCSE (9–1) papers, and exam guidance from examiner reports and grade indicators informed by ResultsPlus.

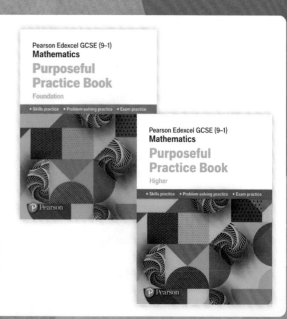

11 Multiplicative reasoning

Prior knowledge

11.1 Growth and decay

- Find an amount after repeated percentage changes.
- Solve growth and decay problems.
- Solve problems using an iterative process.

Active Learn
Homework

Warm up

1 **Fluency** Write these percentages as decimals.

   a 150%   **b** 15%   **c** 1.5%   **d** 0.15%

2 Copy and complete.

   a $1.3 \times 1.3 \times 1.3 \times 1.3 = 1.3^{\square}$ **b** $100 + 1.3 = \square$ **c** $101.3 \div 100 = \square$

3 $n = 4$
What is the value of $2000 \times (0.5)^n$?

4 Work out the multiplier as a decimal for these increases and decreases.
The first two are started for you.

   a an increase of 20%   **b** a decrease of 14%
       $100\% + 20\% = 120\%$ $100\% - 14\% = \square\%$
       $120\% = 1.\square$ $\square\% = 0.\square\square$

   c an increase of 7.2%   **d** a decrease of 2.5%

5 **Problem-solving** A bill is £33.60 including VAT (value added tax).
VAT is 20%.
What is the bill before VAT is added?

6 Wes buys a car for £6500. It loses 35% of its value in the first year. Work out

   a the multiplier to find the value of the car at the end of the first year

   b the value of the car at the end of the first year

   It loses 15% of its value in the second year. Work out

   c the multiplier to find the value of the car at the end of the second year

   d the value of the car at the end of the second year

   e Copy and complete the diagram to show the single decimal multiplier that the original value of
the car can be multiplied by to find its value at the end of two years.

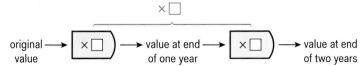

   f Use your diagram to check your answer to part **d**.

7 Work out the decimal multiplier that represents

 a a decrease of 20% followed by a decrease of 15%

 b an increase of 5% followed by an increase of 3%

 c an increase of 9% followed by a decrease of 6%

 8 **Future skills** Abdul has a job with an annual (yearly) salary of £35 000.
At the end of the first year he is given an increase of 2%.
At the end of the second year he is given an increase of 3.5%.
Work out Abdul's salary at the end of two years.

> **Q8 hint**
> Use a single decimal multiplier.

 9 **Future skills** Tristan buys a flat for £135 000.
In the first year, the value of the flat appreciates by 12%.
In the second year, the value of the flat depreciates by 3%.
Work out the value of the flat after the 2 years.

> **Q9 hint**
> In financial terms, 'appreciate' means to gain value and 'depreciate' means to lose value.

 10 **Reasoning** Becky says, 'An decrease of 15% followed by an increase of 22% is the same as an increase of 7%.'
Is Becky correct? Explain.

 11 Work out the decimal multiplier that represents

 a an increase of 12% for 3 years

 b a decrease of 15% for 4 years

 c **Reflect** How could you write each of these multipliers as a power rather than a decimal?

Exam-style question

12 Jon bought a campervan for £16 500.
The value, £V, of Jon's campervan at the end of n years is given by the formula
$$V = 16\,500 \times (0.82)^n$$

 a By what percentage does the campervan depreciate each year? **(1 mark)**

 b At the end of how many years is the value of Jon's campervan first less than 50% of the value of the campervan when it was new?
You must show your working. **(2 marks)**

> **Q12b hint**
> Work out 50% of the value of the campervan when bought new.
> Then try $n = 1, n = 2, n = 3, ...$
> in the formula.

Key point

Iteration means carrying out a process repeatedly.

13 Bella invests £1000 in a savings account.
The savings account pays 2% per annum (yearly) on Bella's original amount plus any interest she has earned.
At time t years, Bella has an amount A_t.

a Copy and complete this iterative process to show how to work out how much Bella has after 5 years.

$$A_0 = 1000$$
$$A_1 = 1000 \times \square$$
$$A_2 = 1000 \times \square \times \square$$
$$A_3 = 1000 \times \square \times \square \times \square$$
$$A_4 = 1000 \dots$$
$$A_5 = 1000 \dots$$

Q13a hint

A_1 is the amount Bella has at the end of year 1: her original investment (A_0) increased by 2%. A_2 is the amount Bella has at the end of year 2: her amount at the end of year 1 (A_1) increased by 2%.

b **Reflect** How could you write the calculation to work out A_5 using powers?

c Work out how much Bella has after 5 years.

> **Key point**
>
> Most interest rates are **compound interest** rates. In compound interest the interest earned each year is added to money in the account and earns interest the next year.

14 **Future skills** £3500 is invested for 3 years at 1.8% per annum compound interest.
Work out the total amount in the account after 3 years.

15 **Reasoning** Anthony says that you can work out compound interest using the formula

$$\text{Amount after } n \text{ years} = \text{initial amount} \times \left(\frac{100 + \text{interest rate}}{100}\right)^n$$

Show that this formula works by using it to answer **Q13c** and **Q14**.
Do you get the same answers when using the formula?

> **Key point**
>
> You can calculate an amount after n years' compound interest using the formula
>
> $$\text{Amount} = \text{initial amount} \times \left(\frac{100 + \text{interest rate}}{100}\right)^n$$

16 **Future skills** £3000 is invested for 2 years at 1.6% per annum compound interest.
Work out the **total interest** earned over the 2 years.

Q16 hint

Total interest = amount in the account at the end of the investment – amount invested

> **Exam-style question**
>
> **17** Marie wants to invest £3500 for 3 years.
> She finds these two savings accounts.
>
> | **Cash savings account** |
> | --- |
> | Compound interest |
> | 1.9% per annum |
>
> | **Bonus savings account** |
> | --- |
> | 2.5% for the first year |
> | 0.7% for each extra year |
>
> Which account will give Marie more interest at the end of 3 years?
> You must show your working.
>
> **(3 marks)**

18 **Future skills** Laura invests £3600 in a savings account for 2 years.
The account pays 3.52% compound interest per annum.
Laura has to pay 40% tax on the interest earned each year.
The tax is taken from the account at the end of each year.
How much is in the account at the end of 2 years?

19 **Reasoning** Ian invests £6000 for 1 year in an account paying
R% compound interest per annum.
At the end of the year, Ian's investment is worth £6084.
Use the compound interest formula to work out the value of R.

Q19 hint

Substitute the values you know into the formula. Divide both sides of the equation by 6000.

Q20 hint

First work out the interest before Ruth paid 20% tax.

Exam-style question

20 A savings accounts pays interest at a rate of x% per year.
Ruth invests £4500 in the account one year.
At the end of the year, Ruth pays 20% tax on the interest.
After paying tax, she gets £46.80.
Work out the value of x. **(3 marks)**

Exam tip

Check your answer by re-reading the question and making sure your value of x works.

Key point

You can use the compound interest formula to work out **growth** (increases) and **decay** (decreases) of quantities other than money.

21 **Future skills** The level of activity of a radioactive source decreases by 5% per hour.
The activity is 1400 counts per second at one point.

a What will it be 2 hours later?

b After how many complete hours will the count be less than 1200 counts per second?

22 **Future skills** In 2020 a fast-food chain has 180 outlets in the UK.
The number of outlets is growing at a rate of 9% each year.

a How many outlets will it have in 2026?

b **Reflect** What is an appropriate degree of accuracy to give for this question? Why?

23 **Problem-solving / Reasoning** A population of insects
increases by 25% per day.
At the end of one week there are 2146 insects.
How many insects were there at the beginning of the week?

Q23 hint

Use x for the number of insects at the start of the week.

11.2 Compound measures

- Calculate rates.
- Convert between metric speed measures.
- Use a formula to calculate speed and acceleration.

Active Learn
Homework

Warm up

1 Fluency Copy and complete these rates.

  a £60 for a 6-hour day   £☐ per hour

  b 300 km on 20 litres of petrol   ☐ km per ☐

2 $a = \dfrac{b}{c}$

Find

  a a when $b = 18$, $c = 3$

  b b when $a = 15$, $c = -2$

  c c when $a = -4$, $b = 0.5$

3 a Karl cycles 48 km in 3 hours.
      What was his average speed?

  b Andy cycles with an average speed of 15 km/h for 2 hours.
      What distance did he travel?

  c Shakil cycles 42 km at an average speed of 14 km/h. How long does it take him?

  d Jean cycles 52 km in 5 hours.

    i Estimate her average speed.

    ii Is your estimate an overestimate or an underestimate?

> **Q3 hint**
>
> Speed = $\dfrac{\text{distance}}{\text{time}}$

4 Convert these times to hours and minutes.

  a 450 minutes

  b 6.2 hours

5 Write **i** the lower bound and **ii** the upper bound of each measurement.

  a 5 km (correct to the nearest km)

  b 1 hour (correct to the nearest minute)

> **Q5b hint**
>
> 1 hour = 60 minutes

6 Problem-solving George works a 35-hour week and some overtime on Saturdays.
He is paid £8.50 an hour for this work.
George is paid time and a half for each hour he works on a Saturday.
One week, George works a 35-hour week and some hours on Saturday.
He is paid £335.75 for this week.
How many hours did George work on Saturday?

> **Q6 hint**
>
> How much more is George paid than for his 35-hour week?

7 **Problem-solving** Water is leaking from a water butt at a rate of 4.5 litres per hour.

   a Work out how much water leaks from the water butt in

     i 20 minutes

     ii 50 minutes

   b Initially there are 180 litres of water in the water butt.
Work out how long it takes for all the water to leak from the water butt.

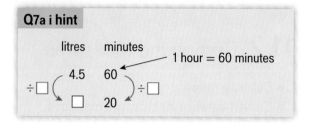

Q7a i hint

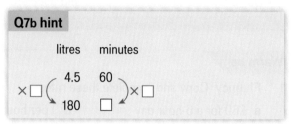

Q7b hint

8 **Reasoning** A car travels 320 km and uses 20 litres of petrol.

   a Work out the average rate of petrol usage in kilometres per litre.

   b **Reflect** Write a sentence explaining why the question asks for the 'average rate' rather than the 'exact rate'.

> **Key point**
>
> **Compound measures** combine measures of two different quantities.
> For example, speed is a measure of distance travelled and time taken.
> It can be measured in metres per second (m/s), kilometres per hour (km/h) or miles per hour (mph).

9 A man walks at an average speed of 5.4 km/h

   a **Reflect** Does he travel more or fewer metres than kilometres in 1 hour?

   b Convert his speed from km/h to m/h.

   c **Reflect** Does he travel more or fewer metres in a minute than an hour?

   d Convert his speed from m/h to m/min.

   e Convert his speed from m/min to m/s.

   f **Reflect** Write a sentence explaining how you know when to multiply and when to divide.

Q9 hint

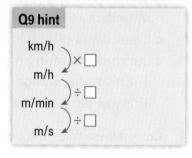

10 Convert these speeds to the units given.

   a 6500 m/h to km/h

   b 256 km/h to m/h

   c 12 m/s to m/h

   d 3600 m/h to m/s

   e 81 km/h to m/s

   f 20 m/s to km/h

Q10e hint

First convert to m/h.

Q10f hint

First convert to m/min then m/h.

11 Problem-solving A commercial aeroplane has a cruising speed of 900 km/h.
Work out

   a the number of seconds it takes the plane to travel 900 km

   b the number of seconds it takes the plane to travel 1 km

Exam-style question

12 A plane travels at a speed of 237 miles per hour.

   a Work out an estimate for the number of seconds the
      plane takes to travel 1 mile. **(3 marks)**

   b Is your answer to part **a** an underestimate or an
      overestimate?
      Give a reason for your answer. **(1 mark)**

Exam tip

When asked to work out an estimate, you can round values so that the numbers are easier to work with. For example, round to the nearest 10.

---

13 a Reasoning A car travels at x km/h.
      Write an expression for this speed in m/s.

   b A cheetah runs at y m/s.
      Write an expression for this speed in km/h.

---

14 Problem-solving Paul swims 750 metres in 25 minutes.
What is his average speed in km/h?

Q14 hint

First work out Paul's speed in m/min.

Exam-style question

15 A car travels a distance of 307 km in 4 hours.
The distance is measured correct to the nearest kilometre.
The time is measured correct to the nearest minute.
By considering bounds, work out the average speed, in km/minute, of the car correct to 1 decimal place.
You must show your working. **(5 marks)**

Q15 hint

Work out the lower and upper bounds for distance, time and speed.

Lower bound for speed = ____ bound for distance / ____ bound for time

Upper bound for speed = ____ bound for distance / ____ bound for time

---

16 Problem-solving Karl travels 35 miles in 45 minutes then
65 km in $1\frac{1}{2}$ hours.
5 miles = 8 kilometres
What is his average speed for the total journey in km/h?

17 Problem-solving Orla drove 155 km in 2 hours.
Then she drove 48 km at a speed of 60 km/h.
Work out Orla's average speed for the whole journey.

Q16 and Q17 hint

Work out the total distance in km and the total time in hours.
Then use the formula
Average speed for total journey = total distance / total time

18 Antony's average speed for long distance cycling is 19.8 km/hour. Antony is planning to cycle 875 miles from Land's End to John o'Groats.
He hopes to cycle for 8 hours every day.

   a Estimate how many days it will take Antony to cycle from Land's End to John o'Groats. **(3 marks)**

   b Antony trains for the cycle ride.
   He is aiming for an average speed of 28 km/hour.
   He wants to complete the ride in 5 days.
   On average, for how long must he cycle each day?
   Give your answer correct to the nearest hour and minute. **(3 marks)**

Exam tip

A good start is to write down the information you are given.
Then write down the information you are trying to find.

Q18a hint

Distance = 875 miles = $\square$ km
Speed = 19.8 km/h $\approx$ $\square$ km/h
Now work out the time.

Key point

These are kinematics formulae:

$$v = u + at$$
$$s = ut + \tfrac{1}{2}at^2$$
$$v^2 = u^2 + 2as$$

In exam questions you will need to decide which formula to use.

where a is constant acceleration, u is initial velocity, v is final velocity, s is displacement from the position when $t = 0$, and t is time taken.
Velocity is speed in a given direction; possible units are m/s.
Initial velocity is speed in a given direction at the start of the motion.
Acceleration is the rate of change of velocity, i.e. a measure of how the velocity changes with time; possible units are m/s^2.

19 A car starts from rest and accelerates at 5 m/s^2 for 200 m.

   a Write down the values of u, a and s.

   b Work out the final velocity in m/s.

Q19 hint

When the car starts from rest, the initial velocity $u = 0$. Use the formula that contains u, a, s and v but not t.

20 **Problem-solving** A tram has an initial velocity of 300 m/minute.
It travels a distance of 0.5 km in 20 seconds.
What is the acceleration of the tram in m/s^2?

21 **Problem-solving** A bus travels with an acceleration of 2 m/s^2 and reaches a speed of 45 km/h in 5 seconds.
What was the initial velocity of the bus in m/s?

11.3 More compound measures

Active Learn
Homework

- Solve problems involving compound measures.

Warm up

1 Fluency What is the formula for
   a the area of a circle   **b** the volume of a prism?

2 Copy and complete.
   a $7.5\,kg = \square\,g$ **b** $1\,g = \square\,mg$
   c $1\,m^2 = \square\,cm^2$ **d** $62\,500\,cm^2 = \square\,m^2$
   e $1\,m^3 = \square\,cm^3$ **f** $95\,000\,cm^3 = \square\,m^3$

3 Solve
   a $\dfrac{m}{5} = 6$ **b** $\dfrac{8}{v} = 0.5$

4 Write **i** the lower bound and **ii** the upper bound of each measurement.
   a 4.1 cm, correct to the nearest mm
   b 410 g, correct to the nearest 5 g
   c 41.1 kg, correct to the nearest 100 g

5 Copy and complete the table.

Metres per second	Kilometres per hour
	54
	72
30	
45	

6 Problem-solving / Reasoning A Formula 1 racing car has a top speed of 350 km/h.
   A peregrine falcon is the fastest bird, with a speed of 108 m/s.
   Which is faster? Explain your answer.

7 Problem-solving A swallow flies for
   40 minutes at an average speed of 11 m/s.
   How far does the swallow fly in kilometres?

> **Q7 hint**
>
> Time is in minutes and speed is in metres per second.
> Change one of them so that they both use the same
> measure of time.
> Then use the formula
> $$\text{speed} = \frac{\text{distance}}{\text{time}}$$

Key point

Density is the mass of substance (g) contained in a certain volume (cm³).
It is often measured in grams per cubic centimetre (g/cm³).

$$\text{Density} = \frac{\text{mass}}{\text{volume}} \text{ or } D = \frac{M}{V}$$

8 **Problem-solving** A sample of brass has a mass of 2 kg
and a volume of 240 cm³.
What is its density in g/cm³?

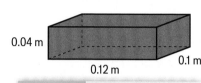

Q8 and Q9 hint

Make sure mass is in g and
volume is in cm³ before working
out the density.

9 **Problem-solving** A cubic metre of concrete has a mass
of 2400 kg.
What is the density of the concrete in g/cm³?

10 The diagram shows a block of wood in the shape of
a cuboid.
The density of wood is 0.6 g/cm³.

a Work out the volume of the block of wood in cm³.

b Work out the mass of the block of wood.

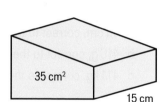

0.04 m   0.12 m   0.1 m

Q10b hint

Substitute into the formula $D = \frac{M}{V}$.
Then solve to find M.

11 **Problem-solving** The area of the cross-section of this plastic
prism is 35 cm².
Its length is 15 cm. The plastic has a density of 3.9 g/cm³.
What is the mass of the prism?

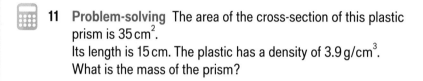

35 cm²   15 cm

12 **Problem-solving** Iron has density 8 g/cm³.
The mass of a piece of iron is 5.4 kg.
What is the volume?

Q12 hint

Substitute into the formula $D = \frac{M}{V}$.
Then solve to find V.

13 **Problem-solving** The density of aluminium is 2.70 g/cm³.
What is the density of aluminium in kg/m³?

Q13 hint

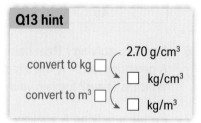

2.70 g/cm³

convert to kg ☐ ⟵ ☐ kg/cm³

convert to m³ ☐ ⟵ ☐ kg/m³

14 **Reasoning** A metal has density x g/cm³.
Write an expression for its density in kg/m³.

15 **Reasoning** 1 m³ of lithium has mass 530 kg.
60 cm³ of sodium has mass 58.2 g.
Is lithium or sodium denser? You must show your working.

16 Problem-solving / Reasoning A block of metal is in the shape of a cuboid.
The length is 8.6 cm, correct to the nearest mm.
The width is 7.4 cm, correct to the nearest mm.
The height is 20.3 cm, correct to the nearest mm.
The mass is 10.3 kg, correct to the nearest 100 g.
By considering bounds, work out the maximum and minimum density for the block of metal.

Exam-style question

17 The density of juice is 1.1 grams per cm^3.
The density of water is 1 gram per cm^3.
270 cm^3 of drink is made by mixing 40 cm^3 of juice with
230 cm^3 of water.
Work out the density of the drink. **(4 marks)**

Q17 hint

Work out the mass of juice and the mass of water.
Then use the formula $D = \dfrac{M}{V}$.

Key point

Pressure is the force in newtons applied over an area in cm^2 or m^2.
It is usually measured in newtons (N) per square metre (N/m^2) or per square centimetre (N/cm^2).

$$\text{Pressure} = \frac{\text{force}}{\text{area}} \text{ or } P = \frac{F}{A}$$

18 Copy and complete the table.
Give your answers correct to 3 significant figures.

	Force	Area	Pressure
a	60 N	2.6 m^2	☐ N/m^2
b	☐ N	4.8 m^2	15.2 N/m^2
c	100 N	☐ m^2	12 N/m^2

d Write your answer to part **a** in N/cm^2.

19 Problem-solving A cylindrical bottle of water has a flat, circular base with a diameter of 0.1 m.
The bottle is on a table and exerts a force of 12 N on the table.
Work out the pressure in N/cm^2. Give your answer correct to 3 significant figures.

20 Problem-solving The pressure between a car's tyres and the road is 99 960 N/m^2.
The car tyres have a combined area of 0.12 m^2 in contact with the road.
What is the force exerted by the car on the road? Give your answer correct to 3 significant figures.

21 Problem-solving / Reasoning Jamie sits on a chair with four identical legs.
Each chair leg has a flat square base measuring 2 cm by 2 cm.
Jamie has a mass of 75 kg and the chair has a mass of 5 kg.

a Weight is a force on an object due to gravity and is measured in newtons.
Use $F = mg$ to work out the combined weight of Jamie and the chair, where $g = 9.8 \, m/s^2$.

b Work out the pressure on the floor, in N/cm^2, when only the four chair legs are in contact with the floor.

c The area of Jamie's trainers is 0.04 m^2.
Work out the pressure Jamie exerts on the floor when he is standing up.
Give your answer in N/m^2.

d Does Jamie exert a greater pressure on the floor when he is standing up or when he is sitting on the chair?

11.4 Ratio and proportion

*Active*Learn
Homework

- Use relationships involving ratio.
- Use direct and inverse proportion.

Warm up

1 Fluency Match each equation to a graph.

a $y = \dfrac{1}{x}$ **b** $y = x$ **c** $y = 2x + 3$

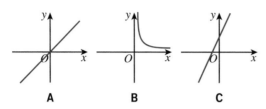

A   **B**   **C**

2 a Draw a graph for each table of values.

i

x	2	4	6	8
y	8	16	24	32

ii

x	1	2	3	4
y	1	4	9	16

b Which graph in part **a** shows direct proportion? Explain.

c Work out the equation of the graph that shows direct proportion.

3 Write the ratio $3 : 4$ in the form $1 : n$.

4 Copy and complete these. The first one is started for you.

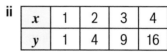

a $A : B = 3 : 5$

$\dfrac{A}{B} = \dfrac{3}{5}$

$A = \square B$

b $P : Q = 7 : 4$

$\dfrac{P}{Q} = \dfrac{\square}{\square}$

$P = \square Q$

c $5X = 9Y$

$\dfrac{X}{Y} = \dfrac{\square}{\square}$

$X : Y = \square : \square$

5 The table shows some lengths in both miles and kilometres.

Miles	5	10	15	20
Kilometres	8	16	24	32

a What is the ratio miles : kilometres in the form $1 : n$?

b Draw a line graph for these values.
Plot miles on the horizontal axis and kilometres on the vertical axis.

c Reasoning Are miles and kilometres in direct proportion? Explain.

d What is the gradient of the line?

e Write a formula that shows the relationship between miles and kilometres.

f Reflect Write a sentence explaining the connection between your answers to parts **a**, **d** and **e**.

Q5e hint

$K = \square M$

6 **Problem-solving** The table shows the distance, s miles, travelled by a car over a period of time, t minutes.

Distance, s (miles)	8	16	24	32	40
Time, t (minutes)	10	20	30	40	50

    a **Reasoning** Is s in direct proportion to t? Explain.

    b What is the relationship between distance (s) and time (t)?

    c Work out the distance travelled after 25 minutes.

Q6b hint

Write a formula.

Key point

When x and y are in **direct proportion** then
- $y = kx$, where k is the gradient of the graph of y against x
- $\dfrac{y}{x} = k$, a constant

Example

A is directly proportional to x. $A = 5$ when $x = 10$.

a Sketch a graph of A against x.

b Use your graph to work out a formula for A in terms of x.

c Use your formula to work out the value of A when $x = 100$.

a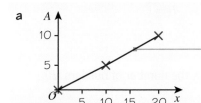

> A sketch does not have to be drawn on graph paper. 'Graph of A against x' means that A is on the vertical axis.

> When A and x are in direct proportion, the graph must go through the origin and, as A doubles, so does x.

b $A = kx$ so $k = \dfrac{A}{x}$

$k = \dfrac{5}{10} = \dfrac{1}{2}$ or 0.5 ◄─── Substitute $A = 5$ and $x = 10$.

$A = 0.5x$

c $A = 0.5 \times 100$ ◄─── Substitute $x = 100$ into the formula $A = 0.5x$.

$A = 50$

---

7 P is directly proportional to Q. $P = 4.5$ when $Q = 3$.

    a Sketch a graph of P against Q.

    b Write a formula for P in terms of Q.

    c Use your formula to work out the value of P when Q is 10.

 8 **Problem-solving** The cost of buying 20 litres of petrol is £26.

    a **Reasoning** Show that the cost, £C, of buying the fuel is directly proportional to the amount, x litres, of fuel bought.

    b What is the relationship between C and x?

    c Work out the cost of 55 litres of fuel.

Q8 hint

Draw a table.

x	0	5	10	20
C		6.5		26

Plot a graph of C against x.

9 The time, T seconds, it takes a water heater to boil some water is directly proportional to the mass of water, m kg, in the water heater.
When $m = 225$, $T = 495$.
Find T when $m = 300$.

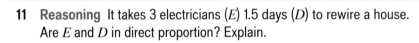

Exam-style question

10 The pressure, P, of water on an object (in bars) is directly proportional to its depth, d (in metres).
When the object is at a depth of 8 metres the pressure on the object is 0.8 bars.
A diver's watch has been guaranteed to work at a pressure up to 8.5 bars.
A diver takes the watch down to 75 m. Will the watch still work?
Give a reason for your answer. **(3 marks)**

11 Reasoning It takes 3 electricians (E) 1.5 days (D) to rewire a house.
Are E and D in direct proportion? Explain.

Key point

When x and y are in **inverse proportion**, y is proportional to $\frac{1}{x}$.
As one doubles ($\times 2$) the other halves ($\div 2$).

12 Problem-solving It takes 5 builders (B) 8 hours (H) to build a wall.

Q12a hint
$$H = \frac{k}{B}$$

   a Write a formula for H in terms of B.

   b How long will it take

     i 6 builders   **ii** 3 builders   **iii** 9 builders?

   c Reflect Multiply each exact answer in hours (H) in part **b** by the number of builders (B). What do you notice?

13 Problem-solving It takes 3 typists (T) 5 hours (H) to type a report.

   a Reasoning Are T and H in direct proportion? Explain.

   b Write a formula for H in terms of T.

   c How long would it take 7 typists to type a report?
     Give your answer correct to the nearest minute.

   d Reflect State at least one assumption you made in working out your answer to part **c**.

Key point

When x and y are in inverse proportion then
• $x \times y = $ a constant
• $xy = k$, so $y = \frac{k}{x}$

14 Reasoning A and B are in inverse proportion.
Work out the values of W, X, Y and Z.

A	10	20	14	Y	Z
B	14	W	X	70	28

15 Reasoning Does each equation represent direct proportion, inverse proportion or neither?

a $y = 3x$ **b** $y = \dfrac{5}{x}$ **c** $x + y = 9$

d $xy = 10$ **e** $\dfrac{y}{x} = 4$

Q15c hint

Rearrange the equation to $y = \underline{\quad}$

16 Problem-solving It takes 120 minutes to fill a swimming pool using 4 hoses.
How long will it take to fill the pool if 5 hoses are used?

Q16 hint

Check whether quantities in problems like this are in direct or inverse proportion by asking yourself, 'If one quantity doubles, does the other quantity double too, or does it halve?' For example, if the number of hoses doubles, does the time double or halve?

17 Problem-solving In a circuit, the resistance, R ohms, is inversely proportional to the current, I amps. When the resistance is 12 ohms, the current in the circuit is 8 amps.
Find the current when the resistance in the circuit is 6.4 ohms.

18 Problem-solving r is inversely proportional to t.
$r = 15$ when $t = 0.3$.

a Write a formula for r in terms of t.

b Calculate the value of r when $t = 4$.

19 a Copy and complete the table for $y = \dfrac{10}{x}$.

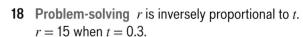

x	1	2	5	10
y				

b Use the table to sketch a graph of y against x.

c Work out the value of y when $x = 20$.

d Does your value of y in part **c** fit the shape of your graph?

e Work out the value of y when $x = 0.5$. Is your answer consistent with your sketch?

Exam-style question

20 The time, T seconds, it takes a water heater to boil a constant mass of water is inversely proportional to the power, P watts, of the water heater.
When $P = 900$, $T = 560$.
When P is 1700, will the water heater boil this mass of water in less than 5 minutes?
You must show your working. **(3 marks)**

Exam tip

When told that two quantities are directly or inversely proportional, write a formula including the constant k.

11 Check up

Active Learn
Homework

Percentages

1 Darika bought a car for £10 000.
The value of the car depreciated by 20% in the first year.
Its value depreciated by another 10% in the second year.
Work out the value of Darika's car at the end of 2 years.

2 £3500 is invested for 3 years at 3.4% compound interest.
Work out the total amount in the account after 3 years.

3 The number of bees in a hive decreases by 3% each year.
There are 7500 bees in the hive at the beginning of 2020.
How many bees will there be in the hive at the end of 2026?

4 Gavin invests £4500 at a compound interest rate of 4.2% per annum.
How many years before the investment has grown to £5527.78?

Compound measures

5 There are 40 litres of water in a barrel.
The water flows out of the barrel at a rate of 125 millilitres per second.
1 litre = 1000 millilitres
Work out the time it takes for the barrel to empty completely.

6 The mass of this plastic cuboid is 2208 g.
Work out the density of the plastic in grams per cm^3.

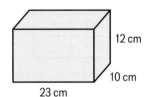

12 cm
10 cm
23 cm

7 A solid cube of steel has sides of length 5 cm.
The density of steel is 8.05 g/cm^3.
 a Convert 8.05 g/cm^3 to kg/m^3.   **b** Work out the mass of the cube.

8 A force of 30 N is applied to an area of 3200 cm^2. Work out the pressure in N/m^2.

9 The greatest recorded speed of Usain Bolt is 12.3 m/s.
The greatest speed of a great white shark is 40 km/h.
Which is faster? Explain your answer.

Ratio and proportion

10 The table shows a comparison of costs in British pounds (P) and euros (E).

Cost in British pounds	1	2	5	10	15
Cost in euros	1.3	2.6	6.5	13	19.5

 a Are British pounds and euros in direct proportion? Explain.
 b What is the relationship between British pounds (P) and euros (E)?
 c Convert £25 to euros.

11 c is directly proportional to d.
$c = 18$ when $d = 2.25$.

  a Write a formula for c in terms of d.

  b Calculate the value of c when $d = 12$.

12 In a circuit, the resistance, R ohms, is inversely proportional to the current, I amps.
When the resistance is 14 ohms, the current in the circuit is 9 amps.
Find the current when the resistance is 12 ohms.

13 **Reflect** How sure are you of your answers? Were you mostly

Just guessing   Feeling doubtful   Confident

What next? Use your results to decide whether to strengthen or extend your learning.

Challenge

14 You can use arrow diagrams to convert between compound measures.
For example, here is an arrow diagram to convert miles per hour to inches per minute.

Distance   Time

```
           1 mile    1 hour
×5280
           ☐ feet    1 hour      1 hour = 60 minutes
  ×12
           ☐ inches  1 hour
  ÷☐
           ☐ inches  1 minute            ÷☐
```

  a Copy and complete the arrow diagram.

  b Draw arrow diagrams to convert

    i km/hour to mm/minute

    ii litres/min to ml/hour

    iii mg/m^3 to kg/cm^3

  c Try some conversions of your own using arrow diagrams.

15 The times, distances and speeds of athletes in a 5 km race
and a 10 km race have got mixed up.
Sort them into the 5 km and 10 km races and then compile
a leader board for each race.

    Allia: 18 minutes 38 seconds, 5 km
    Billie: 1364 seconds, 13.2 km/h
    Chaya: 20 minutes 50 seconds, 4 m/s
    Daisy: 10 km, 4.1 m/s
    Ellie: 10 km, 13 km/h
    Fion: 2105 seconds, 17.1 km/h
    Gracie: 45 minutes 3 seconds, 3.7 m/s
    Hafsa: 5 km, 3.9 m/s

> **Q15 hint**
>
> Use the formula connecting distance, speed and time to work out the missing times and distances for each person.

11 Strengthen

Active Learn
Homework

Percentages

1 Write down the multiplier for each percentage increase as a decimal.

   a 20%   **b** 9%

   c 12%   **d** 3.7%

Q1 hint

Original = 100%
After increase = 100% + ☐%
        = ☐%
Write ☐% as a decimal.

2 Write down the multiplier for each percentage decrease as a decimal.

   a 23%   **b** 6%

   c 8%   **d** 7.5%

Q2 hint

Original = 100%
After decrease = 100% − ☐%
        = ☐%
Write ☐% as a decimal.

3 Use your answers to **Q1a** and **Q1b** to write down the multiplier for an increase of 20% followed by an increase of 9%.

Q3 hint

☐ × ☐ = ☐

4 Use your answers to **Q2a** and **Q2b** to write down the multiplier for a decrease of 23% followed by a decrease of 6%.

5 Write down the multiplier for an increase of 12% followed by a decrease of 8%.

6 Harry invests £400 at 3% **compound interest**. Copy and complete the table.

Year	Amount at start of year	Amount plus 3% interest	Total amount at end of year
1	£400	400×1.03	£412
2	£412	$412 \times 1.03 = 400 \times 1.03^2$	£424.36
3	£424.36	$424.36 \times 1.03 = 400 \times 1.03^3$	£437.09
4	£437.09	$437.09 \times 1.03 =$	
5			
6			

7 **Problem-solving** A company buys a van for £15 000.
Each year the value of the van depreciates by 20%.
Copy and complete the table to work out the value of the van at the end of 4 years.

Year	Value of van (£)	Value subtract 20% depreciation	Value at end of year
0	15 000	$15\,000 \times$ ☐	☐
1	☐		

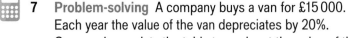

8 **Problem-solving** Molly invests £2000 at a compound interest rate of 2.3% per annum.
After how many years will she have more than £2200?

Q8 hint

Look for when the total amount is more than £2200.

9 **Problem-solving** A population of ants increases at a rate of 30% per day.
At the end of one week there are 3500 insects.
How many insects were there at the beginning of the week?

Q9 hint

Draw a table.

Day	Number of ants
6	↖ ÷□
7	3500 ↙

Compound measures

1 **Reasoning** A bucket holds 12 litres of water.
Water flows out at a rate of 50 ml per second.

  a Copy and complete the diagram to work out the amount of water flowing out per minute.

  b Work out the time it will take the bucket to empty.

×□ (50 ml in 1 second
    ↘ □ ml in 60 seconds) ×□
÷□ ((1 minute)
    ↘ □ litres in 1 minute

2 You can use these triangles to represent the formulae for speed, density and pressure.

$$\text{Speed} = \frac{\text{Distance}}{\text{Time}} \qquad \text{Density} = \frac{\text{Mass}}{\text{Volume}} \qquad \text{Pressure} = \frac{\text{Force}}{\text{Area}}$$

Cover the quantity you want to find.

$$\text{Mass} = \text{Density} \times \text{Volume} \qquad \text{Volume} = \frac{\text{Mass}}{\text{Density}} \qquad \text{Density} = \frac{\text{Mass}}{\text{Volume}}$$

 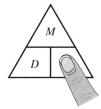 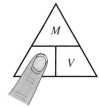

Use this method to copy and complete these formulae.

  a Speed = _____   **b** Time = _____   **c** Distance = _____

  d Force = _____   **e** Area = _____   **f** Pressure = _____

3 Use the method in **Q2** to copy and complete these tables.
Give your answers correct to 3 significant figures.

a

Metal	Mass (g)	Volume (cm³)	Density (g/cm³)
copper		122	8.96
lead	450		11.3
mercury	110	8.15	

b

Force (N)	Area (cm²)	Pressure (N/cm²)
	13	8
48	12	
65		13

4 Copy and complete.

a $\square$ g = 1 kg

b $\square$ cm^2 = 1 m^2

c $\square$ cm^3 = 1 m^3

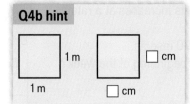

Q4b hint

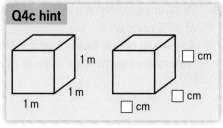

Q4c hint

5 Copy and complete the diagrams to convert

a 270 kg/m^2 to g/cm^2

b 50 g/cm^3 to kg/m^3

270 kg per m^2

$\square$ g per m^2 $)\times\square$

$\square$ g per cm^2 $)\div\square$

50 g per cm^3

$\square$ kg per cm^3 $)\div\square$

$\square$ kg per m^3 $)\times\square$

6 Copy and complete this table to convert from km/h to m/s.

km/h	m/h	m/min	m/s
18			
			10
24			
			16

Q6 hint

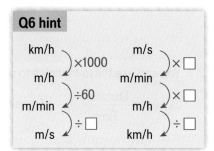

Ratio and proportion

1 A and B are in direct proportion.

This means that $\dfrac{A}{B}$ = constant and always has the same value.

Work out the values of W, X, Y and Z.

A	5	12	X	15	Z
B	10	W	45	Y	36
$\dfrac{A}{B}$	0.5	0.5	0.5	0.5	0.5

2 The time, T seconds, it takes a kettle to boil some water is directly proportional to the volume of water, V cm^3, in the kettle.
When $V = 300$ cm^3, $T = 120$ seconds.
How long will it take to boil 500 cm^3 of water?

Q2 hint

Draw a table.

T	V	$\dfrac{T}{V}$

3 A and B are in inverse proportion.
This means that $A \times B$ = constant and always has the same value.

Work out the values of W, X, Y and Z.

A	6	8	X	12	Z
B	8	W	16	Y	12
$A \times B$	48	48	48	48	48

4 For a constant force, the pressure, P (in N/m^2) is inversely proportional to the area, A (in m^2) the force acts on.
When the area is 2 m^2, the pressure is 16 N/m^2.
Work out the pressure when the area is 0.5 m^2.

Q5c hint

Draw a table.

P	A	PA
16		

11 Extend

 1 In a spring, the tension (T newtons) is directly proportional to its extension (x cm).
When the tension is 150 newtons, the extension is 6 cm.

   a Calculate the tension, in newtons, when the extension is 15 cm.

   b Calculate the extension, in cm, when the tension is 600 newtons.

 2 **Problem-solving** The diagram shows a piece of plastic cut into the shape of a trapezium.
A force is exerted evenly over the trapezium.
Work out the force required to create a pressure of $20\,\text{N/cm}^2$ on the trapezium.

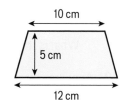

 3 Use the kinematics formulae in these questions.

   a Rafael hits a tennis ball at 20 m/s.
It hits the net, 8 m away, at a speed of 12 m/s.
What was its deceleration (negative acceleration) in m/s^2?

   b A car starts from rest and accelerates to v m/s in 20 seconds.
Find the acceleration in terms of v.

> **Q3 hint**
>
> Kinematics formulae:
> $$v = u + at$$
> $$s = ut + \tfrac{1}{2}at^2$$
> $$v^2 = u^2 + 2as$$

 4 The average distance between Earth and Jupiter is 7.78×10^8 km.

   a If a signal travels at 3.0×10^5 km/s, how long would it take to get to Jupiter?
Give your answer in minutes and seconds.

   b If the speed of the signal was actually less than 3.0×10^5 km/s, how would this affect your answer to part **a**?

 5 George invests £4500 at a compound interest rate of 5% per annum.
At the end of n complete years the investment has grown to £5469.78.
Find the value of n.

 Exam-style question

6 Evie invests £4750 for 5 years.
The investment gets compound interest of $x\%$ per annum.
At the end of 5 years, the investment is worth £5155.02.
Work out the value of x. **(3 marks)**

Exam tip

Show each part of your working on a separate line, so that it is clear.

 7 **Problem-solving** Plastic block **A** has a mass of 1.2 kg. Plastic cube **B** is made from plastic with a density 40% greater than plastic block **A**.
Work out the mass of plastic cube **B**.
Give your answer in grams correct to 3 significant figures.

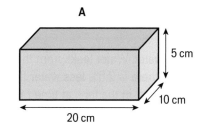

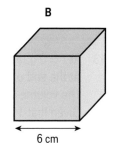

8 **Problem-solving** Carrie buys a motorbike. The value of the motorbike depreciates by 10% each year. Write down the letter of the graph which best shows how the value of Carrie's motorbike changes with time.

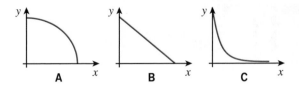

9 **Reasoning** The size of an exterior angle of a regular polygon is inversely proportional to the number of sides of the polygon.

a Sam thinks this means that if one polygon has twice the number of sides of another polygon, the size of the exterior angle is half the size of the original exterior angle.
Is Sam correct? Explain.

b The size of an exterior angle of a regular hexagon is 60°.
What is the size of the exterior angle of a 20-sided polygon?

10 Amy drove from Oxford to Maidenhead and then to London.
Her average speed from Oxford to Maidenhead was 72 km/h.
Her average speed from Maidenhead to London was 84 km/h.
Amy says, 'I can find my average speed from Oxford to London by working out the mean of 72 km/h and 84 km/h.'
If Amy is correct, what does this tell you about the two parts of Amy's journey?

Exam-style question

11 A train travelled along a track in 125 minutes, correct to the nearest 5 minutes.
The track is 280 km long, measured to the nearest 10 km.

a Could the average speed of the train have been greater than 136 km/h?
You must show how you get your answer. **(4 marks)**

b The exact length of the track is 276 km. Does this affect your decision in part **a**?
Give a reason for your answer. **(1 mark)**

> **Exam tip**
> Always write down the formula you are using, for example, Speed = ____

Exam-style question

12 The petrol consumption of a car, in litres per 100 kilometres, is given by the formula

$$\text{Petrol consumption} = \frac{100 \times \text{number of litres of petrol used}}{\text{number of kilometres travelled}}$$

Jonny's car travelled 152 kilometres, correct to 3 significant figures.
The car used 11.2 litres of petrol, correct to 3 significant figures.
Jonny says, 'My car used less than 7 litres of petrol per 100 kilometres.'
Could Jonny be wrong? You must show how you get your answer. **(3 marks)**

> **Exam tip**
> When a question includes rounded numbers (e.g. correct to 3 significant figures) they have an upper and a lower bound.

13 At $t = 0$ a tank is full of water. Water leaks from the tank.
At the end of every hour there is 2.5% less water than at the start of the hour.
The volume of water, in litres, in the tank at time t hours is V.
Given that $V_0 = 1200$ and $V_{t+1} = kV_t$

a write down the value of k

b work out the volume of water in the tank after 4 hours

11 Test ready

Summary of key points

To revise for the test:

- Read each key point, find a question on it in the mastery lesson, and check you can work out the answer.

- If you cannot, try some other questions from the mastery lesson or ask for help.

Key points

1. **Iteration** means carrying out a process repeatedly. → **11.1**

2. Most interest rates are **compound interest** rates. The interest earned each year is added to the money in the account and earns interest the next year. → **11.1**

3. You can calculate an amount after n years' compound interest using the formula

$$\text{Amount} = \text{initial amount} \times \left(\frac{100 + \text{interest rate}}{100}\right)^n$$

→ **11.1**

4. You can use the compound interest formula to work out **growth** (increases) and **decay** (decreases) of quantities other than money. → **11.1**

5. **Compound measures** combine measures of two different quantities. For example, speed is a measure of distance travelled and time taken. It can be measured in metres per second (m/s), kilometres per hour (km/h) or miles per hour (mph). → **11.2**

6. These are kinematics formulae: $v = u + at$ $\quad s = ut + \frac{1}{2}at^2$ $\quad v^2 = u^2 + 2as$

where a is constant acceleration, u is initial velocity, v is final velocity, s is displacement from the position when $t = 0$ and t is time taken.

Velocity is speed in a given direction; possible units are m/s.
Initial velocity is speed in a given direction at the start of the motion.
Acceleration is the rate of change of velocity, i.e. a measure of how the velocity changes with time; possible units are m/s^2. → **11.2**

7. **Density** is the **mass** of substance (g) contained in a certain **volume** (cm^3) and is often measured in grams per cubic centimetre (g/cm^3).

$$\text{Density} = \frac{\text{mass}}{\text{volume}} \text{ or } D = \frac{M}{V}$$

→ **11.3**

8. **Pressure** is the **force** in newtons applied over an **area** in cm^2 or m^2. It is usually measured in newtons (N) per square metre (N/m^2) or per square centimetre (N/cm^2).

$$\text{Pressure} = \frac{\text{force}}{\text{area}} \text{ or } P = \frac{F}{A}$$

→ **11.3**

9. When x and y are in **direct proportion**

$y = kx$, where k is the gradient of the graph of y against x, and so $\frac{y}{x} = k$, a constant. → **11.4**

10. When x and y are in **inverse proportion**, y is proportional to $\frac{1}{x}$.
As one doubles ($\times 2$) the other halves ($\div 2$). → **11.4**

11. When x and y are in inverse proportion

$$x \times y = \text{a constant} \qquad xy = k, \text{ so } y = \frac{k}{x}$$

→ **11.4**

Sample student answers

Exam-style question

1   Amy and Richard cycled along the same 30 km path. Amy took $1\frac{1}{4}$ hours to cycle the 30 km.
    Richard started to cycle 3 minutes after Amy started to cycle.
    Richard caught up Amy when they had both cycled 6 km.
    Amy and Richard both cycled at constant speeds.
    Work out Richard's speed.   **(5 marks)**

Student A

Amy

$$S = \frac{30}{1.25} \overset{\times 4}{\underset{\times 4}{=}} \frac{120}{5}$$

$S = 24$ km/h

At 24 km/h:

6 km

$T = \frac{6}{24} = \frac{1}{4}$ hour

$\quad = 15$ minutes

Richard

$15 - 3 = 12$

$S = \frac{6}{12} = 0.5$

$0.5 \times 60 = 30$ km/h

Student B

$Speed = \dfrac{distance}{time}$

Amy

$D = 30$ km

$T = 1.25$ hours

$S = \dfrac{30}{1.25} = \dfrac{3000}{125}$

$\quad = 24$ km/h

6 km at 24 km/h

$Time = \dfrac{distance}{speed}$

$\quad = \dfrac{6}{24} = \dfrac{1}{4}$ hour

$\quad = 15$ minutes

Richard

Richard takes 3 minutes less to go 6 km

$15 - 3 = 12$ minutes

$\quad\quad\quad = 0.2$ hours

$Speed = \dfrac{6}{0.2} = \dfrac{60}{2}$

$\quad\quad\quad = 30$ km/h

---

a   Students A and B work out $\frac{30}{1.25}$ differently.

   i   Why does Student A multiply the numerator and the denominator by 4?

   ii   Write a sentence explaining Student B's method for working out 30 divided by 1.25.

   iii   Which student's method do you prefer? Why?

b   Give two reasons why Student B's working is better.

11 Unit test

Active Learn
Homework

1 A leopard travels 100 metres in 7.19 seconds, and another 50 metres in 4.2 seconds.
What is its average speed

  a in m/s   **(2 marks)**

  b in km/h?   **(2 marks)**

Give your answers correct to 3 significant figures.

2 A cylindrical glass has a circular base with radius 4 cm.
The glass exerts a force of 7 N on the table.
Work out the pressure in N/m^2.   **(3 marks)**

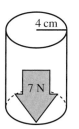

3 The diagram shows a solid prism made from metal.

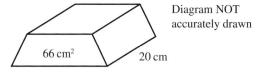

Diagram NOT accurately drawn

The cross-section of the prism is a trapezium with area $66\,cm^2$.
The length of the prism is 20 cm.
The density of the metal is $5\,g/cm^3$.
Calculate the mass of the prism.
Give your answer in kilograms.   **(4 marks)**

4 When travelling at constant speed the distance, D, in km covered by a walker is directly proportional to the time taken, t, in minutes.
When $t = 20$ minutes, $D = 1.6\,km$.

  a Find a formula for D in terms of t.   **(1 mark)**

  b Calculate the value of D when $t = 2$ hours.   **(2 marks)**

  c Calculate the value of t when $D = 11\,km$.
     Give your answer correct to 3 significant figures.   **(2 marks)**

5 Anna cycles 1 km in 3 minutes.
The distance is measured correct to the nearest metre.
The time is measured correct to the nearest second.
Work out the maximum speed in m/s at which she could be travelling.
Give your answer correct to 2 decimal places.   **(3 marks)**

6 A television loses 4% of its value every month.
It was bought for £950 at the beginning of January.
How much will it be worth at the end of June?   **(3 marks)**

7 When a constant force is applied, the resulting pressure, P, is inversely proportional to the area, A.
When $A = 8$, $P = 6$.

   a Find a formula for P in terms of A. **(3 marks)**

   b The pressure must not reach more than a limit of $25\,\text{N/m}^2$.
When the area, A, is $2\,\text{m}^2$, will the pressure be over the limit? **(2 marks)**

8 Nico invests £8000 for 3 years at 1.5% per annum compound interest.
Calculate the value of his investment at the end of the 3 years. **(3 marks)**

9 Emma invests £4000 in an account for 2 years.
The account pays 1.8% compound interest.
Emma has to pay 40% tax on the interest earned each year.
The tax is taken from the account at the end of each year.
How much will Emma have in the account at the end of 2 years? **(4 marks)**

10 Josh invests a sum of money for 25 years at 2.9% per annum compound interest.
Write down the letter of the graph which best shows how the value of Josh's
investment changes over the 25 years. **(1 mark)**

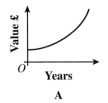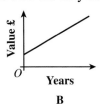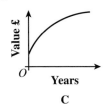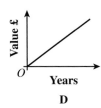

     A   B   C   D

(TOTAL: 35 marks)

11 Challenge Previously you have learned about simple interest.
Simple interest is calculated only on the original amount invested.
Antonia invests £2400 for 5 years at an interest rate of 1.35%.

   a Which is better for Antonia: simple interest or compound interest? Explain.

   Return on investment (ROI) is a measure of how well an investment performs.
ROI is calculated by dividing the profit (or interest) gained by the initial investment.
Then multiply by 100 to express the ROI as a percentage.

   b Assume that Antonia's interest is compound. Work out Antonia's ROI as a percentage.

12 Reflect 'Multiplicative' means involving multiplication or division.
'Reasoning' is being able to explain why you have done some maths a certain way.
This unit is called 'Multiplicative reasoning'.
List three ways you have used multiplication or division in this unit.
Why is it good to reason in mathematics?

13 Reflect Previously you have learned about simple interest.
This is interest calculated only on the original amount invested.
In this unit you have learned about compound interest.
What is the same, and what is different, about simple interest and compound interest?

12 Similarity and congruence

12.1 Congruence

Prior knowledge

Active Learn
Homework

- Show that two triangles are congruent.
- Know the conditions of congruence.

Warm up

1 Fluency What do angles in a triangle add up to?

2 Work out the sizes of angles a, b and c.
Give reasons for your answers.

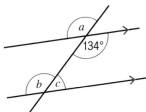

3 a Which of these triangles are congruent to ABC?

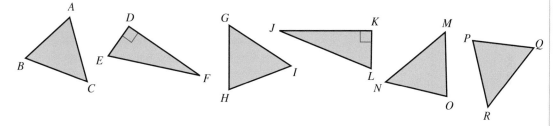

b Triangles DEF and JKL are congruent.
Which of these statements are true?

A Side DE and side KL are corresponding sides.

B Side DF and side JL are corresponding sides.

C Angle DEF and angle KLJ are corresponding angles.

4 In triangle DEF in **Q3**, DE is 8 cm and angle DFE is 28°.

a Use the sine ratio to work out the length of EF.

b Use Pythagoras' theorem to work out the length of DF.

Give your answers correct to 1 decimal place.

> **Q4 hint**
>
> Sketch the triangle.

5   **Reasoning** The two triangles in each pair are congruent.
     Give a reason why (SSS, SAS, AAS or RHS).

a 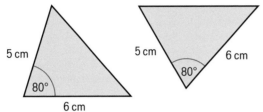   b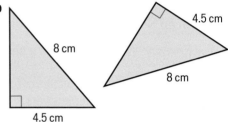

c 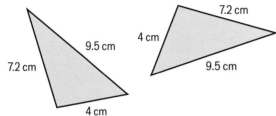   d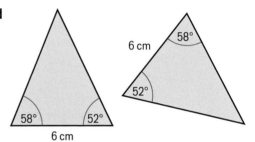

6   **Reasoning** State whether or not the triangles in
     each pair described below are congruent.
     If the triangles are congruent, give the reason and
     write the corresponding vertices in pairs.

     a ABC where $AB = 7$ cm, $BC = 5$ cm, angle $B = 42°$
        PQR where $PQ = 50$ mm, $QR = 7$ cm, angle $Q = 42°$

     b ABC where $AB = 7$ cm, angle $B = 42°$, angle $C = 109°$
        PQR where $PQ = 7$ cm, angle $Q = 109°$, angle $R = 42°$

> **Q6 hint**
>
> Sketch each triangle.

7   **Reasoning** Which of these triangles are congruent to
     triangle ABC?
     Give reasons for your answers.

> **Q7 hint**
>
> Work out the size of angle ACB.

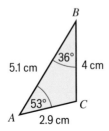   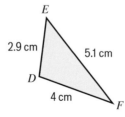   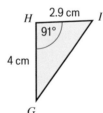   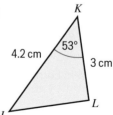

8 **Problem-solving** Are the triangles in each of these pairs congruent? Justify your answers.

Q8 hint

Find a missing side.

a

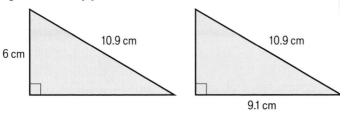

b

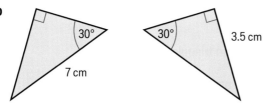

9 **Reasoning** Are all right-angled triangles with one side 5 cm and one side 12 cm congruent? Explain.

10 **Problem-solving** AB and CD are parallel lines, and $AB = CD$.

a Work out the sizes of all the angles.

b **Reasoning** Show that triangle ABE and triangle CED are congruent.

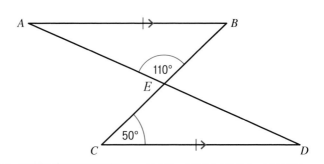

Exam-style question

11 In this arrowhead, angle $JKL = 26°$ and angle $KJL = 35°$.

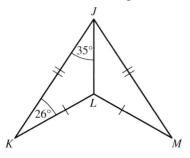

Explain why triangles JKL and JLM are congruent.

(3 marks)

Exam tip

When a diagram is made up of two or more shapes, it can help to sketch each shape separately and then mark on all the information you know.

12.2 Geometric proof and congruence

- Prove shapes are congruent.
- Solve problems involving congruence.
- Use geometric sketching to help solve congruency problems.

Active Learn
Homework

Warm up

1 Fluency What are the conditions for congruence in triangles?

2 Sketch each shape. Mark all the equal sides and angles.

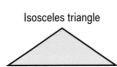 

Rhombus

Isosceles triangle

Equilateral triangle

Isosceles trapezium

3 M is the midpoint of DE. M is also the midpoint of FG.

  a Which length is the same as

    i DM

    ii GM?

  b Which angle is the same as angle DMF?

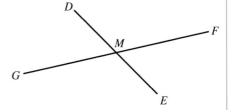

Key point

Conditions of congruence (SSS, SAS, AAS, RHS) can be used for geometric proof to show that shapes are the same.

Example

$ABCD$ is a parallelogram.
Prove that triangle ABC is congruent to ADC.

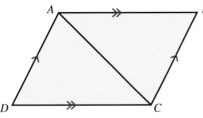 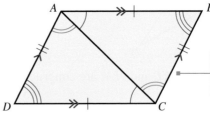

Mark all equal angles and sides.

Length AB = length CD because opposite sides in a parallelogram are equal.

State why $AB = CD$.

Length BC = length AD because opposite sides in a parallelogram are equal.

State why $BC = AD$.

Length AC is common to both triangles.
So triangle ABC is congruent to triangle ADC (SSS).

State the condition used to prove congruence.

4 *WXYZ* is a rectangle.

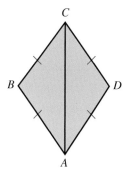

  a Copy the rectangle. Mark all equal sides and angles.

  b Prove that triangle *WXY* is congruent to triangle *XYZ*.

  c Which angle is the same as angle *XWY*?

Q4b hint

Write each statement of your proof on a new line. Give a reason for every statement you make. Then give a reason for congruency.

5 **Reasoning** In the diagram, $AB = BC = CD = AD$.
Prove that triangle *ABC* is congruent to triangle *ACD*.

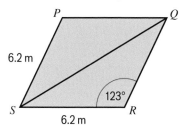

6 **Reasoning** The diagram shows a rhombus *PQRS*.

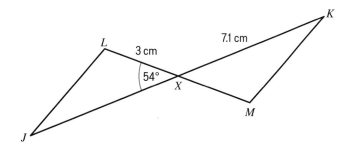

Q6 hint

What do you know about opposite sides in a rhombus?

  a Prove that triangle *PQS* and triangle *QRS* are congruent.

  b Find the size of

   i angle *QPS* **ii** angle *RQS*

  c Betty says, 'I used the fact that the angles are the same in the two triangles to prove that they are congruent.'
  Show Betty's proof.

7 **Reasoning** In the diagram,
X is the midpoint of *JK*.
X is also the midpoint of *LM*.
Prove that the triangles *JLX* and *KMX* are congruent.

Key point

When given a description of a shape, it helps to sketch it.

---

8 **Reasoning** *FGH* is an equilateral triangle.
Point *E* lies on *FH*. *EG* is perpendicular to *FH*.
Prove that triangle *FGE* is congruent to triangle *GHE*.

9 PQRS is an isosceles trapezium.
PQ = RS
Angle PQR = angle QRS
Prove that PR = QS. **(4 marks)**

10 Reasoning RST is an isosceles triangle such that RS = ST. Use congruent triangles to prove that the line SM which cuts the base of the triangle at right angles also bisects the base.

11 Reasoning Prove that triangles PQR and RST are congruent.
Hence, prove that R is the midpoint of PT.

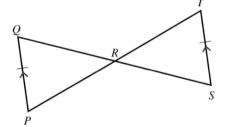

12 Reasoning CDEF is a rhombus.
CD = DE = EF = CF
CD is parallel to EF.
DE is parallel to CF.
Prove that

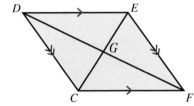

a triangles DEG and CFG are congruent
b triangles CDG and EFG are congruent
c G is the midpoint of both CE and DF and hence that the diagonals of a rhombus intersect at right angles

13 Reasoning ABCD is a square.
AC and BD are the diagonals of the square, which cross at point E.
a Draw the square, showing both diagonals.
b Mark all the equal angles on your diagram.
c Which triangles are congruent in your diagram?
d What can you say about all the angles at point E?
e Using your answers to parts **b** to **d**, show that lines AC and BD bisect at point E, at right angles.

14 XYZ is an isosceles triangle with XY = XZ.
A and B are points on XY and XZ such that AX = BX.
Prove that triangle XAZ is congruent to triangle XBY.
(3 marks)

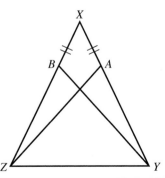

12.3 Similarity

- Use the ratio of corresponding sides to work out scale factors.
- Find missing lengths on similar shapes.
- Use geometric sketching to help solve similarity problems.

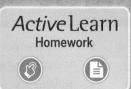

Active Learn
Homework

Warm up

1 Fluency What is the same and what is different in these two regular pentagons?

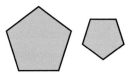

2 What is the scale factor of the enlargement from

a A to B **b** B to A?

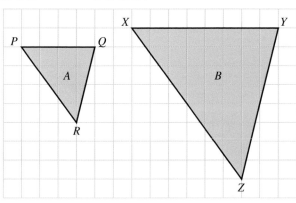

3 a Corresponding sides map on to each other in an enlargement.
Write the pairs of corresponding sides for triangles A and B in **Q2**.

b Measure the angles and sides in the triangles in **Q2**.

c What do you notice about the angles?

d Write the ratios of corresponding sides as fractions.

e What do you notice about the ratios?

> **Q3d hint**
> $$\frac{PQ}{\square} = \frac{\square}{\square}$$

4 Copy and complete to simplify

$$\overset{\times 10}{\overbrace{\frac{10}{22.5}}} = \frac{100}{225} = \frac{\square}{\square}$$
$$\underset{\times 10}{}$$

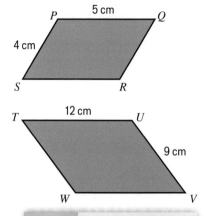

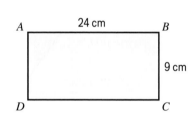

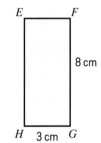

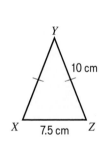

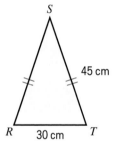

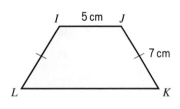

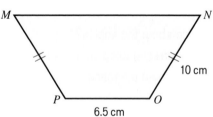

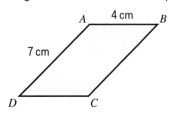

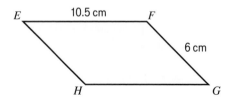

Key point

Shapes are **similar** when one shape is an enlargement of the other.
Corresponding angles are equal and corresponding sides are all in the same ratio.

5 Here are two parallelograms. Angles *SPQ* and *TUV* are the same size.

   a Which side in *TUVW* corresponds to

     i *PQ*   **ii** *PS*?

   b Work out the ratio

     i $\dfrac{PS}{UV}$ **ii** $\dfrac{PQ}{TU}$

   c **Reasoning** Use your answers to parts **a** and **b** to state why the parallelograms are not similar.

6 **Reasoning** Which pairs of shapes are similar?
Use ratios of corresponding sides to explain your answers.

Q6 hint

The shapes may not be in the same orientation.

a *XYZ* and *RST* are isosceles triangles.
Angles *XYZ* and *RST* are equal.

b *ABCD* and *EFGH* are rectangles.

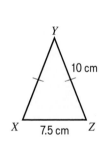

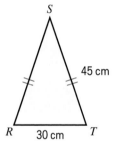

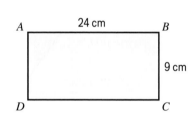

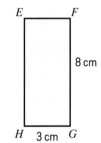

c *IJKL* and *MNOP* are isosceles trapezia.
Angles *IJK* and *NOP* are equal.

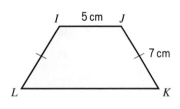

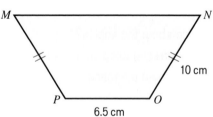

d *ABCD* and *EFGH* are parallelograms.
Angles *ABC* and *FGH* are equal.

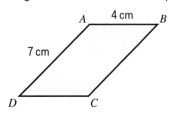

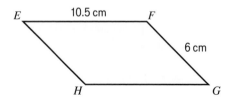

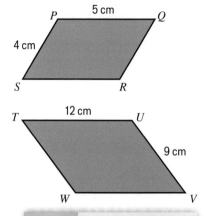

7 Reasoning Show that pentagon $ABCDE$ is similar to pentagon $VWXYZ$.

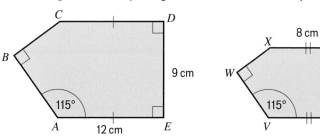

Example

These two rectangles are similar.
Find the missing length x in the smaller rectangle.

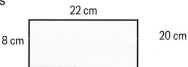

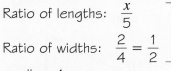

Ratio of lengths: $\dfrac{x}{5}$

Ratio of widths: $\dfrac{2}{4} = \dfrac{1}{2}$

Write the ratio $\dfrac{\text{small}}{\text{large}}$ for the lengths and the widths.

$\dfrac{\text{small}}{\text{large}} = \dfrac{1}{2} = \dfrac{x}{5}$

$2x = 5$

Write an equation to solve for x.

$x = \dfrac{5}{2} = 2.5\,\text{cm}$

8 Problem-solving These two rectangles are similar.
Find the missing side length L in the larger rectangle.

22 cm
8 cm
L cm
20 cm

9 Problem-solving A small photograph has a length of 15 cm and a width of 10 cm.
Shez enlarges the small photograph to make a large photograph.
The large photograph has a length of 27 cm.
The two photographs are similar rectangles.
Work out the width of the large photograph.

Q9 hint

Sketch the rectangles.
Mark the width of the large photograph as x.

10 Problem-solving A small can of soup has a height of 10.5 cm and a diameter of 6 cm.
A large can of soup is similar to the small can.
It has a diameter of 8 cm.
Find the height of the large can of soup.

Q10 hint

Sketch the cans.

11 Problem-solving An aerial photograph shows a campsite with an outdoor swimming pool.
In the photograph, the pool measures 5 cm by 2 cm.
The real pool is 25 m long.

a How wide is the pool?

b Reflect How did you write your ratio?
Does it matter which value is the numerator?

Q11a hint

$25\,\text{m} = \square\,\text{cm}$

12 Triangle *CDE* is similar to triangle *FGH*.

∠*CED* = ∠*FHG*

∠*CDE* = ∠*FGH*

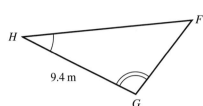

Calculate the length of *FG*.

Give your answer in metres, correct to the nearest centimetre. **(3 marks)**

13 **Reasoning** **a** Use corresponding angles to show that triangles *A* and *B* are similar.

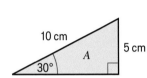
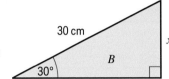
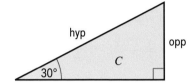

b Work out length *x* in triangle *B*.

c Show that triangle *C* is similar to triangles *A* and *B*. Explain.

d Write down the value of $\dfrac{\text{opposite}}{\text{hypotenuse}}$ hypotenuse for these triangles.

e What is another name for the ratio $\dfrac{\text{opposite}}{\text{hypotenuse}}$?

14 **Reasoning** Are the triangles in each pair similar? Explain.

a

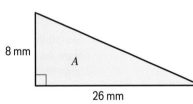
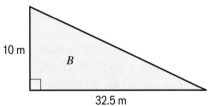

b

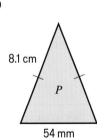
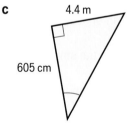

c

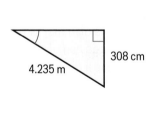

15 **Reasoning** Here are two regular hexagons.

a Are they similar? Explain.

b **Reflect** Are all regular hexagons similar? Explain.

12.4 More similarity

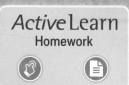

- Use similar triangles to work out lengths in real life.
- Use the link between linear scale factor and area scale factor to solve problems.

Warm up

1 Fluency What is the ratio 5 : 7.5 in the form $1 : n$?

2 Rectangle A has length 5 cm and width 2 cm.
It is enlarged by scale factor 3 to give rectangle B.
What is the ratio of
a corresponding sides of rectangle A to rectangle B
b the perimeter of rectangle A to rectangle B
c the area of rectangle A to rectangle B?

3 What is the scale factor of the enlargement that maps
a triangle Y on to triangle Z **b** triangle Z on to triangle Y?

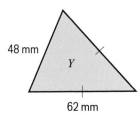

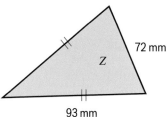

4 Give reasons why angles a, b, c and d are also 62°.

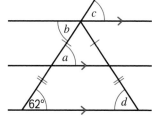

Key point

Angles on parallel lines can be used to show that shapes are similar.

5 a What size are the angles in triangle AEB?
Give reasons for your answers.
b Are triangles ABE and CDE similar?
Explain.

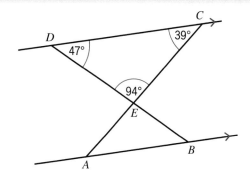

6 **Reasoning**

a Show that triangles *PQR* and *RST* are similar.

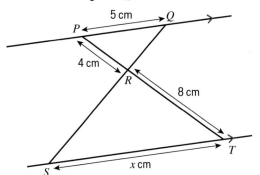

Q6a hint

Find equal angles.

b Find the missing length *x*.

Key point

Similar shapes that are inside each other sometimes share angles and/or sides.

---

7 **Reasoning**

a Explain why triangles *FGH* and *FJK* are similar.

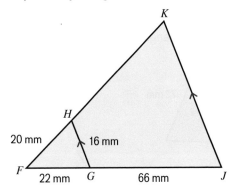

Q7a hint

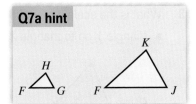

b Calculate the length of *HK*.

c Calculate the length of *JK*.

8 **Reasoning** *LMP* and *LNQ* are straight lines.
MN is parallel to *PQ*.
LN = 22 cm
MN = 16 cm
PQ = 32 cm
ML = 18 cm

a Show that triangle *LMN* is similar to triangle *LPQ*.

b Work out the length of *LQ*.

c Work out the length of *MP*.

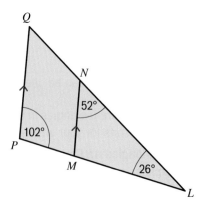

9 The two triangles in the diagram are similar.

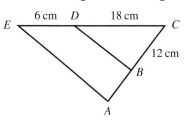

a There are two possible lengths for *AB*.
Work out each of these values. **(4 marks)**

b State any assumptions you made in your working.
(1 mark)

Exam tip

When answering questions about similar shapes, state the corresponding sides.

When a shape is enlarged by linear scale factor k
- perimeter is multiplied by scale factor k
- area is multiplied by scale factor k^2

---

10 The diagram shows two flower beds made in the shape of similar trapezia.

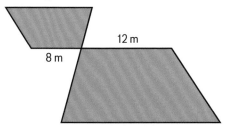

Q10 hint

First work out the linear scale factor of the enlargement (k).

a The perimeter of the small flower bed is 36 m.
Work out the perimeter of the large flower bed.

b The area of the small flower bed is 60 m².
Work out the area of the large flower bed.

11 **Problem-solving** A company makes teddy bears.
The company makes small bears that are 15 cm tall.
A small bear has a surface area of 200 cm².
The same company make giant bears which are 1.8 m tall.
A giant bear is mathematically similar to a small bear.
Work out the surface area of a giant bear.

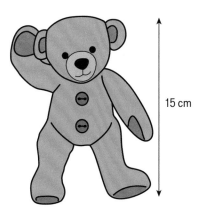

12 Problem-solving Shape A is similar to shape B.
The area of shape A is $4\,cm^2$.
The area of shape B is $36\,cm^2$.

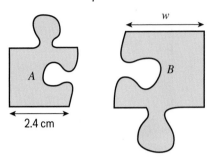

a Write the ratio of the area of shape A to the area of shape B in the form $1 : n$.

b What is the area scale factor?

c Write the ratio of the width of shape A to the width of shape B.

d What is the scale factor of the enlargement?

e The width of shape A is 2.4 cm.
Work out the width, w, of shape B.

Q12c hint

In an enlargement by scale factor k, the area is enlarged by scale factor k^2.

 13 Problem-solving Shape A is similar to shape B.
The area of shape A is $126\,cm^2$.
The area of shape B is $283.5\,cm^2$.
Calculate

a length x

b length y

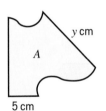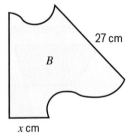

14 Problem-solving A triangle has sides of 4, 5 and 6.4 cm. Its area is $10\,cm^2$.
How long are the sides of a similar triangle that has an area of $90\,cm^2$?

 15 Problem-solving A sheet of A2 paper and a sheet of A4 paper are similar.
The area of a sheet of A2 paper is $2500\,cm^2$ and the area of a sheet of A4 paper is $625\,cm^2$.
The width of a sheet of A2 paper is 42 cm.

a Work out the area scale factor.

b Work out the length scale factor.

c Use the length scale factor to work out the width of a sheet of A4 paper.

16 Problem-solving Two similar triangles have areas of $36\,cm^2$ and $100\,cm^2$ respectively.
The base of the smaller triangle is 3 cm.
Find the length of the base of the larger triangle.

17 Reasoning A painted tile has a surface area of $6\,cm^2$.
What is the area of a similar tile with lengths that are

a twice the corresponding lengths of the first tile

b three times the corresponding lengths of the first tile?

12.5 Similarity in 3D solids

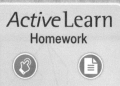

*Active*Learn
Homework

- Use the links between scale factors for length, area and volume to solve problems.

Warm up

1 Fluency Work out the volume and surface area of this cube.

3 cm

2 Each side of the cube in **Q1** is increased by 12.5%.
What decimal multiplier can you use to work out the side lengths of the enlarged cube?

3 Work out

a the cube root of 125

b $\sqrt[3]{\frac{64}{27}}$

4 In each pair of diagrams, solid A is enlarged to make solid B.

i

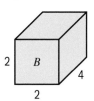

ii

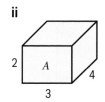

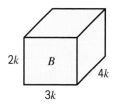

a Copy and complete the table.

	Linear scale factor	Volume A	Volume B	Volume scale factor
i				
ii				

b Reflect What do you notice about the scale factors of volume?

Key point

When the linear scale factor is k
- lengths are multiplied by k
- area is multiplied by k^2
- volume is multiplied by k^3

5 Prisms A and B are similar.
The volume of prism A is 12 cm^3.
Calculate the volume of prism B.

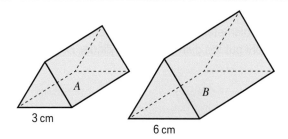

3 cm

6 cm

6 Tetrahedra C and D are similar.
The volume of tetrahedron C is $15\,cm^3$.
Calculate the volume of tetrahedron D.

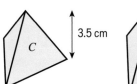 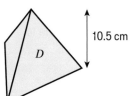

3.5 cm

10.5 cm

C

D

 7 Cones E and F are similar.
The volume of cone E is $202.5\,cm^3$.
Calculate the volume of cone F.

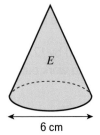

E

F

6 cm

4 cm

 8 **Problem-solving** A can of paint is 18 cm tall.
It holds 5 litres of paint.
A similar bigger can is 30% taller.

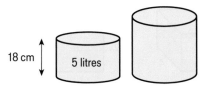

18 cm

5 litres

a Write down the linear scale factor.

b How much paint does the bigger can hold?
Give your answer correct to the nearest litre.

 9 Cylinders G and H are similar.
The volume of G is $32\,cm^3$.
The volume of H is $256\,cm^3$.

a Write the ratio of the volume of cylinder G to the
volume of cylinder H in the form $1 : n$.

b What is the volume scale factor?

c Write the ratio of the diameter of cylinder G
to the diameter of cylinder H.

d What is the scale factor of the enlargement?

e The diameter of cylinder G is 6 cm.
Work out the diameter, d, of cylinder H.

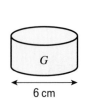 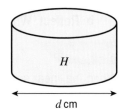

G

H

6 cm

d cm

10 Sphere J is similar to sphere K.
The volume of J is 27 times the volume of K.
Work out the diameter of sphere K.

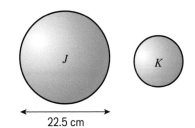

22.5 cm

11 Prisms L and M are similar.
The volume of prism M is 343 times the volume of prism L.
Calculate the value of

a length x

b length y

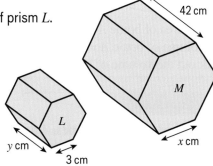

42 cm

M

L

y cm

x cm

3 cm

12 Here are two spheres.

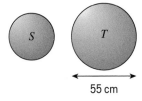

S

T

55 cm

The volume of sphere T is 33.1% more than the volume of sphere S.

a Write the ratio of the volume of sphere S to the volume
of sphere T in the form $1 : n$.

b The diameter of sphere T is 55 cm.
What is the diameter of sphere S?

Q12a hint

Volume of sphere S		Volume of sphere T
100%	:	☐ %
1	:	☐

13 **Reasoning** Pyramid A and pyramid Z are similar.
The volume of pyramid A is 125 times the volume
of pyramid Z.
The surface area of pyramid Z is 60 cm^2.

a Write down the volume scale factor, k^3.

b Work out the linear scale factor, k.

c Work out the area scale factor, k^2.

d Calculate the surface area of pyramid A.

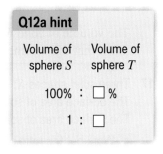

A

Z

14 **Problem-solving** Cylinders B and C are mathematically similar.

The volume of B is $\frac{1}{64}$ of the volume of C.

The surface area of B is 35.2 cm^2.
Work out the surface area of cylinder C.

Exam-style question

15 Here are two similar bottles.

Surface area of bottle A : surface area of bottle $B = 4 : 9$
The volume of bottle $A = 240\,\text{ml}$
Work out the volume of bottle B. **(3 marks)**

Exam-style question

16 The bases of two similar cones, A and B, are $207\,\text{cm}^2$ and $92\,\text{cm}^2$ respectively.

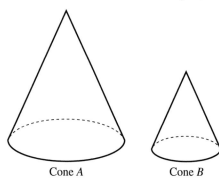

Cone A Cone B

The volume of cone A is $837\,\text{cm}^3$.
Show that the volume of cone B is $248\,\text{cm}^3$. **(5 marks)**

17 Three solid shapes X, Y and Z are similar.
The surface area of solid shape X is $16\,\text{cm}^2$.
The surface area of solid shape Y is $81\,\text{cm}^2$.

a Work out the ratio of the height of solid shape X to the height of solid shape Y.

The ratio of the volume of solid shape Y to solid shape Z is $27 : 1000$.

b Work out the ratio of the height of solid shape Y to the height of solid shape Z.

c Work out the ratio of the height of solid shape X to
solid shape Y to solid shape Z.

12 Check up

Congruence

1 Which of these triangles are congruent? Give reasons for your answer.

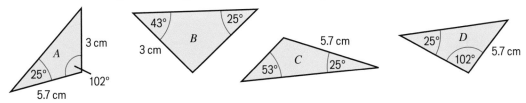

2 *HJKL* is a parallelogram.

    a Prove that triangle *HJK* and triangle *KLH* are congruent. Explain your answer fully.

    b Using a *different* condition for congruence, prove that triangles *HJK* and *KLH* are congruent.

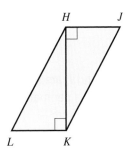

Similarity in 2D shapes

3 Show that the triangles in each pair are similar.

    a

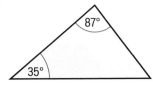

    b

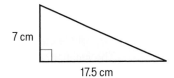

4 Quadrilaterals *ABCD* and *EFGH* are similar.

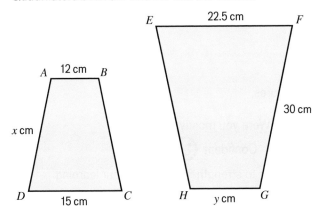

Work out

    a length *AD*   **b** length *GH*

5 **a** Prove that triangle ABE is similar to triangle ACD.

　　b Work out length CD.

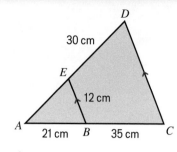

6 **a** Show that PQR and RST are similar triangles.

　　b Work out the missing lengths in the diagram, x and y.

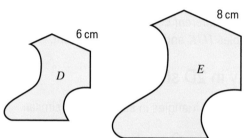

 7 Shapes D and E are similar.
Shape D has a perimeter of 41.4 cm and an area of 112.5 cm^2.
Calculate

　a the perimeter of shape E

　b the area of shape E

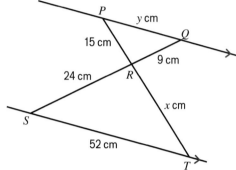

Similarity in 3D solids

 8 Two whole cheeses are mathematically similar in shape.
The smaller cheese has a radius of 8.4 cm and a volume of 665 cm^3.
The larger cheese has a radius of 13.2 cm.
Work out the volume of the larger cheese.
Give your answer correct to 1 decimal place.

9 Pyramids A and B are similar.
The surface area of pyramid A is 1260 cm^2.
The surface area of pyramid B is 180 cm^2.
The volume of pyramid A is 9604 cm^3.
Work out the volume of pyramid B.
Give your answer correct to 3 significant figures.

10 **Reflect** How sure are you of your answers? Were you mostly

Just guessing ☹　Feeling doubtful 😐　Confident ☺

What next? Use your results to decide whether to strengthen or extend your learning.

Challenge

11 There are two possible triangles ABC where $AB = 16$ cm, $BC = 10$ cm and angle $CAB = 40°$.

　a Construct them accurately.

　b Are the two triangles　　　**i** congruent　　**ii** similar?

12 Strengthen

*Active*Learn
Homework

Congruence

1 **Reasoning** Sketch these triangles and work out any missing angles.
Use your sketches to decide which of triangles B, C and D is *not* congruent to triangle A.

Q1 hint

If you rotated triangles B, C and D, which one would not fit exactly on A?

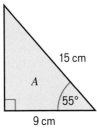

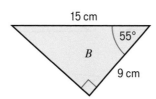

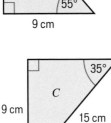

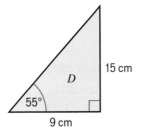

2 Sketch these triangles and work out any missing angles.
Use your sketches to decide which of the triangles are congruent.

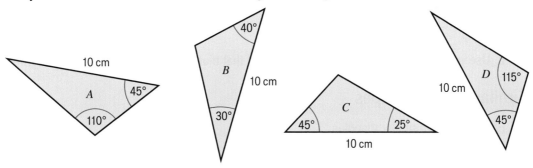

3 Draw two right-angled congruent triangles on paper and cut them out.
What shapes can you make by putting these two triangles together?
You must place the triangles with equal sides touching.

Q3 hint

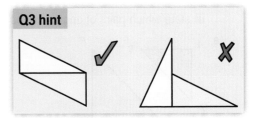

4 **Reasoning** Choose the correct reason for each pair of triangles being congruent.

| SSS | SAS | AAS | RHS |

a

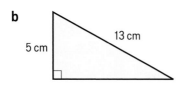

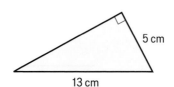

b

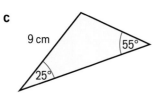

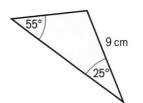

c

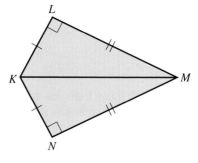

5 **Reasoning** Sam joins two triangles to make a kite.

a Choose a side from the box to copy and complete the start of a proof to show that triangles *KLM* and *KNM* are congruent.

| *KM* | *KN* | *NM* |

____ is a common side.

LM = ____

KL = ____

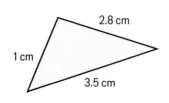

b Use your answer to part **a** to state the reason that triangles *KLM* and *KNM* are congruent.

Similarity in 2D shapes

1 For each pair of similar triangles:

i sketch them in the same orientation (facing the same way)

ii name the three pairs of corresponding sides

iii state which pairs of angles are equal.

a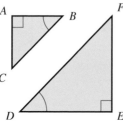

b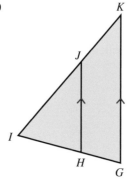

2 Triangles C and D are similar.

a Copy and complete the table showing the pairs of corresponding sides.

C	D	$\dfrac{C}{D}$
3		
5		
y		

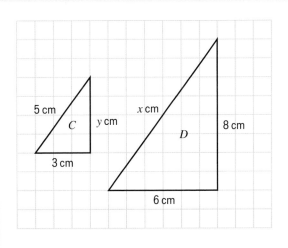

b Use the scale factor to work out x and y.

> **Q2b hint**
>
> $\dfrac{5}{x} = \dfrac{\square}{\square}$

3 **Problem-solving** Find the missing length x in these similar shapes.

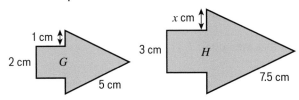

> **Q3 hint**
>
> Draw a table for G and H like the one in **Q2**.

4 **Problem-solving** All these triangles are similar.

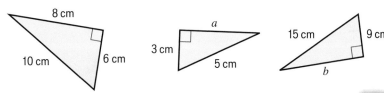

a Sketch the triangles in the same orientation.

b Work out the lengths marked with letters.

> **Q4b hint**
>
> Draw a table.

5 **Reasoning**

a For each triangle, work out $\dfrac{\text{shorter side}}{\text{longer side}}$

Simplify the fraction if possible.

b Use your answer to part **a** to state which of these triangles are similar to triangle A.

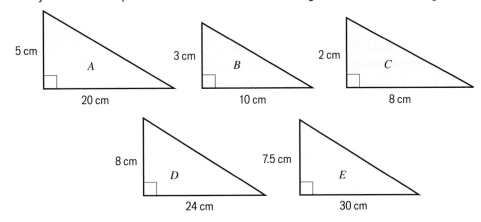

6 **Reasoning** The diagram shows two triangles.

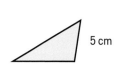

  a Give the reason for

  i $a = f$

  ii $b = e$

  iii $c = d$

  b When three pairs of angles are equal, what does this tell you about the two triangles?

  c Copy and complete the two triangles (sketched separately in the same orientation) with the lengths you know.

  Q6c hint

  Trace each triangle separately and rotate them until they are in the same orientation.

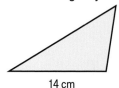

14 cm

5 cm

  d Work out the scale factor and find the missing lengths, x and y.

7 **a** Sketch the diagram. Mark all the angles that are equal.

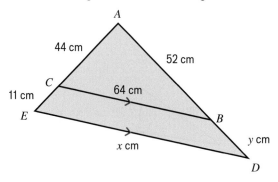

  b Prove that triangles ABC and ADE are similar.

  c Sketch the triangles separately, in the same orientation. Label the lengths of the sides you know.

  d Work out the scale factor and find the missing lengths, x and y.

  Q7b hint

  Show the triangles have the same angles.
  For example,
  angle ACB = angle ____ .
  Give reasons for your answers.

8 **a** Draw a rectangle 2 by 5 on squared paper. Label it A.

  b Draw an enlargement of A with scale factor 2. Label the new shape B.

  c Work out the perimeter of A and the perimeter of B.

  d Copy and complete:

  lengths on A : lengths on B, scale factor is 2.

  perimeter of A : perimeter of B, scale factor is ☐.

  e Enlarge A by scale factor 3. Label the new shape C.

  f Predict the perimeter of C.

  Now work out the perimeter to check your prediction.

9 Triangle A is similar to triangle B.

   a Work out the scale factor from A to B by comparing the given side lengths.

   The perimeter of triangle A is 12 cm.

   b What is the perimeter of triangle B?

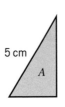

25 cm

5 cm

A

B

10 Draw a 1 cm by 3 cm rectangle on squared paper. Label it A.
Draw an enlargement of A with scale factor 2. Label the new shape B.
How many times will A fit into B?
Copy and complete:
   lengths on A : lengths on B, scale factor is 2
   area of A : area of B, scale factor is $\square = 2^{\square}$

11 The area of triangle A in **Q9** is 6 cm^2. What is the area of triangle B?

Similarity in 3D solids

1 The diagram shows two cubes, with side length 1 cm and side length 2 cm.

   a How many times does A fit into B?

   b Copy and complete:
      lengths on A : lengths on B, scale factor is 2
      volume of A : volume of B, scale factor is $\square = 2^{\square}$

   c Cube A is enlarged to make cube C, so the side length is now 3 cm. What is the scale factor of the lengths for this enlargement?

   d Predict: volume of A : volume of C, scale factor is $\square = \square^{\square}$

   e Draw a sketch to check your prediction.

A

1 cm

B

2 cm

2 Cuboid X and cuboid Y are similar.

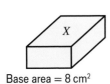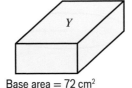

X

Y

Base area = 8 cm^2 Base area = 72 cm^2

The area of the base of cuboid X is 8 cm^2.
The area of the base of cuboid Y is 72 cm^2.

   a Work out the area scale factor to change X to Y.

   b Square root the area scale factor to find the linear scale factor.

   c Find the volume scale factor.

The volume of cuboid X is 12.5 cm^3.

   d Find the volume of cuboid Y.

Q2d hint

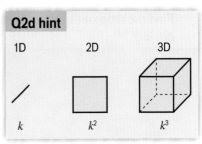

1D   2D   3D

k k^2 k^3

12 Extend

1 **Problem-solving** The vertices of triangle ABC are at $(0.5, 0)$, $(3.5, 0)$ and $(3.5, 4)$.
The vertices of triangle DEF are at $(-2, -2)$, $(5.5, -2)$ and $(5.5, 8)$.
Show that the two triangles are similar.

2 **Problem-solving** $KLMN$ is a parallelogram such that $KN = LM$ and $KL = MN$.
The diagonals intersect at O.
Prove that the point O is the midpoint of each diagonal in the parallelogram.

> **Q2 hint**
>
> Sketch the parallelogram. Use triangles NKO and LMO and show that MO and KO are the same length. Do the same for the other pair of triangles.

3 **Problem-solving** A parallelogram $ABCD$ has sides 8 cm and 15 cm.
One pair of angles is twice the size of the other pair.
The diagonal, AC, joins the bigger angles.

   a What is the size of angle ABC?

   b Prove that triangles ABC and ACD are congruent.

> **Q3a hint**
>
> Sketch a diagram. Label all angles in terms of x.

> **Q3b hint**
>
> Use the diagram you sketched for part **a** and the conditions for congruence.

4 Cylinders Y and Z are similar.
The volume of Y is $6\pi\,\text{cm}^3$.
The volume of Z is $93.75\pi\,\text{cm}^3$.
Calculate the length of the radius of cylinder Z.

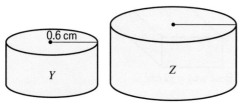

Exam-style question

5 The circumference of circle Y is 80% of the circumference of circle X.
Find the ratio of the area of circle X to the area of circle Y. **(2 marks)**

> **Exam tip**
>
> The question does not ask for the ratio to be in simplified form so your final answer can be a ratio that includes a decimal.

6 **Problem-solving** D and E are regular pentagonal prisms.
They are mathematically similar.
Prism D has cross-sectional area $15.5\,\text{cm}^2$.
The side length of pentagon D is $3\,\text{cm}$.
The side length of pentagon E
is $10.5\,\text{cm}$.
The length of prism D is $20\,\text{cm}$.

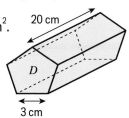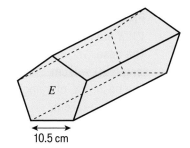

a Work out the volume of prism E.

b How many prisms the same size as prism E could be made from $1\,\text{m}^3$ of plastic?

Exam-style question

7 A statue has a mass of $840\,\text{kg}$.
A similar statue made out of the same material is $\frac{2}{5}$
of the height of the first statue.
What is the mass of the small statue? **(3 marks)**

Q7 hint

Mass = volume × density, so
mass and volume are in direct
proportion.

8 Square A has sides of length $a\,\text{cm}$.
Square B has sides of length $b\,\text{cm}$.
The area of square A is 21% greater than the area of square B.
Work out the ratio $a : b$.

9 **Reasoning** A **frustum** is a cone or pyramid
with the point cut off, parallel to the base.
The diagram shows a frustum.

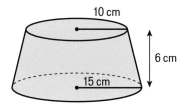

Q9a hint

Imagine the whole cone, before the top was
cut off to make the frustum. Use similar
triangles to work out the heights.

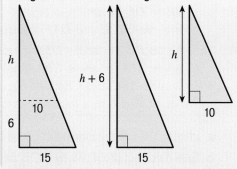

a Find the height of the whole cone.

b Find the volume of the whole cone.

c Find the volume of the smaller cone
(the bit that is missing).

d Hence, find the volume of the frustum.
Give your answer correct to 3 significant
figures.

Q9b hint

Volume of a cone =
$\frac{1}{3}$ × area of base × vertical height
Leave your answer in terms of π.

10 **Problem-solving** This is the frustum of a square-based pyramid.
The length of the base is $18\,\text{cm}$.
The length of the top of the frustum is $7.5\,\text{cm}$.
The vertical height of the frustum is $20.4\,\text{cm}$.
Find the volume of the frustum.
Give your answer correct to 3 significant figures.

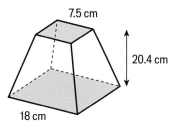

11 Problem-solving Triangles STU and VWX are mathematically similar.
The base, SU, of triangle STU has length $2(x-1)$ cm.
The base, VX, of triangle VWX has length (x^2-1) cm.
The area of triangle STU is $8\,cm^2$.
The area of triangle VWX is $A\,cm^2$.
Prove that $A = 2x^2 + 4x + 2$.

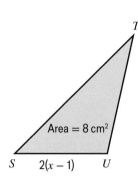
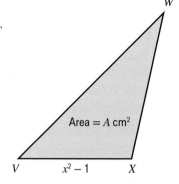

Area = 8 cm² Area = A cm²

S $2(x-1)$ U V x^2-1 X

Working towards A level

12 A student attempted this question.

> $ABCD$ is a parallelogram.
> E and F are points on AB and CD such that $BE = DF$.
>
> **i** Prove that triangles BCE and DAF are congruent.
>
> **ii** Deduce that EC is parallel to AF.

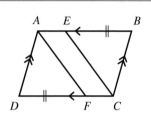

Here is the student's answer.

> **i**   Angle EBC = angle FDA (opposite angles of a parallelogram)
> $BC = DA$ (opposite sides of a parallelogram)
> $BE = DF$ (given)
> Triangles BCE and DAF are congruent (SAS)
>
> **ii** Angle BEC = angle DFA (since the triangles BCE and DAF are congruent)
> Angle AEC = angle CFA (angles on a straight line from angles BEC and DFA)
> Angles AEC and CFA are equal opposite angles of quadrilateral $AECF$
> hence $AECF$ is a parallelogram and so EC must be parallel to AF.

a Identify the false reasoning in part **ii** of the student's answer.

b Give a correct proof of why EC is parallel to AF.

> At A level, every statement needs to be justified and the proof must not contain any false reasoning.

13 In the diagram, AB is parallel to DC.
$AE = BE$ and $AF = BG$.
Prove that triangles ACG and BDF are congruent.

Q13 hint

Always accompany a statement with a reason. Set out your proof line by line and give a reason for congruency.

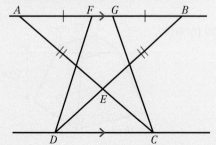

12 Test ready

Summary of key points

To revise for the test:

- Read each key point, find a question on it in the mastery lesson, and check you can work out the answer.

- If you cannot, try some other questions from the mastery lesson or ask for help.

Key points

1 **Congruent** triangles have exactly the same size and shape.
   Their angles are the same and corresponding sides are the same length.
   Two triangles are congruent when one of these conditions of congruence is true.
   - **SSS** (all three sides equal)
   - **SAS** (two sides and the included angle are equal)
   - **AAS** (two angles and a corresponding side are equal)
   - **RHS** (right angle, hypotenuse and one other side are equal)   → **12.1**

2 Conditions of congruence (SSS, SAS, AAS, RHS) can be used for geometric proof, to show that shapes are the same.   → **12.2**

3 When given a description of a shape, it helps to sketch it.   → **12.2, 12.3**

4 Shapes are **similar** when one shape is an enlargement of the other.
   Corresponding angles are equal and corresponding sides are all in the same ratio.   → **12.3**

5 Angles on parallel lines can be used to show that shapes are similar.   → **12.4**

6 Similar shapes that are inside each other sometimes share angles and/or sides.   → **12.4**

7 When a shape is enlarged by linear scale factor k
   - perimeter is multiplied by scale factor k
   - area is multiplied by scale factor k^2 → **12.4**

8 When the linear scale factor is k
   - lengths are multiplied by k
   - area is multiplied by k^2
   - volume is multiplied by k^3 → **12.5**

Sample student answers

Exam-style question

1 Solids *A* and *B* are mathematically similar.
The ratio of the volume of solid *A* to the volume of solid *B* is 64 : 729.
The surface area of solid *A* is 72 cm^2.
Work out the surface area of solid *B*. **(3 marks)**

64 : 729

8 : 27

×9 ⟨ 72 : ⟨243⟩

a Is this student's working clear? Explain.

b The student's working is incorrect. What mistakes has the student made?

Exam-style question

2 *PQR* is an equilateral triangle.
PS is the perpendicular bisector of *QR*.
QX is the angle bisector of angle *PQR*.
Show that triangle *QXS* is similar to
triangle *PRS*. **(2 marks)**

Diagram **NOT** drawn accurately

<u>Triangle *PRS*</u>
The line *PS* is perpendicular to *RS* so ∠*PSR* is a right angle.
Angles in an equilateral triangle are 60°, so ∠*PRS* = 60°
The last angle must be 180 − 90 − 60 = 30°

<u>Triangle *QXS*</u>
The line *XS* is perpendicular to *QS* so ∠*QSX* is a right angle.
QX bisects ∠*PQR*, so ∠*SQX* is 30°
The last angle must be 180 − 90 − 30 = 60°
Therefore all the corresponding angles are equal, so the triangles must be similar.

a How has the student's diagram helped check similarity?

b How else has the student clearly 'shown' that the triangles are similar?

12 Unit test

1 *JKL* and *PQR* are triangles.
Show that the triangles are not similar. **(4 marks)**

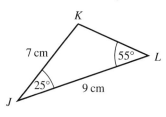
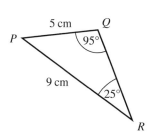

2 Identify two triangles that are congruent and give a reason for your answer. **(2 marks)**

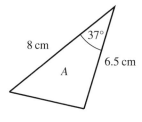
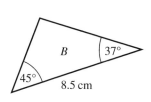

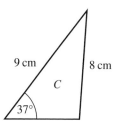
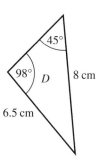

3 *M* and *N* are similar shapes.
Work out the missing lengths, *x* and *y*. **(3 marks)**

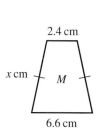
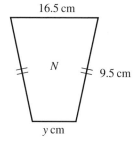

Unit 12 Similarity and congruence   57

4 a Show that triangles *ABC* and *CDE* are similar. **(3 marks)**

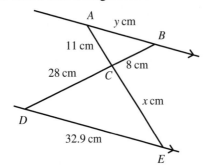

  b Work out the lengths *x* and *y*. **(3 marks)**

5 *ABCD* is a rectangle.
  AC and *BD* are the diagonals of the rectangle, which cross at point *E*.

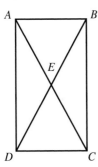

  a Prove that *ABC* and *ABD* are congruent triangles. **(4 marks)**
  b Prove that *E* is the midpoint of *AC* and *BD*. **(2 marks)**

 6 Shapes *B* and *C* are similar.
  The area of *B* is $18\,\text{cm}^2$. The area of *C* is $112.5\,\text{cm}^2$.
  Find the length of side *a*. **(3 marks)**

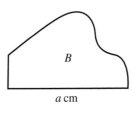 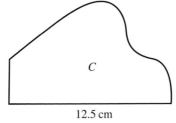

7 a The two triangles in the diagram are similar.

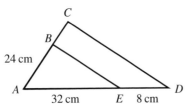

  There are two possible lengths for *BC*.
  Work out each of these values. **(4 marks)**

  b State two assumptions you have made in your working. **(1 mark)**

 8 The circumference of circle *B* is 20% larger than the circumference of circle *A*.
  Write down the ratio of the area of circle *A* to the area of circle *B*. **(2 marks)**

9 *F* and *G* are mathematically similar greenhouses.

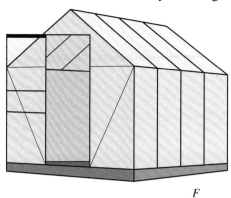

F

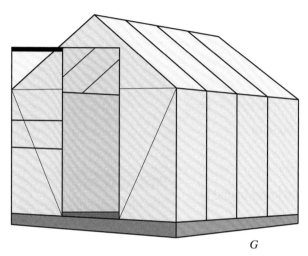

G

F has a volume of $5.6\,\text{m}^3$ and a surface area of $15\,\text{m}^2$.
The volume of *G* is $18.9\,\text{m}^3$.
What is the surface area of *G*? **(4 marks)**

(TOTAL: 35 marks)

 10 Challenge

   a Draw three different right-angled triangles, each with an angle of 45°.
      Are they all similar? Explain.

   b On a calculator, work out the tangent of the 45° angle for each triangle.
      What do you notice?

   c Draw three more right-angled triangles, each with an angle of 60°.
      Are they all similar? Explain.

   d On a calculator, work out the cosine of the 60° angle for each triangle.
      What do you notice?

11 Reflect

   a Are similar shapes congruent? Write a sentence to explain.

   b Are congruent shapes similar? Write a sentence to explain.

13 More trigonometry

Prior knowledge

13.1 Accuracy

Active Learn
Homework

- Understand and use upper and lower bounds in calculations, especially involving trigonometry.

Warm up

1 Fluency The height of a book is measured as 15.4 cm, correct to 1 decimal place. What are the upper and lower bounds of the height of the book?

13.7
15.4

2 Write the possible lengths in **Q1** as an inequality.

$\square$ cm $\leqslant$ length $<$ $\square$ cm

3 $y = 3.6$ to 1 d.p. $z = 9.2$ to 1 d.p. $w = yz$ $x = \dfrac{y}{z}$

  a Find the upper bound and the lower bound of

    i y **ii** z

  b Work out the value of

    i the upper bound of w **ii** the lower bound of w

  c Work out the value of

    i the upper bound of x **ii** the lower bound of x

> **Q3b and c hint**
>
> Which bounds of y and z will give the largest possible value for w? What about for x?

4 ABC and DEF are right-angled triangles.
Work out the missing length on each triangle.
Give your answers to a suitable degree of accuracy.

a

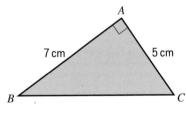

b

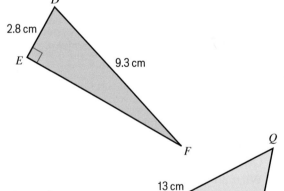

5 PQR is a right-angled triangle.

  a Which trigonometric ratio would you use to work out angle θ?

    [sine] [cosine] [tangent]

  b Work out the size of angle θ.

The upper and lower bounds of a length or angle in a right-angled triangle can be found using Pythagoras' theorem and/or trigonometry.

6 In the diagram, the lengths of AC and BC are given correct to 1 decimal place.

a Find the upper bound for the length of

i AC ii BC

b Use your answers to part **a** to work out the upper bound of x.

c Find the lower bound for the length of

i AC ii BC

d Use your answers to part **c** to work out the lower bound of x.

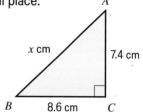

Exam-style Q6.

a i AC $UB = 7.45$ cm

ii BC $UB = 8.65$ cm

b $UB_x = \sqrt{7.45^2 + 8.65^2} = 11.46$ cm

7 Judith

She measures the circumference and the diameter
of a ci b f
She m $UB_x = $
nearest millimetre.
She measures the diameter, d, as 51 mm to the nearest
millimetre.

Judith uses $\pi = \dfrac{C}{d}$ to find the value of π.

Calculate the upper bound and the lower bound for
Judith's value of π. **(4 marks)**

Exam tip

Clearly state in your working
what values you are finding.
For example:
Upper bound for d = ____

8 In this diagram, the measurements are correct to 3 significant figures.

a Find the upper bound for the length of

i AB ii BC

b Find the lower bound for the length of

i AB ii BC

c Copy and complete to write the upper bound of the trigonometric
ratio to find x.

$$\_\_ \, x = \frac{\square}{\square}$$
$$= \square$$

Q8c hint

The upper bound of a fraction $= \dfrac{\text{upper bound of the numerator}}{\text{lower bound of the denominator}}$

For your final answer, write down all the figures on your calculator display.

d Copy and complete to write the lower bound of the trigonometric ratio to find x.

$$\_\_ \, x = \frac{\square}{\square}$$
$$= \square$$

For your final answer, write down all the figures on your calculator display.

e Use your answers to parts **c** and **d** to write the possible values of x as an inequality.

$$\square° \leqslant x < \square°$$

Round the values of x to 3 decimal places.

f **Reflect** Does the upper bound for $\cos x$ give the upper bound for x?

9 In each of these diagrams, the measurements are correct to 3 significant figures.

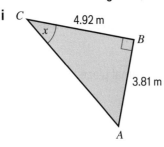
i

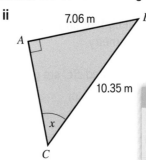
ii

Q9g hint

Round the upper and lower bounds for x to a level of accuracy so that they are the same value.

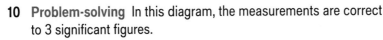

a–e Answer parts **a–e** in **Q8** for each diagram.

f Reflect Do the upper bounds for $\tan x$ and $\sin x$ give the upper bound for x?

g Reasoning For each diagram, give the value of x to a suitable level of accuracy.

10 Problem-solving In this diagram, the measurements are correct to 3 significant figures.

a Find the upper and lower bounds for the value of x, to 3 decimal places.

b Write x to a suitable level of accuracy.

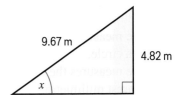

11 Problem-solving In this diagram, the measurements are correct to 2 significant figures.

a Find the upper and lower bounds for the value of x, to 3 decimal places.

b Write x to a suitable level of accuracy.

c Reflect Compare your answers with those for **Q10**.
Write a sentence stating how the upper and lower bounds for x are affected by reducing the accuracy of the measurements to 2 significant figures.

12 In each of these diagrams, the measurements are correct to 2 significant figures.
Find the upper and lower bounds for the value of x.

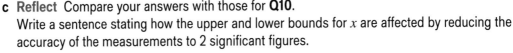

Q12 hint

The angle and the length are correct to 2 significant figures.

a

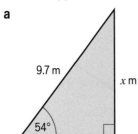

b

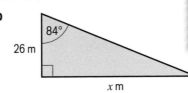

Exam-style question

13 In this diagram, the lengths are correct to the nearest centimetre. The angle is exact. Find the upper and lower bounds for the length l. **(6 marks)**

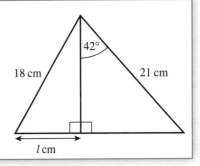

Exam tip

When a diagram includes a right angle and you are asked to find a length, you usually need to use Pythagoras' theorem or trigonometry (or both) to solve the problem.

13.2 Graph of the sine function

Active Learn
Homework

- Understand how to find the sine of any angle.
- Know the graph of the sine function and use it to solve equations.

Warm up

1 **Fluency** What is the exact value of

  a $\sin 30°$ **b** $\sin 45°$ **c** $\sin 60°$ **d** $\sin 90°$?

2 Here is a right-angled triangle.

  a Write down the value of $\sin \theta$.

  b Find the size of angle θ.

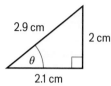

2.9 cm, 2 cm, θ, 2.1 cm

3 Use this graph to estimate two values of x when $y = 0.5$.

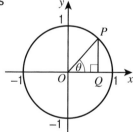

4 The diagram shows a circle drawn on axes with $-1 \leqslant x \leqslant 1$ and $-1 \leqslant y \leqslant 1$. The centre of the circle is at $(0, 0)$.

  a What is the radius of the circle?

  b What is the length OP?

  OPQ is a right-angled triangle.

  c Write the sine ratio for angle θ in terms of PQ.

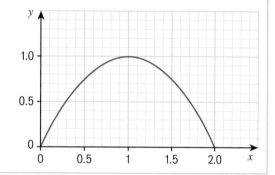

Q4c hint

$\sin \theta = $ ____

5 **Reasoning** The diagram shows a circle of radius 1 unit with centre at $(0, 0)$. Explain how you can use the diagram to work out the value of $\sin 30°$.

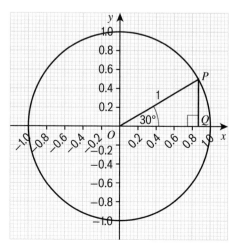

6 **Reasoning** $\sin 30° = 0.5$

a Use the diagram to find an obtuse angle θ such that $\sin \theta = 0.5$.

b For what two values of θ on the diagram does $\sin \theta = -0.5$?

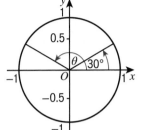

Q6 hint

Use symmetry.

Key point

The length of PQ gives the **sine** of the angle.
This is shown on the vertical axis by the position of the arrow.
You can find the sine of any angle using this method.

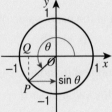

---

7 **Problem-solving** Find the value of $\sin \theta$ on each diagram.

a 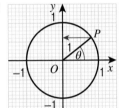 b 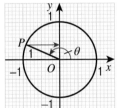 c 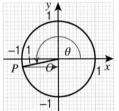 d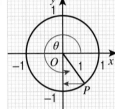

8 **Reasoning**

a What is the largest value that $\sin \theta$ can take?

b What is the smallest value that $\sin \theta$ can take?

c Find two values of θ so that $\sin \theta = 0$.

Q8 hint

Look at the diagrams in **Q7**.

9 **Reasoning** As θ increases from 0° to 90°, $\sin \theta$ increases from 0 to 1.
Copy and complete these statements in the same way.

a As θ increases from 90° to 180°, $\sin \theta$ _____ .

b As θ increases from 180° to 270°, $\sin \theta$ _____ .

c As θ increases from 270° to 360°, $\sin \theta$ _____ .

10 **Problem-solving / Reasoning**
Here is the graph of $y = \sin x$ for $0° \leqslant x \leqslant 180°$.

a Use the graph to i find $\sin 90°$ ii estimate $\sin 75°$

b Copy and complete to describe the symmetry of the curve.
The graph of $y = \sin x$ for $0° \leqslant x \leqslant 180°$ is symmetrical about $x = \square°$

c i Use the graph to find $\sin x$ when $x = 60°$
ii What other value of x on the graph has the same $\sin x$ value as 60°?

d Copy and complete, inserting numbers greater than 90.
  i $\sin 0° = \sin \square°$ ii $\sin 45° = \sin \square°$ iii $\sin 15° = \sin \square°$

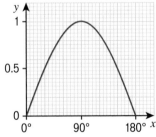

11 Here is the graph of
$y = \sin x$ for $0° \leqslant x \leqslant 360°$.

  a Describe the symmetry of the curve.

  b Use the graph to check your answer to **Q6**.

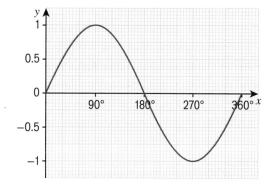

12 The graph of $y = \sin x$ repeats every $360°$ in both directions.
Sketch the graph of $y = \sin x$ for $0° \leqslant x \leqslant 720°$.
Include x values $0°$, $90°$, $180°$, $270°$, $360°$, $450°$, $540°$, $630°$, $720°$ and y values 1, 0.5, 0, -0.5, -1.

13 **Problem-solving** Use your sketch from **Q12** to find **a** $\sin 540°$ **b** $\sin 450°$

14 **a** **Problem-solving** The exact value of $\sin 60°$ is $\dfrac{\sqrt{3}}{2}$.
Use your sketch from **Q12** to write down the exact value of **i** $\sin 420°$ **ii** $\sin 480°$

  b **Reasoning** Explain how you worked out your answers to part **a**.

15 **a** **Problem-solving** Use your sketch from **Q12** to find four values of x such that

   i $\sin x = -0.5$ **ii** $\sin x = -\dfrac{\sqrt{3}}{2}$

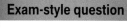

  b Check your answers to part **a** using your calculator.

Exam-style question

16 Here is a sketch of $y = \sin x$.

Write down the coordinates of each of the labelled points. **(4 marks)**

Exam tip

Look for opportunities to check your answers are correct. For example, for this question you can check your answers by making sure $\sin x = y$ on your calculator.

17 **a** Rearrange the equation $5 \sin x = 3$ to make $\sin x$ the subject.

  b Use your answer to part **a** and your calculator to solve the equation $5 \sin x = 3$.

  c Use the sketch to solve the equation $5 \sin x = 3$ for all values of x in the interval $0°$ to $540°$.

Q17a and b hint

$$\sin x = \frac{\square}{\square}$$

$$x = \sin^{-1}\left(\frac{\square}{\square}\right)$$

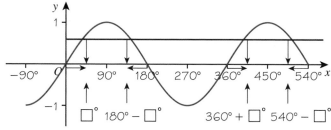

18 **Problem-solving / Reasoning** Solve these equations for all values of x in the interval $0°$ to $720°$.

  a $8 \sin x = 2.5$ **b** $6 \sin x = 5$

13.3 Graph of the cosine function

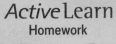

Active Learn
Homework

- Understand how to find the cosine of any angle.
- Know the graph of the cosine function and use it to solve equations.

Warm up

1 Fluency What is the exact value of

    a $\cos 30°$ **b** $\cos 45°$ **c** $\cos 60°$ **d** $\cos 90°$?

2 Here is a right-angled triangle.

    a Write down the value of $\cos \theta$.

    b Find the size of angle θ.

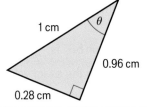

---

3 The diagram shows a circle drawn on axes
with $-1 \leqslant x \leqslant 1$ and $-1 \leqslant y \leqslant 1$.
The centre of the circle is at $(0, 0)$.

    a What is the length OP?

    OPQ is a right-angled triangle.

    b Write the cosine ratio for angle θ in terms of OQ.

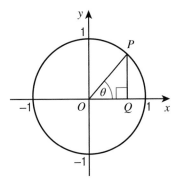

4 Reasoning The diagram shows a circle of radius 1 unit
with centre at $(0, 0)$.

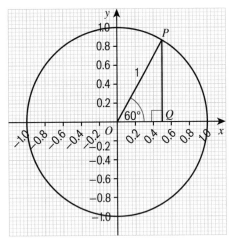

    a Explain how you can use the diagram to work out the value of
$\cos 60°$.

    b Write down the value of $\cos 60°$.

    c Use the diagram to find

        i a reflex angle θ with the same cosine value as $60°$

        ii two angles with the same cosine value as $-\cos 60°$

Q4c hint

Use symmetry.

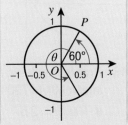

5 **Reasoning** $\cos 30° = \dfrac{\sqrt{3}}{2}$

    a Find a reflex angle θ such that $\cos \theta = \dfrac{\sqrt{3}}{2}$.

    b Find an obtuse angle θ such that $\cos \theta = -\dfrac{\sqrt{3}}{2}$.

Q5 hint

Draw a diagram and make use of symmetry.

> **Key point**
>
> The length of OQ gives the **cosine** of the angle.
> This is shown on the horizontal axis by the position of the arrow.
> You can find the cosine of any angle using this method.

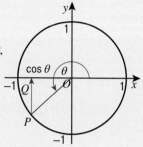

6 **Problem-solving** Find the value of $\cos \theta$ on each diagram.

a 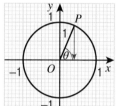   **b** 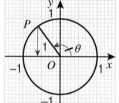   **c** 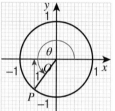   **d**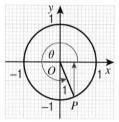

7 **Reasoning** As θ increases from $0°$ to $90°$, $\cos \theta$ decreases from 1 to 0.
Write similar statements for θ increasing from

    a $90°$ to $180°$ **b** $180°$ to $270°$ **c** $270°$ to $360°$

8 **Problem-solving / Reasoning**
Here is the graph of $y = \cos x$ for $0° \leqslant x \leqslant 360°$.

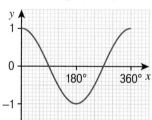

    a Use the graph to
      i find $\cos 180°$ **ii** estimate $\cos 150°$

    b Describe the symmetry of the curve.

    c **i** Use the graph to find $\cos x$ when $x = 60°$.
       ii What other value of x on the graph has the same $\cos x$ value as $60°$?

    d Copy and complete.
      i $\cos 90° = \cos \square°$
      ii $\cos 120° = \cos \square°$
      iii $\cos 0° = \cos \square°$

    e Use the graph to give estimates for both solutions to $\cos x = -0.3$.

9 The graph of $y = \cos x$ repeats every 360° in both directions.
Sketch the graph of $y = \cos x$ for $0° \leqslant x \leqslant 720°$.
Include x values 0°, 90°, 180°, 270°, 360°, ..., 720° and y values 1, 0.5, 0, −0.5, −1.

10 Problem-solving

   a Use your sketch from **Q9** to find

      i $\cos 300°$

      ii $\cos 480°$

   b Find four values of x such that $\cos x = 0.5$.

11 Problem-solving The exact value of $\cos 30°$ is $\dfrac{\sqrt{3}}{2}$.

   a Use your sketch from **Q9** to write down the exact value of

      i $\cos 390°$

      ii $\cos 210°$

   b Find four values of x such that $\cos x = -\dfrac{\sqrt{3}}{2}$.

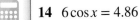 **12** Check your answers to **Q10b** and **Q11b** using a calculator.

 13 Problem-solving / Reasoning $15 \cos \theta = -6.8$

   a Use your calculator to find one value of θ.

   b Use your sketch from **Q9** to solve $15 \cos \theta = -6.8$ for all values of θ in the interval 0° to 720°.

Q13a hint

$\cos \theta = \dfrac{\square}{\square}$

Q13b hint

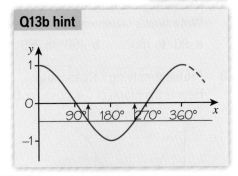

Exam-style question

14 $6 \cos x = 4.86$

   a Use your calculator to find one value of x.   **(2 marks)**

   b Solve $6 \cos x = 4.86$ for all values of x in the interval 0° to 720°.   **(4 marks)**

Exam tip

It always helps to sketch a graph when finding more than one value of x for a trigonometric ratio.

13.4 Graph of the tangent function

Active Learn
Homework

- Understand how to find the tangent of any angle.
- Know the graph of the tangent function and use it to solve equations.

Warm up

1 **Fluency** What is the exact value of

    a $\tan 30°$ **b** $\tan 45°$ **c** $\tan 60°$?

2 Here is a right-angled triangle.

    a Write down the value of $\tan \theta$.

    b Find the size of angle θ.

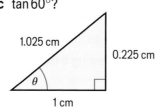

3 Which of these is equivalent to $\dfrac{1}{\sqrt{3}}$?

    $\boxed{\dfrac{3}{\sqrt{3}}}$ $\boxed{\dfrac{\sqrt{3}}{3}}$ $\boxed{\dfrac{3\sqrt{3}}{3}}$

Key point

The diagram shows a circle of radius 1 unit with centre at $(0, 0)$.

A vertical line is drawn, perpendicular to the x-axis and touching the circle only once.

OP is extended to meet the vertical line and gives the value of $\tan \theta$.

$$\tan \theta = \frac{0.8}{1} = 0.8$$

You can find the **tangent** of any angle using this method except for angles of the form $90° \pm 180n°$.

Unlike sine and cosine, the tangent can take *any* value, positive or negative, not just values between -1 and 1.

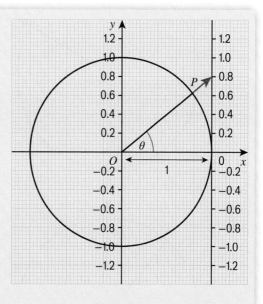

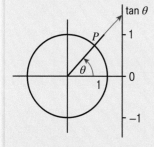 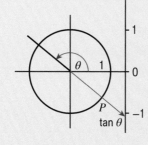

4 **Reasoning** $\tan 60° = \sqrt{3}$

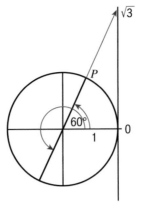

a Use the diagram to find a reflex angle θ such that $\tan \theta = \sqrt{3}$.

b For what two values of θ on the diagram does $\tan \theta = -\sqrt{3}$?

Q4 hint

Use symmetry.

5 **Reasoning** $\tan 45° = 1$

a Find a reflex angle θ such that $\tan \theta = -1$.

b Find an obtuse angle such that $\tan \theta = -1$.

c **Reflect** The hint suggested drawing a diagram to help you answer parts **a** and **b**.
Did it help? How?

Q5a and b hint

Draw a diagram like the one in **Q4** and use symmetry.

6 **Problem-solving** Find the value of $\tan \theta$ in each diagram.

a

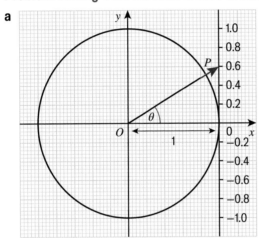

b

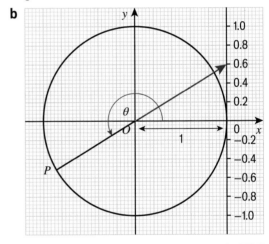

c

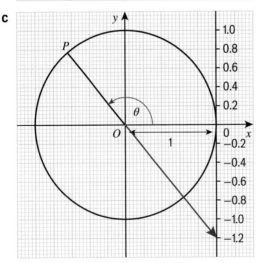

d

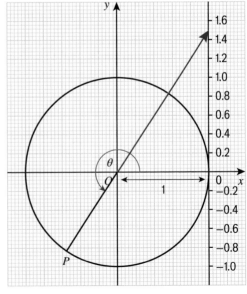

7 **Reasoning** As θ increases from $0°$ to $90°$, $\tan\theta$ increases from 0 to infinity.

    a Write similar statements for

      i θ decreasing from $180°$ to $90°$

      ii θ increasing from $180°$ to $270°$

      iii θ decreasing from $360°$ to $270°$

    b **Reflect** $\tan\theta$ is not defined when θ is exactly $90°$.
    Write a sentence explaining why you think this is.

    c **i** Write at least one other value of θ for which $\tan\theta$ is not defined.

      ii Check your answer to part **c i** by using your calculator to try to find $\tan\theta$ for this value of θ.

Q7 hint

'Infinity' is the word mathematicians use for a value that is greater than any other number. The symbol for infinity is ∞.

8 **Problem-solving / Reasoning** Here is the graph of $y = \tan x$ for $0° \leqslant x \leqslant 360°$.

    a How often does the graph repeat?

    b Use the graph to estimate the value of

      i $\tan 60°$ **ii** $\tan 300°$

    c Describe the symmetry of the curve.

    d Copy and complete, inserting numbers greater than 180.

      i $\tan 60° = \tan\square°$

      ii $\tan 100° = \tan\square°$

      iii $\tan 120° = \tan\square°$

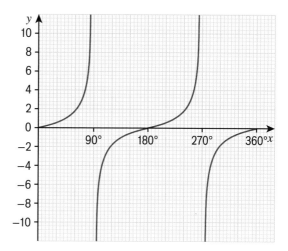

9 Sketch the graph of $y = \tan x$ for $0° \leqslant x \leqslant 720°$.

10 **Problem-solving**

    a Use your sketch from **Q9** to find **i** $\tan 540°$ **ii** $\tan 405°$

    b The exact value of $\tan 60°$ is $\sqrt{3}$. Write down the exact value of **i** $\tan 240°$ **ii** $\tan 120°$

    c **Reasoning** Explain how you worked out your answers to part **b**.

11 **a** **Problem-solving** Use your sketch from **Q9** to find four values of x such that

      i $\tan x = 1$ **ii** $\tan x = -1$

    b Check your answers to part **a** using a calculator.

12 **Problem-solving / Reasoning** $3\tan x = 11$

    a Use your calculator to find one value of x.

    b Use your sketch from **Q9** to solve $3\tan x = 11$ for all values of x in the interval $0°$ to $720°$.

Q12a hint

$\tan x = \dfrac{\square}{\square}$

Exam-style question

13 **a** Sketch the graph of $y = \tan x$ in the interval $0°$ to $720°$.   **(2 marks)**

    b Given that $\tan 30° = \dfrac{1}{\sqrt{3}}$, solve the equation

    $3\tan x = \sqrt{3}$ in the interval $0°$ to $720°$.   **(2 marks)**

Exam tip

You are expected to use your sketch from part **a** to solve the equation in part **b**.

13.5 Calculating areas and the sine rule

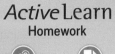

Active Learn

Homework

- Find the area of a triangle and a segment of a circle.
- Use the sine rule to solve 2D problems.

Warm up

1 Fluency Sketch the graph of $y = \sin x$ for $0° \leqslant x \leqslant 360°$.

2 Work out the area of each shape.
Give your answers correct to 1 decimal place where necessary.

a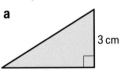
3 cm
6 cm

b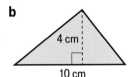
4 cm
10 cm

c
5 cm

d
7 cm

e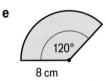
120°
8 cm

3 a Write the equation $\frac{1}{4}x(x+3) = 7$ in the form $ax^2 + bx + c = 0$, where a, b and c are integers.

b Solve the equation to find two possible values for x.

4 Calculate the perpendicular height, h, of this triangle.
Give your answer correct to 2 decimal places.

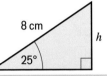
8 cm
h
25°

5 a Write h, the perpendicular height of the triangle, in terms of p and θ.

b Write a formula in terms of p and θ to calculate the area of the triangle.

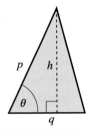
p h
θ
q

> **Q5a hint**
> You need to use trigonometry, as in **Q4**.

Key point

The **area** of this triangle $= \frac{1}{2}ab \sin C$.
C is the given angle.
a is the side opposite angle A.
b is the side opposite angle B.

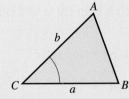
A
b
C a B

---

6 Sketch these triangles. For each triangle

i label vertex C (the given angle)

ii label vertices A and B and their opposite sides a and b
(it doesn't matter which way round you label vertices A and B)

iii calculate its area using the formula Area of a triangle $= \frac{1}{2}ab \sin C$.
Give your answers correct to 3 significant figures.

a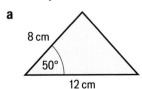
8 cm
50°
12 cm

b
2.3 m 132° 1.8 m

7 The area of each of these triangles is 38.8 cm². For each triangle, work out the value of x.
Give your answers correct to 3 significant figures.

a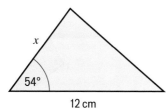

b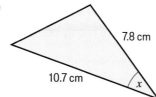

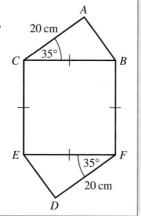

Exam-style question

8 The diagram shows a square *BCEF* and two congruent triangles, *ABC* and *DEF*.

Area of square =
    area of triangle *ABC*
    + area of triangle *DEF*
Work out the length of *BC*.
Give your answer correct to 1 decimal place. **(3 marks)**

Exam tip

Sometimes you are not given all the values to substitute into a formula. For example, write the area of the square in terms of *BC*. Then, for triangle *ABC*, substitute 20, 35° and *AC* into the formula
Area of a triangle $= \frac{1}{2}ab \sin C$.

9 **a** Find the area of triangle *AOB* in this circle.

b Find the area of the sector *AOB*.

c Use your answers to parts **a** and **b** to find the area of the shaded segment of the circle.

Give your answers correct to 2 decimal places.

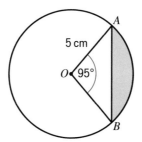

Q9b hint

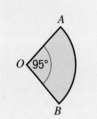

10 **Problem-solving** *ABC* is an arc of a circle centre *O* with radius 60 m.
AC is a chord of the circle.
Angle $AOC = 35°$
Calculate the area of the shaded region.
Give your answer correct to 3 significant figures.

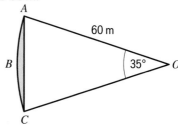

11 **Problem-solving**

a Calculate the size of angle *AOB*.
Give your answer correct to 1 decimal place.

b Work out the area of the shaded segment.
Give your answer correct to 3 significant figures.

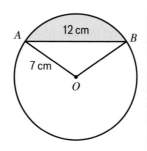

Q11a hint

Sketch the triangle. Split it into two right-angled triangles and mark on it any values you know.

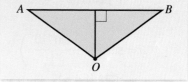

12 Problem-solving The area of triangle ABC is $5\,\text{cm}^2$.

    a Use this information to write an equation in the form $ax^2 + bx + c = 0$.

    b Use the equation in part **a** to find a solution for x.

    c Reasoning Explain why there is only one possible value for x.

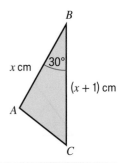

Key point

The **sine rule** can be used in any triangle.

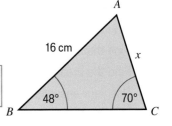

- $\dfrac{a}{\sin A} = \dfrac{b}{\sin B} = \dfrac{c}{\sin C}$ Use this to calculate an unknown *side*.

- $\dfrac{\sin A}{a} = \dfrac{\sin B}{b} = \dfrac{\sin C}{c}$ Use this to calculate an unknown *angle*.

To use the sine rule you need to know one angle and the opposite side. Then:

- if you know another *angle*, you can work out the length of its opposite *side*
- if you know another *side*, you can work out the size of its opposite *angle*.

Example

Find the length of the side labelled x in this triangle.
Give your answer correct to 3 significant figures.

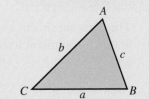

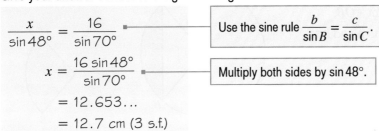

$$\frac{x}{\sin 48°} = \frac{16}{\sin 70°}$$ Use the sine rule $\dfrac{b}{\sin B} = \dfrac{c}{\sin C}$.

$$x = \frac{16\sin 48°}{\sin 70°}$$ Multiply both sides by $\sin 48°$.

$$= 12.653\ldots$$

$$= 12.7\text{ cm (3 s.f.)}$$

---

13 Find the length of the side labelled x in each diagram.
Give your answers correct to 3 significant figures.

a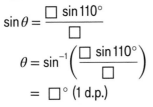

b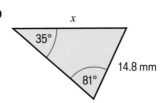

c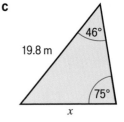

---

14 Copy and complete the sine rule to find the size of angle θ.

$$\frac{\sin\theta}{\square} = \frac{\sin 110°}{\square}$$

$$\sin\theta = \frac{\square\,\sin 110°}{\square}$$

$$\theta = \sin^{-1}\left(\frac{\square\,\sin 110°}{\square}\right)$$

$$= \square°\text{ (1 d.p.)}$$

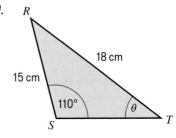

15 Find the size of angle θ in each diagram.
Give your answers correct to 1 decimal place.

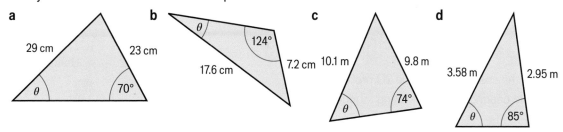

a   29 cm   23 cm θ 70°

b θ 124°   17.6 cm   7.2 cm

c   10.1 m   9.8 m θ 74°

d   3.58 m   2.95 m θ 85°

16 a Work out the length of AC.
Give your answer correct to 3 significant figures.

b Work out the size of angle BAC.
Give your answer correct to 1 decimal place.

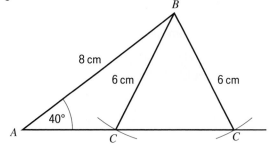

B 95°   7.1 cm C 6.7 cm   33°   67° A D

17 In triangle ABC, $AB = 8$ cm, $BC = 6$ cm and angle $BAC = 40°$.

a Use the sine rule to find the value of $\sin C$.

The diagram shows that there are two
possible triangles with an angle of 40°,
an adjacent side of length 8 cm and an
opposite side of length 6 cm.
Hence there are two possible sizes for the
angle ACB.

b Use the sine graph to work out the two
possible values.
Give each answer correct to 1 decimal place.

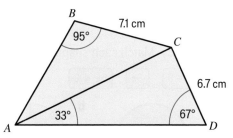

B 8 cm   6 cm   6 cm   40° A C C

18 In triangle XYZ, $XY = 12$ cm, $YZ = 9.5$ cm and angle $YXZ = 50°$.
Work out the size of angle XZY.
There are two possible answers. Give each of them correct to 1 decimal place.

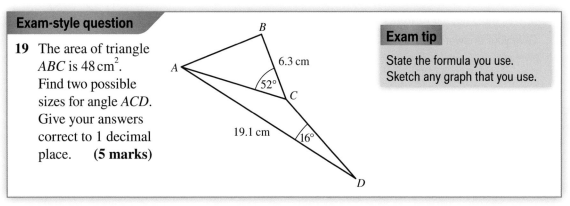

B 6.3 cm A 52° C 19.1 cm   16° D

13.6 The cosine rule and 2D trigonometric problems

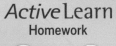

Active Learn
Homework

- Use the cosine rule to solve 2D problems.
- Solve bearings problems using trigonometry.

Warm up

1 Fluency Match each shape below to the correct perimeter.

$2\pi + 8$ 8π $6\pi + 8$

a

4 cm

b

4 cm

c

4 cm
270°

2 In the diagram, what is the bearing of
   a B from A
   b A from B?

Q2b hint

First use the parallel north lines to find this angle

3 Work out the positive value of x when $x^2 = 9^2 + 7^2 - 2 \times 9 \times 7 \times \cos 34°$.
   Give your answer correct to 3 significant figures.

4 a Work out the area of this triangle.
   b Use the sine rule to work out the size of the angle labelled x.
   Give your answers correct to 3 significant figures.

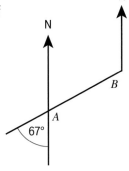

Key point

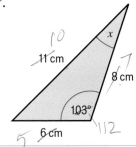

The **cosine rule** can be used in any triangle.
- $a^2 = b^2 + c^2 - 2bc \cos A$ Use this to calculate an unknown *side*.
- $\cos A = \dfrac{b^2 + c^2 - a^2}{2bc}$ Use this to calculate an unknown *angle*.

You can use the cosine rule to find:
- the length of a *side* if you know two sides and the included angle
- the size of an unknown *angle* if you know all three sides.

Work out the length of the side labelled x in this triangle.
Give your answer correct to 3 significant figures.

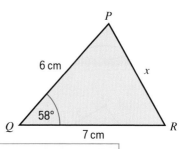

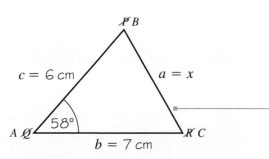

Sketch the triangle. Label the missing side a, and the others b and c.

$a^2 = b^2 + c^2 - 2bc \cos A$

Use the cosine rule to find the side.

$x^2 = 7^2 + 6^2 - 2 \times 7 \times 6 \times \cos 58°$
$ = 40.486...$
$x = \sqrt{40.486...}$
$ = 6.3629...$
$ = 6.36\,\text{cm}$ (3 s.f.)

5 Find the length of the sides marked with letters in these diagrams.
Give your answers correct to 3 significant figures.

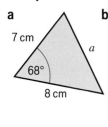

a 7 cm, 68°, 8 cm, a

b b, 11.6 cm, 28°, 16.3 cm

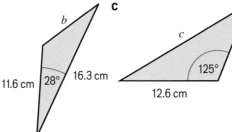

c c, 11.2 cm, 125°, 12.6 cm

d 6.75 m, 102°, 9.26 m, d

6 a Sketch this triangle. Label

    i the given angle A and its opposite side a

    ii the other sides b and c
    (it doesn't matter which way round you label sides b and c)

  b Use the cosine rule to work out the size of angle y.
Give your answer correct to 1 decimal place.

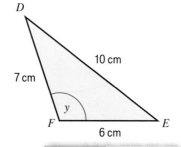

Q6b hint

Use $\cos A = \dfrac{b^2 + c^2 - a^2}{2bc}$.

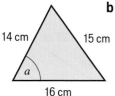

7 Calculate the sizes of the angles marked with letters in these triangles.
Give your answers correct to 1 decimal place.

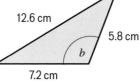

a 14 cm, 15 cm, a, 16 cm

b 12.6 cm, 5.8 cm, b, 7.2 cm

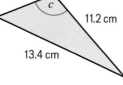

c 5.8 cm, c, 11.2 cm, 13.4 cm

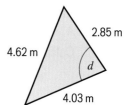

d 2.85 m, 4.62 m, d, 4.03 m

8 In the diagram, O is the centre of the circle of radius 5 cm. AB is a chord of length 8 cm.

Q8 hint

What is the length of OA?

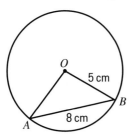

Work out the size of angle AOB.
Give your answer correct to 1 decimal place.

9 Reasoning

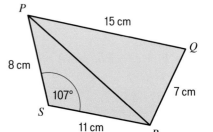

   a Work out the length of PR.
     Give your answer correct to 3 significant figures.

   b Work out the size of angle QPR.
     Give your answer correct to 1 decimal place.

   c Work out the area of quadrilateral $PQRS$.
     Give your answer correct to 3 significant figures.

Q9c hint

Work out the area of each triangle.

10 Problem-solving
This shape is made from a triangle and a sector of a circle, centre O and radius 7 cm.
$AD = 12$ cm
Angle $AOD = 107°$
Angle $OAD = 39°$

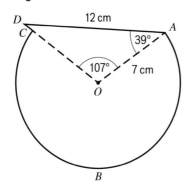

   a Use the cosine rule to work out the length OD.
     Give your answer correct to 3 significant figures.

   b Reasoning Ed says, 'You can use the sine rule to work out the length OD too.'
     Is Ed correct? Explain.

   c Calculate the perimeter of the shape.

11 Reasoning The diagram shows the positions of three towns, A, B and C.

Calculate the bearing of C from A.

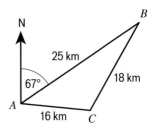

Exam-style question

12 A ship leaves its harbour (H) and sails for 10 km
on a bearing of 054°.
It then sails a further 14 km on a bearing of 148° to reach
port P.

Q12 hint

Sketch the diagram. Join H and
P to create a triangle.

Exam tip

When a diagram includes
parallel lines, first look for angles
that can be found using the
rules of angles on parallel lines.

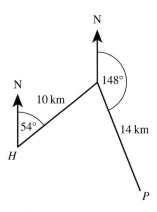

a What is the direct distance between H and P?

(3 marks)

b What is the bearing of P from H?   **(3 marks)**

13 For each triangle, decide whether to use the sine rule or the cosine rule to work out the size of
angle θ. Then work out the angle.

a

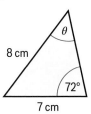

b

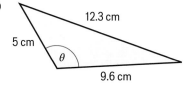

Exam-style question

14 ABC is a triangle.
$BC = 14.3$ cm
Angle $ABC = 75°$
The area of triangle ABC is 48 cm².

Exam tip

Sometimes you need to use
more than one rule involving
sine or cosine to find a missing
length or angle on a triangle.

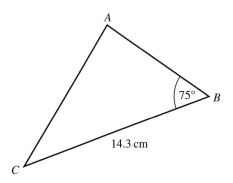

Work out the length of AC.
Give your answer correct to 3 significant figures.

(6 marks)

13.7 Solving problems in 3D

Active Learn
Homework

- Use Pythagoras' theorem in 3D.
- Use trigonometry in 3D.

Warm up

1 Fluency Which rule would you use to work out the value of x in each of these diagrams?

a

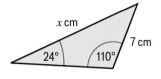
x cm
7 cm
24°
110°

b

x cm
15.8 cm
130°
17.5 cm

2 Find the size of a in each triangle.
Give your answers correct to 1 decimal place.

a

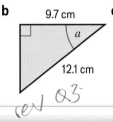
a 5.4 cm
11.6 cm
rev Q2

b

9.7 cm
a
12.1 cm
rev Q3

c

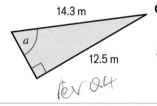
14.3 m
a
12.5 m
rev Q4

d

a
8.7 cm
11.5 cm
rev Q5

Key point

A **plane** is a flat surface. For example, the surface of your desk lies in a horizontal plane; the surface of a wall in your classroom lies in a vertical plane.
In the diagram
- BC is perpendicular to the plane $WXYZ$
- triangle ABC is in a plane perpendicular to the plane $WXYZ$
- θ is the angle between the line AB and the plane $WXYZ$.

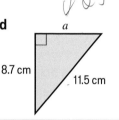

3 $ABCDEFGH$ is a cuboid.
Base $EFGH$ is in a horizontal plane.
A diagonal in the plane $EFGH$ joins vertex E to its opposite vertex, G.
A diagonal in the cuboid $ABCDEFGH$ joins vertex A to its opposite vertex, G.
Triangle AEG is in a vertical plane.
The angle that AG makes with $EFGH$ is θ.

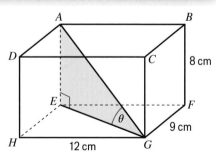

8 cm
9 cm
12 cm

a Sketch the right-angled triangle EGH. Label the lengths that you know.

b Work out the length of diagonal EG.

c Use your answer to part **b** to work out the length of diagonal AG.

d Work out the angle the AG makes with $EFGH$, θ.
Give your answer correct to 1 decimal place.

Q4c and d hint

Look at the right-angled triangle AEG. Label EG with the length you have just found.

4 **Reasoning** *ABCDEFGH* is a cuboid.

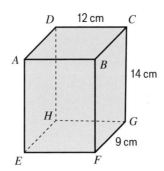

D   12 cm   C

A   B   14 cm

H   G   9 cm

E   F

a Calculate the length of diagonals

  i *FH*

  ii *BH*

  iii *FC*

  iv *CE*

b Find the angle between the diagonal *DF* and the plane *EFGH*.

c Find the angle between the diagonal *GA* and the plane *ABCD*.

d Find the angle between the diagonal *CE* and the plane *AEHD*.

Q4 hint

Sketch separate right-angled triangles using information from the cuboid.

Exam-style question

5 The diagram shows a triangular prism.
The base *CDEF* of the prism is a rectangle with length 12 cm and width 8 cm.
Angle *AED* and angle *AEF* are right angles.

Exam tip

When finding lengths and angles in 3D solids, sometimes you have to work with more than one right-angled triangle.

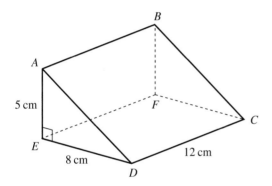

B

A

5 cm   F   C

E

8 cm   12 cm

D

Calculate the angle that the diagonal *AC* makes with the plane *CDEF*.   **(4 marks)**

6 **Reasoning** *ABCDE* is a square-based pyramid with side length 16 cm.
The base *BCDE* lies in a horizontal plane.
AB = AC = AD = AE = 24 cm
AM is perpendicular to the base.

a Calculate the length of

  i *CE*

  ii *CM*

  iii *AM*

Give your answers correct to 3 significant figures.

b Calculate the angle that *AB* makes with the base, correct to the nearest degree.

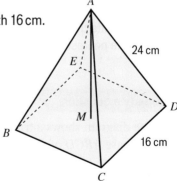

A

24 cm

E

M   D

B

16 cm

C

7 **Reasoning** In the diagram, *ABCD* is a tetrahedron.
$AB = 11\,cm$, $BC = 8\,cm$, angle $ABC = 63°$, angle $CAD = 37°$ and angle $ADC = 68°$.

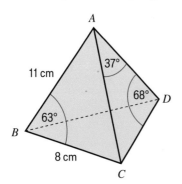

Q7 hint

Sketch each triangle separately. Put as many lengths/angles as possible on your sketch to help you answer the questions.

a Work out the length of *AC*.

b Work out the length of *CD*.

c Given that $BD = 12$ cm, calculate the size of angle *BCD*.

8 The diagram shows a square-based pyramid *ABCDE*.
$BC = 14\,cm$ and $AD = 26\,cm$.
Each triangular face is an isosceles triangle.

a Calculate the size of the angle *ABD*.
Give your answer correct to 1 decimal place.

b Calculate the area of triangle *ABD*.

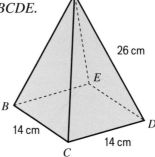

Q8 hint

Sketch the triangle *ABD*. Label it with all the information you know.

9 **Problem-solving** *ABCDE* is a pyramid with a rectangular base.
$AB = AC = AD = AE = 28\,cm$
AM is perpendicular to the base.

a Calculate the size of the angle between *AM* and the face *ACD*, correct to the nearest degree.

b Calculate the size of angle *BAD*, correct to the nearest degree.

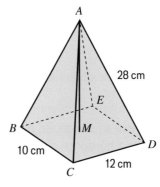

Q9a hint

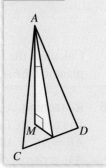

Exam-style question

10 The diagram shows a triangular prism.
The base *ABCD* of the prism is a square of side length $12\,cm$.
The cross-section is an equilateral triangle.
M is the point on *BC* such that $BM : MC = 1 : 2$.

Calculate the size of the angle between *FM* and the base of the prism.
Give your answer correct to 1 decimal place. **(4 marks)**

Exam tip

Always begin with a sketch of the 3D solid. Label it with all the information you know. Then draw a separate sketch of each triangle that you are going to use to find the final answer.

13.8 Transforming trigonometric graphs 1

- Recognise how changes in a function affect trigonometric graphs.

Warm up

1 Fluency What are the coordinates of the image of the point $(3, 5)$ under each of these transformations?

   a reflection in the x-axis

   b reflection in the y-axis

   c rotation through $180°$ about $(0, 0)$

2 Write down the exact value of

   a $\sin 60°$ **b** $\tan 30°$ **c** $\cos 45°$

   d $\sin 0°$ **e** $\cos 90°$ **f** $\tan 60°$

3 Sketch these graphs for values of x from $0°$ to $360°$.

   a $y = \sin x$ **b** $y = \cos x$ **c** $y = \tan x$

4 Here is the graph of $y = \sin x$ for $-180° \leqslant x \leqslant 180°$.

   a Copy and complete the table.

   i Write in the values of x and $\sin x$ at points A to H on the graph.

   ii For each x-value, write in the value of $-\sin x$.

   b Sketch the graph of $y = -\sin x$ for $-180° \leqslant x \leqslant 180°$.

   c Describe how the graph of $y = \sin x$ is transformed to give the graph of $y = -\sin x$

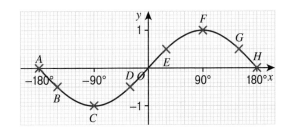

	x	$\sin x$	$-\sin x$
A	$-180°$	0	0
B	$-150°$	-0.5	0.5
C			

Key point

The graph of $y = -f(x)$ is the reflection of the graph of $y = f(x)$ in the x-axis.

5 a Use your table from **Q4**. Add a column for $\sin(-x)$.
   Find the sine values from the graph to fill in the $\sin(-x)$ column.

   b Sketch the graph of $y = \sin(-x)$ for $-180° \leqslant x \leqslant 180°$.

   c Describe the transformation that turns the graph of $y = \sin x$ into the graph of $y = \sin(-x)$.

Key point

The graph of $y = f(-x)$ is the reflection of the graph of $y = f(x)$ in the y-axis.

6 Here is the graph of $y = \tan x$
for $-180° \leqslant x \leqslant 180°$.
Sketch the graph of $y = \tan(-x)$
for $-180° \leqslant x \leqslant 180°$.

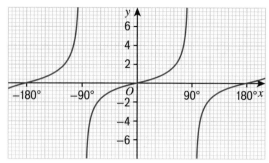

7 a Look at your graph of $y = -\sin x$ for $-180° \leqslant x \leqslant 180°$ in **Q4**.
What transformation will turn the graph of $y = -\sin x$ into the graph of $y = -\sin(-x)$?
  b Sketch the graph of $y = -\sin(-x)$. What do you notice?

> **Key point**
>
> The graph of $y = -f(-x)$ is a reflection of the graph of $y = f(x)$ in the x-axis and then the y-axis,
> or vice versa. These two reflections are equivalent to a rotation of $180°$ about the origin.

8 Reasoning Explain why the graph of $y = -\sin(-x)$
is the same as the graph of $y = \sin x$.

> **Q8 hint**
>
> Sketch the graphs.

9 a Sketch the graph of $y = \cos x$ for $-180° \leqslant x \leqslant 180°$.
  b Sketch the graph of $y = -\cos(-x)$.

10 a Describe the transformation that maps the graph of $y = \cos x$ to the graph of $y = -\cos x$.
  b Sketch the graphs of $y = \cos x$ and $y = -\cos x$ for the interval $-180°$ to $180°$.

11 a Describe the transformation that maps the graph of $y = \tan x$ to the graph of $y = -\tan(-x)$.
  b Sketch the graphs of $y = \tan x$ and $y = -\tan(-x)$ for the interval $-180°$ to $180°$.

> **Exam-style question**
>
> **12** Here is a sketch of the graph of $y = -\sin x$.
>
>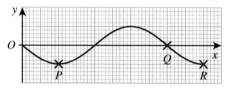
>
> > **Exam tip**
> >
> > When asked for coordinates of
> > points labelled with letters, make
> > sure you write the letter as well
> > as the coordinates.
> > For example, $P(\square, \square)$.
>
> **a** Write down the coordinates of each of the labelled points. **(3 marks)**
> **b** Which of the labelled points are the same on the graph of $y = -\sin(-x)$? **(1 mark)**

13 Reasoning Which pairs of these sections of graphs are the
same shape?

> **Q13 hint**
>
> Draw sketch graphs of
> $y = \cos x, \quad y = -\cos x,$
> $y = \cos(-x)$ and $y = -\cos(-x)$.

  A $y = \cos x$ for $90° \leqslant x \leqslant 450°$
  B $y = \cos(-x)$ for $-180° \leqslant x \leqslant 180°$
  C $y = -\cos x$ for $-270° \leqslant x \leqslant 90°$
  D $y = \cos x$ for $180° \leqslant x \leqslant 540°$
  E $y = -\cos(-x)$ for $90° \leqslant x \leqslant 450°$
  F $y = \cos(-x)$ for $-270° \leqslant x \leqslant 90°$

13.9 Transforming trigonometric graphs 2

Active Learn
Homework

• Recognise how changes in a function affect trigonometric graphs.

Warm up

1 Fluency Match each equation to one of the sketch graphs below.

a $y = \sin x$ **b** $y = \cos x$ **c** $y = \tan x$

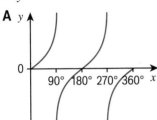 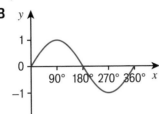 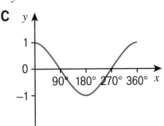

2 Point P is at $(30, 0.5)$. What are the coordinates of the new points when P is translated by

     a $\begin{pmatrix} 0 \\ 2 \end{pmatrix}$ **b** $\begin{pmatrix} -15 \\ 1 \end{pmatrix}$

3 a Copy the graph of $y = \sin x$ for $-180° \leqslant x \leqslant 180°$ from **Q4** in lesson 13.8.

     b Add 0.5 to the y-coordinate at each of the labelled points.

     c Draw and label the sine graph that passes through the new points.

     d Describe the transformation from the graph of $y = \sin x$ to this graph.

     e Now subtract 0.5 from the y-coordinate at each of the labelled points on the original graph.

     f Repeat parts **c** and **d**.

> **Q3c hint**
>
> $y = \sin x + \square$

Key point

The graph of $y = f(x) + a$ is the translation of the graph of $y = f(x)$ by $\begin{pmatrix} 0 \\ a \end{pmatrix}$.

Exam-style question

4 Here is the graph of $y = \sin x$ for $-180° \leqslant x \leqslant 180°$.

Copy the graph. Then on the same grid sketch

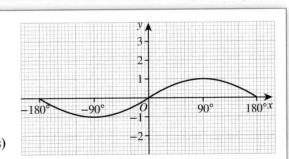

     a $y = \sin x + 2$ for $-180° \leqslant x \leqslant 180°$

                        (2 marks)

     b $y = \sin x - 1$ for $-180° \leqslant x \leqslant 180°$

                        (2 marks)

> **Exam tip**
>
> Look for points where the graph passes through integer coordinates and transform these points carefully.

5 Write down the equation of each graph.

a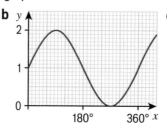

b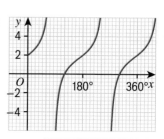

c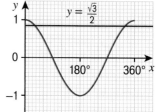

6 Here is the graph of $y = \cos x$ for $0° \leqslant x \leqslant 360°$.

a Use the graph to copy and complete this table of values for $\cos(x+30°)$.

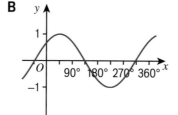

x	$0°$	$30°$	$60°$	$90°$
$\cos(x+30°)$	$\cos 30° = \square$			

b Sketch the graph of $y = \cos(x+30°)$.

c Describe the transformation that takes the graph of $y = \cos x$ to the graph of $y = \cos(x+30°)$.

7 Describe the transformation of the graph of $y = \cos x$ to make the graph with equation

a $y = \cos(x+60°)$ b $y = \cos(x+20°)$ c $y = \cos(x-30°)$

8 Describe the transformation of the graph of $y = \tan x$ to make the graph with equation

a $y = \tan(x+40°)$ b $y = \tan(x+30°)$ c $y = \tan(x-60°)$

9 Match each equation to one of the graphs below.

a $y = \tan(x-45°)$ b $y = \sin(x+45°)$ c $y = \cos(x+45°)$

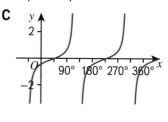

10 **Reasoning** Sarah says, 'The graphs of $y = \sin(x+30°)$ and $y = \cos(x-60°)$ are the same.'
Is Sarah correct? Explain.

11 Sketch the graph of $y = \sin(x+30°)+3$ for $-180° \leqslant x \leqslant 180°$.

12 Here is a sketch of the curve
$y = \sin(x+a°)+b$ for $0° \leqslant x \leqslant 360°$.
Find the values of a and b.   **(2 marks)**

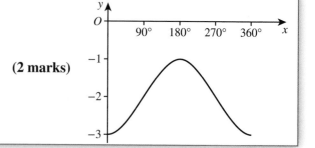

13 Check up

Active Learn
Homework

Accuracy and 2D problem-solving

 1 Work out the area of this triangle.

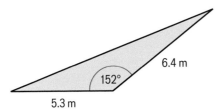

6.4 m

152°

5.3 m

 2 Calculate the length of the side labelled x in each diagram.

a

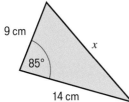

9 cm

x

85°

14 cm

b

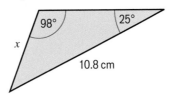

98°   25°

x

10.8 cm

 3 Find the size of the acute angle θ in each triangle.

a

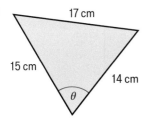

17 cm

15 cm

14 cm

θ

b

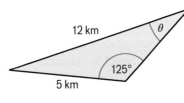

12 km

θ

125°

5 km

 4 The measurements in this diagram are given correct to 2 significant figures.
Find the upper and lower bounds for the value of x in the diagram.

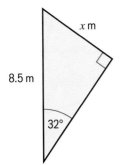

x m

8.5 m

32°

Write x to a suitable degree of accuracy.

Trigonometric graphs

5 Sketch the graph of $y = \tan \theta$ for $-360° \leqslant \theta \leqslant 360°$.

6 **a** Sketch the graph of $y = \sin \theta$ for $0° \leqslant \theta \leqslant 360°$.
   b Use your sketch to find $\sin 270°$.

Unit 13 More trigonometry   87

7 Here is the graph of $y = \cos x$ for $0° \leqslant x \leqslant 360°$.

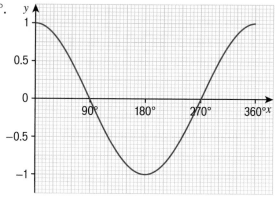

a Use the graph to solve $\cos x = 0.4$.

b Which of these graphs is the same as $y = \cos x$ for $0° \leqslant x \leqslant 360°$?

   A $y = -\cos x$ for $0° \leqslant x \leqslant 360°$

   B $y = \cos(-x)$ for $0° \leqslant x \leqslant 360°$

   C $y = -\cos(-x)$ for $0° \leqslant x \leqslant 360°$

8 Match each equation to a translation of the graph of $y = \cos x$.

a $y = \cos(x - 10°)$

b $y = \cos x + 10$

c $y = \cos x - 10$

d $y = \cos(x + 10°)$

  A $\begin{pmatrix} -10 \\ 0 \end{pmatrix}$ **B** $\begin{pmatrix} 0 \\ 10 \end{pmatrix}$ **C** $\begin{pmatrix} 10 \\ 0 \end{pmatrix}$ **D** $\begin{pmatrix} 0 \\ -10 \end{pmatrix}$

9 Solve the equation $3 \sin x = 1$ for $0° \leqslant x \leqslant 720°$.

3D problem-solving

10 *ABCDEFGH* is a cuboid.
Calculate the length of the diagonal *AG*.
Give your answer correct to 1 decimal place.

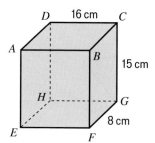

11 *ABCD* is a tetrahedron.
$AB = 14$ cm, $BC = 12$ cm,
angle $ABC = 71°$, angle $CAD = 42°$, angle $ADC = 64°$
Calculate the length of

a *AC*

b *CD*

Give your answers correct to 1 decimal place.

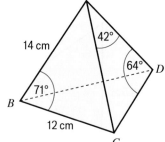

12 Reflect How sure are you of your answers? Were you mostly

Just guessing 😞   Feeling doubtful 😐   Confident 🙂

What next? Use your results to decide whether to strengthen or extend your learning.

Challenge

13 The diagram shows a sequence of isosceles triangles *A*, *B*, *C*, ...
Continue the sequence and find the lengths of the sides
of triangle *D*.

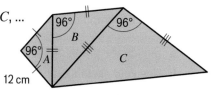

13 Strengthen

Accuracy and 2D problem-solving

1 a Copy the triangle. Label
  i the vertex at the given angle C, and the other two vertices A and B (it doesn't matter which way round you label vertices A and B)
  ii the sides a (opposite A) and b (opposite B).
b Use Area $= \frac{1}{2}ab\sin C$ to find the area of the triangle. Give your answer correct to 3 significant figures.

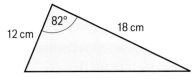

2 Find the areas of these triangles correct to 3 significant figures.

a

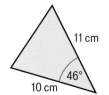

b

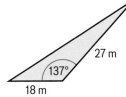

> **Q2 hint**
>
> Use the method in **Q1**.

3 a Copy the triangle. Label
  i the vertices A, B and C (it doesn't matter which letter you use for each vertex)
  ii the opposite sides a, b and c
b Substitute the values from the diagram into the sine rule:
$$\frac{a}{\sin A} = \frac{b}{\sin B} = \frac{c}{\sin C}$$
c Use two parts of the sine rule with values in them to find x. Give your answer correct to 3 significant figures.

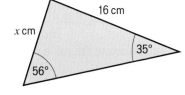

4 Use the method in **Q3** to find the value of x in each triangle. Give your answers correct to 3 significant figures.

a

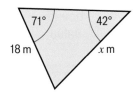

b

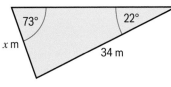

5 a Copy the triangle. Label the vertices A, B and C and the opposite sides a, b and c.
b Substitute the values from the diagram into the sine rule:
$$\frac{\sin A}{a} = \frac{\sin B}{b} = \frac{\sin C}{c}$$
c Use two parts of the sine rule with values in them to find θ. Give your answer correct to 1 decimal place.

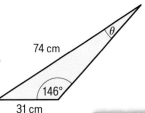

> **Q5c hint**
>
> $$\frac{\sin\theta}{\square} = \frac{\square}{\square}$$

6 Use the method in **Q5** to find the size of the acute angle θ in each triangle.
Give your answers correct to 1 decimal place.

a
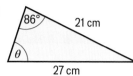
86° 21 cm
θ
27 cm

b
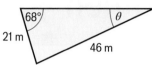
68° θ
21 m
46 m

7 **a** Copy the triangle. Label the x side a, and the other sides b and c.
Label the vertices A, B and C.

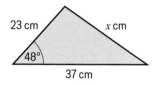
23 cm x cm
48°
37 cm

b Substitute the values from the triangle into the cosine rule:
$$a^2 = b^2 + c^2 - 2bc \cos A$$

c Solve the equation to find x, correct to 3 significant figures.

8 Use the method in **Q7** to calculate the value of x in each triangle.
Give your answers correct to 3 significant figures.

a
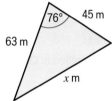
76° 45 m
63 m
x m

b
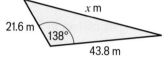
x m
21.6 m
138°
43.8 m

9 **a** Copy the triangle. Label the vertex and angle θ, A.
Label the other vertices and the other sides.

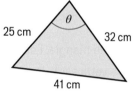
θ
25 cm 32 cm
41 cm

b Substitute the values from the triangle into the cosine rule:
$$\cos A = \frac{b^2 + c^2 - a^2}{2bc}$$

c Calculate the value of θ correct to 1 decimal place.

10 Use the method in **Q9** to calculate the size of angle θ in each triangle.
Give your answers correct to 1 decimal place.

a
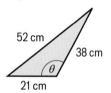
52 cm
38 cm
θ
21 cm

b
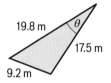
19.8 m θ
17.5 m
9.2 m

11 All the measures in this equation are given to 2 significant figures.
$$x = \frac{5.7}{\sin 23°}$$

a Copy and complete the table for the upper and lower bounds.

b Find sin(upper bound for 23) and sin(lower bound for 23).

c Which upper and lower bound values give the lower bound
for x?

d Which upper and lower bound values give the upper bound
for x?

	Upper bound value	Lower bound value
5.7		5.65
23	23.5	

Q11c hint

Which values – one from each
row of the table – give the
smallest possible value of x?

12 All the measures in this equation are given to
2 significant figures.
$$x = 14 \tan 36°$$
Find the upper and lower bounds for the value of x.

Q12 hint

Use the same method as in **Q11**.

Trigonometric graphs

1　**a** Copy and complete the table for $y = \sin x$.

x	0°	10°	20°	30°	40°	50°	60°	70°	80°	90°
$\sin x$										

b Draw the graph of $y = \sin x$ for $0° \leqslant x \leqslant 90°$.

The sine graph is symmetrical about the line $x = 90°$.

c Sketch the graph of $y = \sin x$ for $0° \leqslant x \leqslant 180°$.

The graph of $y = \sin x$ has rotational symmetry about the point $(180°, 0)$.

d Extend your sketch from part **c** to cover the interval $0° \leqslant x \leqslant 360°$.

e $\sin x = -1$. Use your graph to work out the value of x.

> **Q1e hint**
>
> $y = \sin x$ so read across from -1 on the y-axis to your graph. Then read up to the x-axis.

2　**a** Copy and complete the table for $y = \tan x$.

x	0°	10°	20°	30°	40°	50°	60°	70°	80°
$\tan x$									

b Draw the graph of $y = \tan x$ for $0° \leqslant x \leqslant 80°$.

c What happens to the value of $\tan x$ as x increases from 80° towards 90°?

d Sketch the graph of $y = \tan x$ for $0° \leqslant x \leqslant 90°$.

The graph of $y = \tan x$ has rotational symmetry about the point $(90°, 0)$.

e Extend your sketch from part **d** to cover the interval $0° \leqslant x \leqslant 180°$.

The graph of $y = \tan x$ repeats every 180°.

f Extend your sketch from part **e** to cover the interval $0° \leqslant x \leqslant 360°$.

g $\tan x = -1$. Use your graph to work out two possible values of x.

3 Here is the graph of $y = \cos x$ for $0° \leqslant x \leqslant 360°$.

Find -0.6 on the y-axis and read off the corresponding values from the curve to estimate a solution to $\cos x = -0.6$. There are two values in the interval $0° \leqslant x \leqslant 360°$.

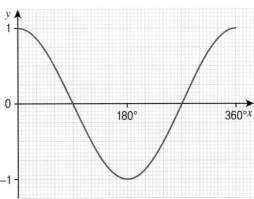

4 The graph of $y = \cos x$ repeats every 360° in both directions.

Sketch the graph of $y = \cos x$ for $-360° \leqslant x \leqslant 720°$.

> **Q4 hint**
>
> Sketch the graph in **Q3**.

5　**a** The graph of $y = -\cos x$ is a reflection of $y = \cos x$ in the x-axis.
　　Is the graph of $y = -\cos x$ the same as the graph of $y = \cos x$?

> **Q5 hint**
>
> Use your sketch from **Q4**.

　　b The graph of $y = \cos(-x)$ is a reflection of $y = \cos x$ in the y-axis.
　　Is the graph of $y = \cos(-x)$ the same as the graph of $y = \cos x$?

　　c The graph of $y = -\cos(-x)$ is a rotation of $y = \cos x$ by 180° about the origin.
　　Is the graph of $y = -\cos(-x)$ the same as the graph of $y = \cos x$?

6 Here are sketch graphs of $y = \cos x° + 4$ and $y = \cos(x + 4°)$.

a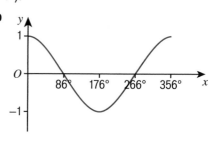

b

Compare the graphs with the graph in **Q3**. For each graph

    i has the graph been translated in the x or y direction

    ii by how much is it translated?

    iii Write the translation as a column vector.

Q6 iii hint

 7 **a** Rearrange the equation $4\cos x = 3$ to make $\cos x$ the subject.

    b Sketch the graph of $y = \cos x$ for $0° \leqslant x \leqslant 720°$.

    c Use your calculator to find one value of x that satisfies the equation.

    d Use your sketch to solve the equation $4\cos x = 3$ for $0° \leqslant x \leqslant 720°$.

3D problem-solving

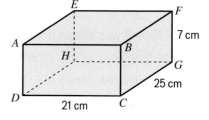

 1 **Reasoning** The diagram shows a cuboid $ABCDEFGH$.

    a Sketch triangle CDG. Label

      i its right angle

      ii the information shown on the diagram

    b Calculate the length of DG.
      Give your answer correct to 3 significant figures.

    c Sketch triangle DFG. Label

      i its right angle

      ii the information shown on the diagram
         and your answer to part **b**

    d Calculate the angle that the diagonal DF
      makes with the plane $DCGH$.

Q1d hint

The angle between the diagonal DF and the plane $DCGH$ is angle FDG on your sketch for part **c**. Use your *unrounded* answer to part **b**.

2 **Reasoning** In the diagram, $ABCD$ is a tetrahedron.
    $AB = AC = 11$ cm, $BD = 12$ cm, angle $BAC = 56°$

    a Sketch triangle ABC.
      Label the triangle with the information shown on the diagram.

    b Calculate the length of BC.
      Give your answer correct to 3 significant figures.

    c Sketch triangle BCD.
      Label the triangle with the information shown on
      the diagram and your answer to part **b**.

    d Calculate angle BCD.

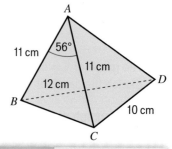

Q2d hint

Use your *unrounded* answer to part **b**.

13 Extend

1 **a** Write an expression for the area of this triangle using angle C.

b Write two more expressions for the area of the triangle using angles B and A.

c Using your answers to parts **a** and **b**, show that
$$\frac{a}{\sin A} = \frac{b}{\sin B} = \frac{c}{\sin C}$$

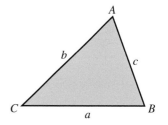

 2 The diagram shows a cuboid.
The diagonal of the cuboid has length d.
Write d^2 in terms of a, b and c.

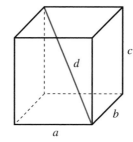

3 **a** In triangle BCD, write a^2 in terms of h, b and x and expand the brackets.

b In triangle ABD, write c^2 in terms of h and x.

c Use your answers to parts **a** and **b** to write a^2 in terms of b, c and x.

d Show that $a^2 = b^2 + c^2 - 2bc \cos A$.

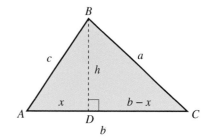

4 The diagram shows the triangle XYZ.
$XY = 1\,\text{cm}$, $YZ = \sqrt{3}\,\text{cm}$
The area $A = 0.75\,\text{cm}^2$.
Show that $\theta = 60°$.

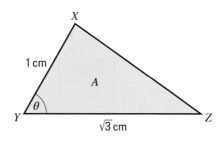

Q4 hint

Use
Area of a triangle $= \frac{1}{2}ab \sin C$
and your knowledge of the exact values of the sine of some angles.

5 **Problem-solving** Find the area of the shaded segment in terms of r.

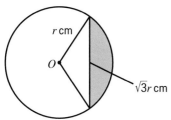

Exam-style question

6   OSR is a sector of a circle with centre O and radius 13 cm. P is the point on OS and Q is the point on OR, such that OPQ is an equilateral triangle of side 8 cm.

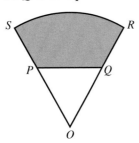

Calculate the area of the shaded region as a percentage of the area of the sector OSR.
Give your answer correct to 1 decimal place.   **(5 marks)**

Exam tip

Sketch the shape. Label your sketch with all the information you are given.

Exam-style question

7   Here are two triangles, A and B.

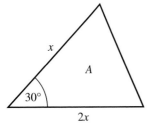

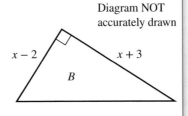

Diagram NOT accurately drawn

The lengths of the sides are in centimetres.
The area of triangle A is equal to the area of triangle B.
Work out the value of x.   **(4 marks)**

Exam tip

Consider the different ways you know of finding the area of triangles.

8   **a** Sketch the graph of $y = \sin x$ for $0° \leqslant x \leqslant 180°$.

b Copy and complete this table of values for $y = \sin\left(\dfrac{x}{2}\right)$.

x	$0°$	$60°$	$90°$	$120°$
$\sin\left(\dfrac{x}{2}\right)$			$\dfrac{1}{\sqrt{2}} \approx 0.7$	

c Sketch the graph of $y = \sin\left(\dfrac{x}{2}\right)$ on the same axes as your graph of $y = \sin x$.

9   **Problem-solving** The depth, d metres, of water in a harbour at a time t hours after midnight is $d = 12 + 5\sin(30t)$, where $0 \leqslant t \leqslant 24$.
Sketch the graph of d against t.

10 **Problem-solving** $ABCDE$ is a square-based pyramid.
Calculate the volume of the pyramid.
Give your answer correct to 3 significant figures.

11 Problem-solving

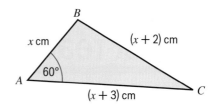

a Find the value of x in triangle ABC.

b The exact value of the area of triangle ABC is $\sqrt{3}\,k$ cm^2. Find the value of k.

12 ABD and CBE are congruent triangles.

$AB = BC = 8$ cm

$BD = BE = x$ cm

Angle CAD = angle $ACE = 90°$

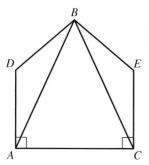

Diagram NOT accurately drawn

Exam tip

When dealing with more than one triangle in a single diagram, draw them separately and write the information you know on them, or beside them. For example

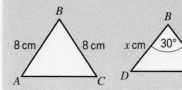

$AC = DE$

Given that angle $DBE = 30°$,

prove that $\cos ABC = 1 - \dfrac{(2-\sqrt{3})}{128}x^2$.

(5 marks)

13 In triangle ABC, $AB = (x+2)$ cm,

$BC = (x+4)$ cm and angle $ABC = 120°$.

Given that the area of triangle ABC is $11\sqrt{3}$ cm^2,

a show that x satisfies the equation $x^2 + 6x - 36 = 0$

b work out the exact value of x, giving your answer in the form $x = p\sqrt{q} + r$, where p, q and r are integers.

14 Here is the graph of $y = \cos(x - k°) + 1$, where $k > 0$.

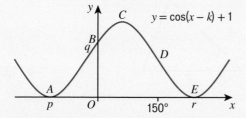

In A level you might have to make links between topics. This question connects the formula for the area of a triangle, the exact value of $\sin 120°$, quadratic equations and giving answers in surd form.

Q14a hint

Use the given point to find the value of k. Now use the equation of the graph and your knowledge of the cosine graph to answer the other parts.

The point $(150°, 1)$ lies on the graph.

The graph touches the x-axis at A and E, crosses the y-axis at B and has a maximum value at C.

Work out

a the smallest value of k **b** the coordinates of C

c the values of p and r **d** the value of q

In A level you might have to adapt your knowledge to solve problems. In this question you have to adapt your knowledge to a cosine graph that has undergone two transformations.

13 Test ready

Summary of key points

To revise for the test:

- Read each key point, find a question on it in the mastery lesson, and check you can work out the answer.

- If you cannot, try some other questions from the mastery lesson or ask for help.

Key points

1 The upper and lower bounds of a length or angle in a right-angled triangle can be found using Pythagoras' theorem and/or trigonometry. → **13.1**

2 The **sine** graph repeats every 360° in both directions.

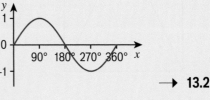

→ **13.2**

3 The **cosine** graph repeats every 360° in both directions.

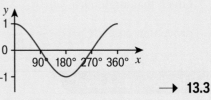

→ **13.3**

4 The **tangent** graph repeats every 180° in both directions.

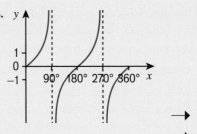

→ **13.4**

5 $\tan x$ is not defined for angles of the form $90 \pm 180n°$. → **13.4**

6 The **area** of this triangle $= \frac{1}{2}ab\sin C$.
   C is the given angle.
   a is the side opposite angle A.
   b is the side opposite angle B.

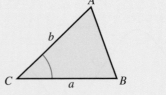

→ **13.5**

7 The **sine rule** can be used in any triangle.

- $\dfrac{a}{\sin A} = \dfrac{b}{\sin B} = \dfrac{c}{\sin C}$ Use this to calculate an unknown *side*.

- $\dfrac{\sin A}{a} = \dfrac{\sin B}{b} = \dfrac{\sin C}{c}$ Use this to calculate an unknown *angle*.

→ **13.5**

8 The **cosine rule** can be used in any triangle.

- $a^2 = b^2 + c^2 - 2bc\cos A$ Use this to calculate an unknown *side*.

- $\cos A = \dfrac{b^2 + c^2 - a^2}{2bc}$ Use this to calculate an unknown *angle*.

→ **13.6**

Key points

9 A **plane** is a flat surface. In the diagram

- BC is perpendicular to the plane $WXYZ$
- triangle ABC is in a plane perpendicular to the plane $WXYZ$
- θ is the angle between the line AB and the plane $WXYZ$

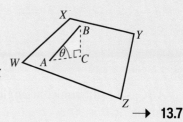

→ **13.7**

10 The graph of $y = -f(x)$ is the reflection of the graph of $y = f(x)$ in the x-axis. → **13.8**

11 The graph of $y = f(-x)$ is the reflection of the graph of $y = f(x)$ in the y-axis. → **13.8**

12 The graph of $y = -f(-x)$ is a reflection of the graph of $y = f(x)$ in the x-axis and then the y-axis, or vice versa. These two reflections are equivalent to a rotation of $180°$ about the origin. → **13.8**

13 The graph of $y = f(x) + a$ is the translation of the graph of $y = f(x)$ by $\begin{pmatrix} 0 \\ a \end{pmatrix}$. → **13.9**

14 The graph of $y = f(x + a)$ is the translation of the graph of $y = f(x)$ by $\begin{pmatrix} -a \\ 0 \end{pmatrix}$. → **13.9**

Sample student answers

Exam-style question

The diagram shows a circle and an equilateral triangle. Two sides of the equilateral triangle are radii of the circle. The circle has a circumference of $8\,\text{cm}$.

Show that the area of the triangle is $\dfrac{4\sqrt{3}}{\pi^2}\ \text{cm}^2$.

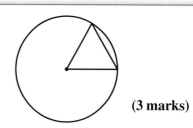

(3 marks)

Here are the starts of two students' answers.

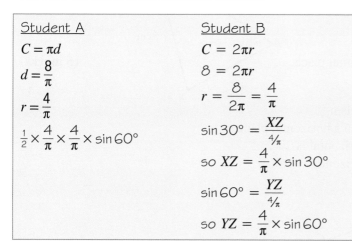

Student A

$C = \pi d$

$d = \dfrac{8}{\pi}$

$r = \dfrac{4}{\pi}$

$\dfrac{1}{2} \times \dfrac{4}{\pi} \times \dfrac{4}{\pi} \times \sin 60°$

Student B

$C = 2\pi r$

$8 = 2\pi r$

$r = \dfrac{8}{2\pi} = \dfrac{4}{\pi}$

$\sin 30° = \dfrac{XZ}{4/\pi}$

so $XZ = \dfrac{4}{\pi} \times \sin 30°$

$\sin 60° = \dfrac{YZ}{4/\pi}$

so $YZ = \dfrac{4}{\pi} \times \sin 60°$

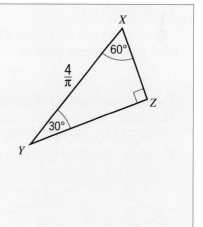

a Both students began by using the circumference to find r, but then they each used a different method. Is each student's method correct? Explain.

b Each student's next step was to work out an answer on a calculator. Why is this not a good idea?

c Continue each student's working to fully answer the exam question.

13 Unit test

 Active Learn
Homework

 1 *ABC* is a right-angled triangle.
Angle *ACB* is 90°.
AB is length 11.2 m correct to 3 significant figures.
Angle *BAC* is 48° correct to 2 significant figures.
Find the length of *BC* to a suitable degree of accuracy. **(4 marks)**

2 The diagram shows the triangle *PQR*.
PQ = 3*x* cm
PR = 4*x* cm
Angle *QPR* = 30°
The area of triangle *PQR* = *A* cm².
Show that $x = \sqrt{\dfrac{A}{3}}$. **(4 marks)**

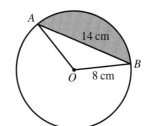

 3 The diagram shows a circle with centre *O*
and radius 8 cm.
The chord *AB* has length 14 cm.
Calculate the area of the shaded segment. **(3 marks)**

 4 *PQR* is a triangle.
QR = 7.6 cm
Angle *PQR* = 47°
Angle *RPQ* = 64°
Calculate the area of triangle *PQR*.
Give your answer correct to 1 decimal place. **(5 marks)**

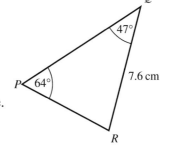

5 The diagram shows a triangular prism.
The base *CDEF* of the prism lies in a horizontal plane.
The angle between *AD* and the horizontal is 20°.
The angle between *AC* and the horizontal is 16°.

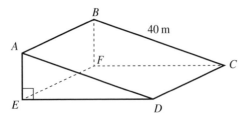

Calculate the length of *AC*.
Give your answer correct to 3 significant figures. **(5 marks)**

6 In triangle ABC, $AB = 8$ cm, $BC = 6$ cm and angle $BAC = 35°$.
Find the two possible sizes of angle BCA.
Give your answers correct to 1 decimal place. **(5 marks)**

7 The diagram shows the positions of towns A, B and C.

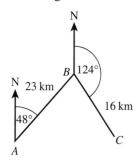

   a Calculate the direct distance from A to C. **(3 marks)**
   b Calculate the bearing of C from A. **(3 marks)**

8 Sketch the graph of $y = \sin(x - 90°)$ for $0° \leqslant x \leqslant 360°$. **(3 marks)**

9 **a** The graph of $y = \cos x$ is reflected in the x-axis.
     What is the equation of the new graph? **(1 mark)**

   b The graph of $y = \sin x$ is reflected in the y-axis.
     What is the equation of the new graph? **(1 mark)**

10 Here is a sketch of the curve $y = \cos(x - 90°) + 1$.

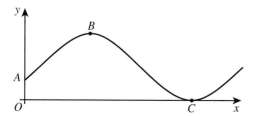

   Write down the coordinates of points A, B and C. **(3 marks)**

 11 **a** Show that **one** solution to $5\tan\theta = 7$ is $54.5°$ correct to 1 decimal place. **(2 marks)**
   b Solve the equation $5\tan\theta = 7$ for all values of θ in the interval $0°$ to $720°$. **(3 marks)**

   (TOTAL: 45 marks)

12 **Challenge** The angles that satisfy an equation in the interval $0° \leqslant x \leqslant 720°$ are $30°$, $210°$, $390°$ and $570°$.
Write down a possible equation.

13 **Reflect** For each statement **A**, **B** and **C**, choose a score:
1 – strongly disagree;   2 – disagree;   3 – agree;   4 – strongly agree
A I always try hard in mathematics
B Doing mathematics never makes me worried
C I am good at mathematics
For any statement where you scored lower than 3, write down two things you could do so
that you agree more strongly in the future.

1 **Reasoning** Here are some graphs.

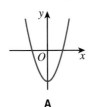

A

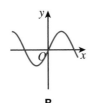

B

C

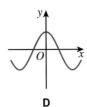

D

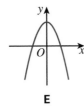

E

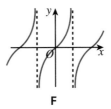

F

Here are some equations.

i $y = \sin x$ **ii** $y = \dfrac{1}{x}$ **iii** $y = \tan x$

iv $y = x^2 - 4$ **v** $y = 4 - x^2$ **vi** $y = \cos x$

Match each equation to the letter of its graph.

2 **Reasoning** ABC is a triangle.

D is a point on AB and E is a point on BC.
DE is parallel to AC.
State whether each statement is true or false.

a $\dfrac{BD}{BA}$ is equivalent to $\dfrac{DE}{BC}$ **b** $\dfrac{BD}{AB}$ is equivalent to $\dfrac{DE}{AC}$

c $\dfrac{AC}{DE}$ is equivalent to $\dfrac{BC}{BE}$ **d** $\dfrac{BE}{EC}$ is equivalent to $\dfrac{DE}{AC}$

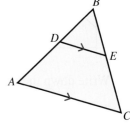

3 **Problem-solving** PQR is a triangle.

X is a point on PQ and Y is a point on QR.
XY is parallel to PR.
Calculate the perimeter of the trapezium $XYRP$.

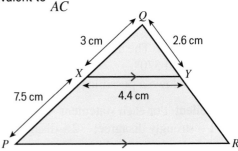

4 A force of 90 newtons acts on an area of $40\,\text{cm}^2$.
The force is increased by 20 newtons.
The area is increased by $20\,\text{cm}^2$.
Adam says, 'The pressure decreases by 40%.'
Is Adam correct? You must show how you get your answer. **(4 marks)**

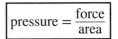

$$\text{pressure} = \frac{\text{force}}{\text{area}}$$

5 **Reasoning** $PQRS$ is a parallelogram.

Prove that triangle PQR is congruent to triangle RSP.

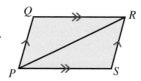

6 **Problem-solving** Asma drove to her gran's house.

At 11.30 am the satnav said she was 55 miles from her destination.

Asma drove at a steady (constant) speed. She arrived at her gran's house at 12.45 pm.

Work out the average speed of the car from 11.30 am to 12.45 pm in km/h.

Use 1 mile $\approx$ 1.6 km.

You must show all your working.

Exam-style question

7 Sam says, '53 kilometres per hour is faster than 15 metres per second.'

Is Sam correct? You must show how you get your answer. **(2 marks)**

8 **Reasoning** Here is a table of values.

x	2	3.5	8	11
y	16	28	64	88

a What type of proportion is shown between the values x and y?

b Which formula describes the relationship between x and y, where k is a constant?

    A $y = kx$ **B** $xy = k$ **C** $y = \dfrac{k}{x}$ **D** $k = xy$

Exam-style question

9 Zaid invests £4550 in a savings account for 3 years.

He is paid 1.8% per annum compound interest for each of the first 2 years.

He is paid $R\%$ interest for the third year.

Zaid had £4819.01 in his savings account at the end of the 3 years.

Work out the value of R.

Give your answer correct to 1 decimal place. **(3 marks)**

10 **Problem-solving** A rectangular table top measures 85 cm by 160 cm.

The measurements of the table are given to the nearest centimetre.

A force of 24 newtons is applied to the top of the table.

The force is given to the nearest newton.

Work out the upper and lower bounds of the pressure.

Give your answers to 3 significant figures.

Exam-style question

11 Liquid X and liquid Y are mixed in the ratio 3 : 7 by volume to make liquid Z.

Liquid X has density 1.16 g/cm³

Liquid Y has density 1.05 g/cm³

A cylindrical container is filled completely with liquid Z.

The cylinder has radius 4 cm and height 20 cm.

Work out the mass of the liquid in the container.

Give your answer correct to 3 significant figures.

You must show all your working. **(4 marks)**

12 **Problem-solving** 8 typists work in an office.
When all typists are working, they complete their work in 5 hours.
After 4 hours, 2 of the typists leave the office.
The remaining typists carry on working until the work is finished.
How many hours do each of the remaining typists work in total?

13 P and Q are two similar cylindrical containers.
The ratio of the surface area of container P : the surface
area of container Q is 9 : 25.
Ethan fills container P with water.
He then pours all the water into container Q.
Ethan repeats this and stops when container Q is full of water.
Work out the number of times that Ethan fills container P with water.
You must show all your working.　　　　　　　　　　　　**(4 marks)**

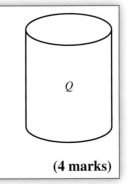

14 **Reasoning** Here are some graphs.

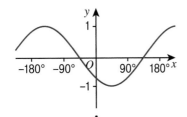

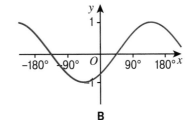

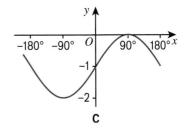

A　　　　　　　　　　B　　　　　　　　　　C

Here are some equations.

i　$y = \sin x + 2$　　　　　ii　$y = \sin x - 1$　　　　　iii　$y = -\sin x - 1$

iv　$y = \sin(x + 45°)$　　　v　$y = -\sin(x + 45°)$　　　vi　$y = \sin(x - 45°)$

a　Match each graph to its equation.

b　Sketch the graph of each equation not used in part **a**.

15 The diagram shows triangle XYZ.

P is the point on XY such that
the size of angle PZX : the size of angle $PXZ = 3 : 2$.

Calculate the length PX.
Give your answer correct to 3 significant figures.
You must show all your working.　　　　　　　　　　**(5 marks)**

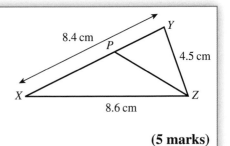

16 **Problem-solving** The diagram shows triangle XYZ.
Calculate the area of triangle XYZ.
Give your answer correct to 3 significant figures.

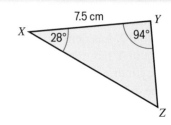

17 Reasoning Here are two graphs.

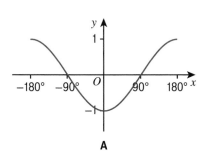

A   B

Write two possible equations for each graph.

Exam-style question

18 The diagram shows a frustum of a cone.

The diagram shows that the frustum is made by removing a cone with height 4 cm from a solid cone with height 8 cm and base diameter 6 cm.

The frustum is joined to a solid hemisphere of diameter 6 cm to form the solid D shown.

The density of the frustum is 3.2 g/cm^3
The density of the hemisphere is 5 g/cm^3
Calculate the average density of solid D.   **(5 marks)**

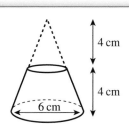

Volume of sphere $= \frac{4}{3}\pi r^3$
Volume of cone $= \frac{1}{3}\pi r^2 h$

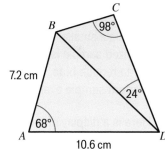

19 Problem-solving The diagram shows the quadrilateral $ABCD$.
Work out the length of BC.
Give your answer correct to 3 significant figures.

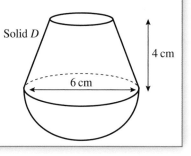

20 Problem-solving The diagram shows a pyramid.
The base of the pyramid is a square with sides of length 8 cm.
The other faces of the pyramid are equilateral triangles with sides of length 8 cm.

a Calculate the volume of the pyramid.
Give your answer correct to 3 significant figures.

b Find the size of angle PTR.

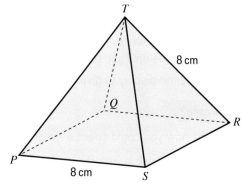

14 Further statistics

14.1 Sampling

Prior knowledge

- Use random numbers to select a random sample.
- Understand the assumptions made when using a sample to predict results for a population.
- Use the Petersen capture–recapture method.

*Active*Learn
Homework

Warm up

1 **Fluency a** Copy and complete. $\dfrac{9}{20} = \dfrac{\square}{140}$ **b** Find the value of N when $\dfrac{20}{N} = \dfrac{4}{11}$

2 John wants to find out about adults' shopping habits in his local town.
   He plans to ask the first 30 adults he sees in the street outside the supermarket.

   a Is this sample likely to be representative of the population? Explain.

   b Janna suggests that he pick 30 people at random from the Electoral Register.
   Is this sample likely to be representative of the population? Explain.

   c Write at least two reasons why surveys often use a sample instead of the whole population.

Key point

A **population** is the set of items that you are interested in.
A **census** is a survey of the whole population.
A **sample** is a smaller number of items from the population.
A good-sized sample is at least 10% of the population.
In order to reduce **bias**, a sample must represent the whole population.
In a **random** sample every item is equally likely to be chosen.

3 Here is a display of random numbers.

   8613607878488056990932660231779879509790513199 2...

   a Follow these steps to get seven
   random numbers between 0 and 50.
   Start with 86 and write the digits in pairs: 86, 13, ...
   Cross out any over 50 and any repeats.
   Continue until you have seven numbers.

   b Use the display to give six random numbers between 0 and 99.

Example

Describe how you could select a random sample of size 15 from a population of 90 people.

List them in alphabetical order of their last names.
Number the list from 1 to 90. ●────────────── Explain how to list the population.

Use a calculator to generate 15 random numbers
between 1 and 90. Ignore any repeated numbers
or numbers greater than 90. ●────── Explain how to use random numbers to choose the sample from the population list.

Choose the people with these numbers from the list.

4 **Reasoning** A company with 60 employees needs to try out a new flexi-time scheme.
It decides on a random sample of 8 employees.
Explain how it could use this table of random numbers to select a sample.
Write down the numbers of the employees who will be in the sample.

4612671248069924148378376573394...

5 **Reasoning** 50 people were asked how many pets they had.
The results are shown in this frequency table.

Number of pets	Frequency
0	10
1	14
2	13
3	4
4	3
5	2
6	1
7	2
8	1

a Calculate the summary statistics for this population.

b Maisie used random numbers to select this sample of
5 pieces of data from the original set:
0, 6, 2, 1, 7
Calculate the summary statistics of Maisie's sample.

c Ruby used random numbers to select a sample of
20 pieces of data.
The summary statistics for Ruby's sample are:
    Mean 1.95   Median 2   Mode 2   Range 7
The best estimate is closest to the actual value
for the population.
Which sample, Maisie's or Ruby's, gives the better estimate for the summary statistics for
the population?

6 Theo asks a sample of 40 sixth-form students what
type of lunch they would like to buy from the
school cafeteria.
Each student chooses one option.
The table shows his results.

There are 240 students in the sixth form.

Lunch	Number of students
sandwiches	20
salad	13
pizza	2
curry	5

a Explain why you think Theo surveyed a sample of
students instead of carrying out a census.

b What proportion of the students in the sample chose salad?

c Work out how many of the 240 students you think will choose salad.

7 Mark and Dan answered **Q6**. They were asked what assumptions they had made.
Here are their answers:

Mark   Dan

I assumed the sample
was representative of
the whole population.

I assumed the 40 students
were a random sample
from the 240 students.

Explain why their answers to **Q6c** may be wrong if their assumptions are not correct.

Exam-style question

8   Last year Alexis sold holidays in Greece to 2400 people.
He takes a sample of 300 of these people.
He asks each person to choose their favourite type of holiday in Greece.
The table shows the results.

Type of holiday	Number of people
city hotel	63
beach hotel	120
city self catering	52
beach self catering	65

Exam tip

For part **b** you need to write one assumption and **explain** how it affects your answer.

This year Alexis expects to sell 3000 holidays in Greece.

a   Work out how many of the 3000 holidays you think will be beach hotel holidays. (**2 marks**)

b   State any assumptions you make **and** explain how this may affect your answer. (**1 mark**)

9   **Reasoning**   A scientist wishes to find out how many fish are in a lake.
He catches 40 fish and marks them with a small tag.
Two weeks later, he returns to the lake and catches another 40 fish.
5 of the fish he catches have been marked with his tag.

a   What fraction of the fish he catches in the second sample are tagged?

b   Assume the fraction tagged in the sample is the same as the fraction tagged in the lake.
Estimate how many fish are in the lake.

Q9b hint

Let f be the total number of fish in the lake. Write an expression for the fraction of tagged fish in the lake.

Key point

To estimate the size of the population N of an animal species:

• Capture and mark a sample size n.
• Recapture another sample of size M. Count the number marked (m).

$$\frac{n}{N} = \frac{m}{M}$$
So, $N = \frac{n \times M}{m}$

This is the **Petersen capture–recapture method**.
Assumptions:

• The population has not changed between the release and recapture times.
• The probability of being captured is the same for all individuals.
• Marks or tags are not lost.

Exam-style question

10   Lara wants to estimate the number of ants in a nest.

On Monday Lara catches 60 ants and marks each one.
She then puts all 60 ants back in the nest.

On Tuesday Lara catches 40 ants.
7 of these ants have a mark on them.

a   Work out an estimate for the total number of ants in the nest. (**2 marks**)

b   State any assumptions you have made. (**2 marks**)

Exam tip

Set out your calculation for part **a** clearly, so the examiner can see the method you used.

14.2 Cumulative frequency

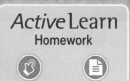

Active Learn
Homework

- Draw and interpret cumulative frequency tables and graphs.
- Work out the median, quartiles and interquartile range from a cumulative frequency graph.

Warm up

1 Fluency a How many lengths of wood are

    i less than or equal to 5 m long

    ii less than or equal to 10 m long

    iii more than 5 m long?

  b Estimate the range.
    Explain why your value is only an estimate.

Length of wood, l (m)	Frequency
$0 < l \leqslant 5$	4
$5 < l \leqslant 8$	6
$8 < l \leqslant 10$	7

2 For 11, 14, 15, 16, 18, 21, 22, 25, 26, 27, 30

  a write down the median

  b which value is **i** $\frac{1}{4}$ of the way into the list   **ii** $\frac{3}{4}$ of the way into the list?

Key point

A **cumulative frequency table** shows how many data values are less than or equal to the **upper class boundary** of each data class.
The upper class boundary is the highest possible value in each class.

3 The frequency table shows the masses of 50 cats.
Copy and complete the cumulative frequency table.

Mass, m (kg)	Frequency
$3 < m \leqslant 4$	4
$4 < m \leqslant 5$	12
$5 < m \leqslant 6$	17
$6 < m \leqslant 7$	10
$7 < m \leqslant 8$	7

Mass, m (kg)	Cumulative frequency
$3 < m \leqslant 4$	4
$3 < m \leqslant 5$	$4 + 12 = \square$
$3 < m \leqslant 6$	
$3 < m \leqslant 7$	
$3 < m \leqslant 8$	

4 This frequency table gives the heights of 70 giraffes.
Draw a cumulative frequency table for this data.

Q4 hint

Start every height group with the shortest height given in the table:
$4.0 < h \leqslant 4.2$, $4.0 < h \leqslant 4.4$, etc.

Height, h (m)	Frequency
$4.0 < h \leqslant 4.2$	2
$4.2 < h \leqslant 4.4$	3
$4.4 < h \leqslant 4.6$	5
$4.6 < h \leqslant 4.8$	8
$4.8 < h \leqslant 5.0$	12
$5.0 < h \leqslant 5.2$	18
$5.2 < h \leqslant 5.4$	15
$5.4 < h \leqslant 5.6$	7

Example

The cumulative frequency table shows the amount of pocket money for 40 teenagers.

Pocket money, x (£)	Cumulative frequency
$1 \leqslant x \leqslant 2$	8
$1 \leqslant x \leqslant 3$	19
$1 \leqslant x \leqslant 4$	32
$1 \leqslant x \leqslant 5$	39
$1 \leqslant x \leqslant 6$	40

a Draw a cumulative frequency graph.

b Use the cumulative frequency graph to find an estimate for the median amount of pocket money.

c Estimate the range.

a

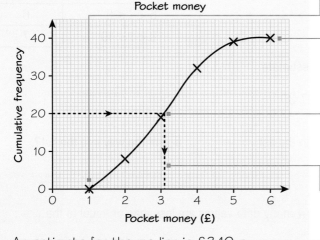

Plot (1, 0). There are no people with less than £1 pocket money.

Plot each frequency at the upper class boundary: (2, 8), (3, 19), etc. Draw a smooth curve through the points.

To estimate the median, draw a line from the halfway cumulative frequency value, $\dfrac{n}{2} = \dfrac{40}{2} = 20$

Draw a line down to the x-axis and read off the value.

b An estimate for the median is £3.10.

This is an estimate because it depends on how the curve is drawn. Also, you don't know the exact values in the class containing the median.

c Lowest possible value = £1
Highest possible value = £6
An estimate for the range is $6 - 1 = £5$

5 a Draw a cumulative frequency graph for the giraffe data in **Q4**.

b Find an estimate for the median height of the giraffes.

Exam-style question

6 This table gives the times taken by 50 students to solve a maths puzzle.

Time, t (minutes)	Frequency
$0 < t \leqslant 2$	3
$2 < t \leqslant 4$	12
$4 < t \leqslant 6$	19
$6 < t \leqslant 8$	10
$8 < t \leqslant 10$	6

a Draw a cumulative frequency graph. **(2 marks)**

b Use your graph to find an estimate for the median time. **(1 mark)**

c Maria says that the range of times is $10 - 0 = 10$.
Explain why the exact range may not be 10. **(1 mark)**

7 **Reasoning** The times taken for 80 runners to complete a 10-kilometre fun run are shown in the table.

Time, t (minutes)	Frequency
$40 < t \leqslant 45$	3
$45 < t \leqslant 50$	17
$50 < t \leqslant 55$	25
$55 < t \leqslant 60$	26
$60 < t \leqslant 65$	8
$65 < t \leqslant 70$	1

a Draw a cumulative frequency graph.

b Estimate the median time taken.

c Draw a line across to the curve from the cumulative frequency value one-quarter of the way up. Draw a line down to the x-axis to estimate the lower quartile of the time taken.

d Estimate the upper quartile.

e What fraction of the times are

   i less than the LQ

   ii less than the UQ?

f Use your answers to parts **c** and **d** to work out an estimate for the interquartile range.

8 **Reasoning** The table shows the masses of 60 hippos.

Mass, m (tonnes)	Frequency
$1.3 < m \leqslant 1.4$	4
$1.4 < m \leqslant 1.5$	7
$1.5 < m \leqslant 1.6$	21
$1.6 < m \leqslant 1.7$	18
$1.7 < m \leqslant 1.8$	10

a Draw a cumulative frequency graph.

b Estimate the median, quartiles and interquartile range.

c Draw a line to the curve from 1.55 on the x-axis. Estimate how many hippos weigh less than 1.55 tonnes.

d Copy and complete.

   40 hippos are estimated to weigh less than ☐ tonnes.

Exam-style question

9 Charlie drives to work. The table gives information about the time (t minutes) it took him to get to work on each of 100 days.

Time, t (minutes)	Frequency
$0 < t \leqslant 10$	16
$10 < t \leqslant 20$	34
$20 < t \leqslant 30$	32
$30 < t \leqslant 40$	14
$40 < t \leqslant 50$	4

Exam tip

For part **c** draw lines on the graph with a ruler to show how you got your answers.

a Draw a cumulative frequency table. **(1 mark)**

b Draw a cumulative frequency graph. **(2 marks)**

c Use your graph to find an estimate for the percentage of days it took Charlie more than 18 minutes to drive to work. **(2 marks)**

14.3 Box plots

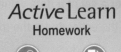

Active Learn
Homework

- Find the quartiles and the interquartile range from stem-and-leaf diagrams.
- Draw and interpret box plots.

Warm up

1 Fluency a In a set of 50 data values, which value is the median?

b The upper quartile is 26 and the lower quartile is 10. What is the interquartile range?

2 The stem-and-leaf diagram gives the ages of the members of a judo club.

a Find the median age.

b Work out the range.

c What fraction of the members are less than the median age?

1	3 5 8 8
2	0 1 1 2 5 6 6 7 9
3	1 7
4	0 5
5	1

Key: 3 | 7 represents 37 years

Key point

A **box plot**, sometimes called a **box-and-whisker diagram**, displays a data set to show the median and quartiles.

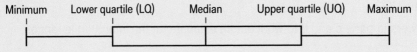

Minimum   Lower quartile (LQ)   Median   Upper quartile (UQ)   Maximum

Summary statistics for a set of data are the averages, ranges and quartiles.

Example

The table shows **summary statistics** from a data set of the lengths of ladybirds.

Minimum	Lower quartile (LQ)	Median	Upper quartile (UQ)	Maximum
3 mm	5 mm	8 mm	9 mm	11 mm

Draw a box plot for the data.

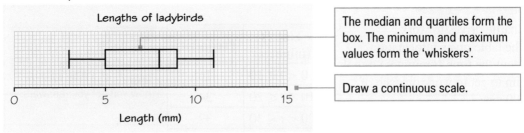

Lengths of ladybirds

Length (mm)

The median and quartiles form the box. The minimum and maximum values form the 'whiskers'.

Draw a continuous scale.

3 Draw a box plot for this data on the masses of tomatoes.

Minimum	LQ	Median	UQ	Maximum
120 g	135 g	140 g	150 g	154 g

---

4 **Reasoning** This data shows the lengths of time, in minutes, it took 11 students to complete an essay.

15, 18, 19, 21, 22, 25, 26, 26, 28, 30, 31

a Write down the median time taken.   b Find the upper and lower quartiles.

c Draw a box plot for the data.

d **Reflect** How accurately can you read values from a graph?

Why do you think you use the $\dfrac{n+1}{4}$ th data value for the lower quartile in a data set but

the $\dfrac{n}{4}$ th value in a cumulative frequency graph? Write a sentence to explain.

5 **Problem-solving** This data shows the heights, in metres, of 15 trees.

4.5, 5.3, 11, 4.8, 6.1, 10.2, 5.8, 7.3, 8, 9.6, 6.3, 8.8, 4.9, 6, 8

a Draw a box plot for the data.

b What percentage of the trees are less than the LQ height?

6 **Reasoning** This stem-and-leaf diagram shows the ages of 37 people on a bus.

a Draw a box plot for the data.

b Work out the interquartile range.

```
1 | 6 7 9 9
2 | 1 2 2 4 6 7 8
3 | 3 6 6 7 8 9 9
4 | 0 0 1 3 4 5 8 8 9 9
5 | 0 1 2 4 4 5 6 7 9
```

Key: 3 | 6 represents 36 years

Exam-style question

7 The table gives some information about the heights of 60 sunflowers.

a Draw a box plot to represent this information.

(**3 marks**)

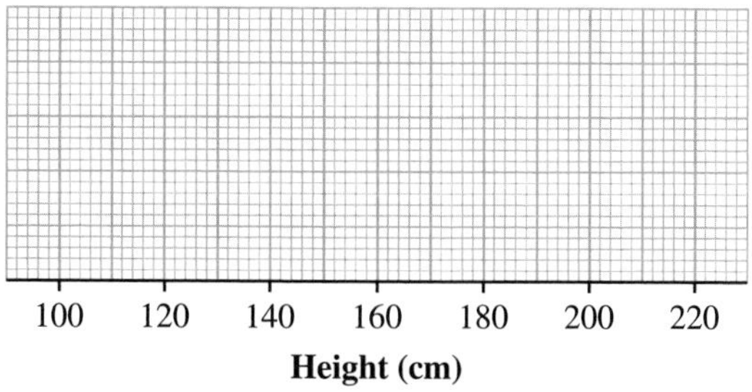

Least height	105 cm
Greatest height	220 cm
Lower quartile	130 cm
Upper quartile	175 cm
Median	160 cm

Height (cm)

Exam tip

Use a sharp pencil and a ruler. Draw the box plot as accurately as possible.

b Work out an estimate for the number of these sunflowers with a height between 105 cm and 130 cm. (**2 marks**)

Josh says, 'There must be some sunflowers with a height between 190 cm and 200 cm.'

c Is Josh right? You must give a reason for your answer. (**1 mark**)

8 **Problem-solving** This frequency table gives the heights of some schoolchildren.

Height, h (m)	Frequency
$1.0 < h \leqslant 1.1$	3
$1.1 < h \leqslant 1.2$	4
$1.2 < h \leqslant 1.3$	6
$1.3 < h \leqslant 1.4$	9
$1.4 < h \leqslant 1.5$	10
$1.5 < h \leqslant 1.6$	4

a Draw a cumulative frequency graph for this data.

b Draw a box plot for the data. Use the same horizontal axis as you used in your cumulative frequency graph.

c Copy and complete this inequality to show the heights of the middle 50% of the schoolchildren

$\square \leqslant h \leqslant \square$

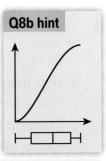

Q8b hint

Key point

Comparative box plots are box plots for two different sets of data drawn in the same diagram.

9 The ages of children in two different after-school clubs are shown in the comparative box plots.

Ages of children in after–school clubs

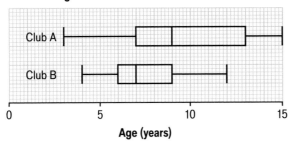

a Which club had the higher median age?

b What is the upper quartile for club A?

c What percentage of the children in club A are less than the UQ age?

d Work out the interquartile range for each club.

e Work out the range for each club.

10 **Future skills** These box plots show the average daily temperature in March and June at a holiday resort.

a Give a reason why a holiday-maker might prefer to go to the resort in March.

b Give a reason why they might prefer to go in June.

Average daily temperature at holiday resorts

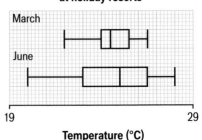

11 Reasoning Summary statistics on the masses, in grams, of two different species of birds are given in this table.

	Minimum	LQ	Median	UQ	Maximum
Species A	45	52	60	65	69
Species B	33	44	65	77	90

a Draw comparative box plots for the two species.
Use the same scale and draw one box plot above the other.

b Which species has the lower median mass?

c Which species has the greater spread of masses?

12 Reasoning The cumulative frequency graph gives information about the masses, in kilograms, of 36 male and 36 female gibbons.

a Use the graph to find the median and quartiles for each gender.

b Draw comparative box plots for the two genders.

c Compare the median masses and the spread of masses for the two genders.

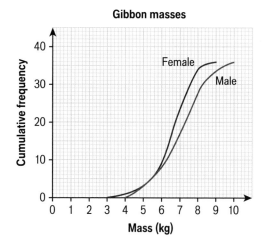
Gibbon masses

Exam-style question

13 The box plot shows information about the distribution of the prices of games consoles on website A.

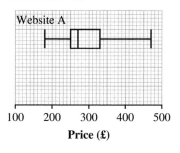
Website A
Price (£)

Exam tip

Plot points carefully. Use the grid and the scale given.

a Work out the interquartile range for the prices on website A. **(2 marks)**

The table shows information about the distribution of the prices of games consoles on website B.

	Smallest	Lower quartile	Median	Upper quartile	Largest
Price (£)	190	270	320	370	450

b Draw a copy of the grid above, and draw a box plot for the information in the table. **(2 marks)**

Kate says,
    'The box plot shows that the prices are higher on website B than on website A.'

c Is Kate correct? Give a reason for your answer. **(1 mark)**

14.4 Drawing histograms

Active Learn
Homework

- Understand frequency density.
- Draw histograms.

Warm up

1 Fluency Work out

  a $32 \div 10$

  b $158 \div 20$

  c $30 \div 0.2$

  d $8 \div 0.05$

2 The masses of 101 birds are recorded in a table.

Mass, m (g)	$10 < m \leqslant 12$	$12 < m \leqslant 14$	$14 < m \leqslant 16$	$16 < m \leqslant 18$	$18 < m \leqslant 20$
Frequency	11	23	31	20	16

  a Draw a frequency diagram for the data.

  b Write down the modal class.

  c What does the height of each bar represent?

Key point

A **histogram** is a type of frequency diagram used for grouped continuous data.
In a histogram for unequal class intervals the area of the bar represents the frequency.
The height of each bar is the **frequency density**.

$$\text{Frequency density} = \frac{\text{frequency}}{\text{class width}}$$

---

3 The heights of 70 trees are recorded in this table.

Height, h (metres)	Frequency	Class width	Frequency density
$20 < h \leqslant 25$	8	5	$\frac{8}{5} = 1.6$
$25 < h \leqslant 30$	12		
$30 < h \leqslant 40$	35		
$40 < h \leqslant 50$	15		

  a Work out each class width.

  b Work out the frequency density for each class.

4 Work out the frequency density for each class in **Q2**.

Example

The lengths of 48 worms are recorded in this table.

Length, x (mm)	$15 < x \leqslant 20$	$20 < x \leqslant 30$	$30 < x \leqslant 40$	$40 < x \leqslant 60$
Frequency	6	14	26	2
Class width	5	10	10	20
Frequency density	$\frac{6}{5} = 1.2$	$\frac{14}{10} = 1.4$	$\frac{26}{10} = 2.6$	$\frac{2}{20} = 0.1$

Draw a histogram to display this data.

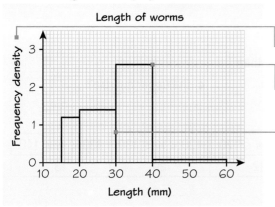

Length of worms

Work out the frequency density for each class.

Label the y-axis 'Frequency density'.

The height of each bar is the frequency density for each class.

Draw the bars with no gaps between them.

---

5 This table shows the times taken for 55 runners to complete a fun run.

  a Copy and complete the table.

  b Draw a histogram for this data.

Time, t (minutes)	Frequency	Class width	Frequency density
$40 < t \leqslant 45$	4		
$45 < t \leqslant 50$	17		
$50 < t \leqslant 60$	22		
$60 < t \leqslant 80$	12		

---

6 **Reasoning** Harry is asked to draw a histogram for the data in the table.
Here is Harry's diagram. It is not a histogram.

Height, h (cm)	Frequency
$120 < h \leqslant 130$	10
$130 < h \leqslant 135$	15
$135 < h \leqslant 140$	11
$140 < h \leqslant 160$	20

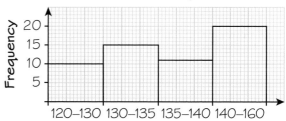

  a Write down two things that are wrong with this diagram.

  b Draw the correct histogram for the data.

  c **Reflect** On a frequency diagram, the tallest bar represents the modal class.
     Is this also true for a histogram? Explain.

Exam-style question

7 This table contains data on the heights of 77 students.
Draw a histogram for this data. **(3 marks)**

Height, h (m)	Frequency
$1.50 < h \leqslant 1.52$	4
$1.52 < h \leqslant 1.55$	18
$1.55 < h \leqslant 1.60$	25
$1.60 < h \leqslant 1.65$	15
$1.65 < h \leqslant 1.80$	15

14.5 Interpreting histograms

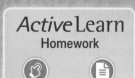
• Interpret histograms.

Warm up

1 Fluency What fraction of a set of data is greater than

  a the median **b** the lower quartile **c** the upper quartile?

2 The masses of 80 apples are recorded in a table.

Mass, m (grams)	$100 < m \leqslant 110$	$110 < m \leqslant 120$	$120 < m \leqslant 130$	$130 < m \leqslant 140$
Frequency	16	22	29	13

  a Work out an estimate for the mean mass.

  b Find the class interval that contains the median.

3 Reasoning The histogram shows the masses of a number of squirrels.

  a What does the area of a bar on a histogram represent?

  b How many squirrels weigh between 150 grams and 200 grams?

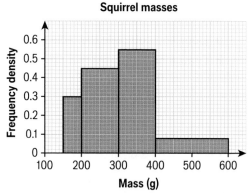

Squirrel masses

  c Copy and complete the frequency table for the masses of squirrels.

  d How many squirrels weigh between 200 grams and 400 grams?

  e How many squirrels are there in total?

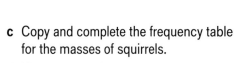

Mass, m (g)	Frequency
$150 < m \leqslant 200$	$50 \times 0.3 = \square$
$200 < m \leqslant 300$	
$300 < m \leqslant 400$	
$400 < m \leqslant 600$	

4 Reasoning The histogram shows the distances a group of football fans have to travel to a match.

  a How many fans travelled less than 5 km?

  b Draw a vertical line at distance 15 km.

   i What are the height and width of the bar between 10 km and 15 km?

   ii Calculate the area of this bar to estimate how many fans travelled between 10 km and 15 km.

  c Estimate how many fans travelled between

   i 25 km and 30 km   **ii** 30 km and 32 km   **iii** 25 km and 32 km

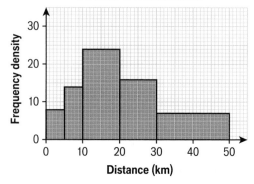

Distance to football match

5 The incomplete table and histogram give some information about distances people travel to work.

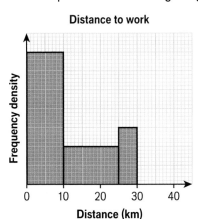

Distance to work

Distance to work, x (km)	Frequency
$0 < x \leqslant 10$	140
$10 < x \leqslant 25$	
$25 < x \leqslant 30$	
$30 < x \leqslant 40$	100

Q5a hint

Area of bar $= 10 \times \square = 140$
Use this to label the frequency density axis.

a Reasoning Use the information in the histogram to complete the frequency table.

b Complete the histogram.

6 Problem-solving The histogram shows the times taken for a number of students to complete an arithmetic test.

a Estimate how many students took at least 19 minutes to complete the test.

b Draw a grouped frequency table for the data.

c Work out an estimate for the mean time taken.

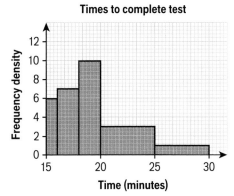

Times to complete test

Example

The histogram shows the masses of pumpkins in a farm shop.

a Work out an estimate for the median mass.

b Explain the assumption you have made.

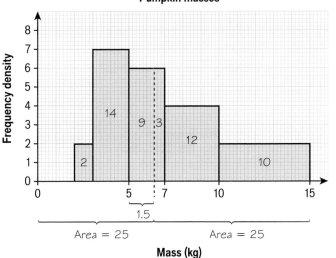

Pumpkin masses

a Total frequency $= 1 \times 2 + 2 \times 7 + 2 \times 6 + 3 \times 4 + 5 \times 2 = 50$

The median is the $\frac{50}{2}$th $= 25$th value.

It lies in the class $5 < m \leqslant 7$

Frequency $=$ area $= 9$

Frequency density $= 6$

Class width $= \frac{9}{6} = 1.5$

An estimate for the median is $5 + 1.5 = 6.5$ kg

b I have assumed that all the values in the class $5 < m \leqslant 7$ are evenly distributed through the class interval.

Add the class width to the lower class boundary.

Work out the areas of all the bars to find the total frequency.

Work out which class contains the median.

Use $\frac{n}{2}$ as it is a large data set.

Use frequency density $= \dfrac{\text{frequency}}{\text{class width}}$ to find class width of class from 5 to median.

7 Work out an estimate for the median of the data in **Q5**.

8 The histogram shows the masses of some elephants.

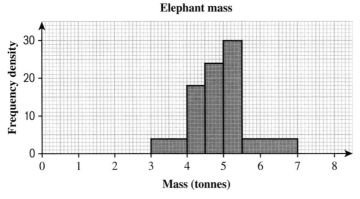

Exam tip

Label the bars with their frequencies as you work them out.

a How many elephants are there in total? **(1 mark)**

b Estimate the median mass. **(2 marks)**

c 40% of the elephants that weigh more than 5.2 tonnes are female.
Work out an estimate for the number of female elephants over 5.2 tonnes. **(3 marks)**

9 **Reasoning** The histogram shows the heights of some students.

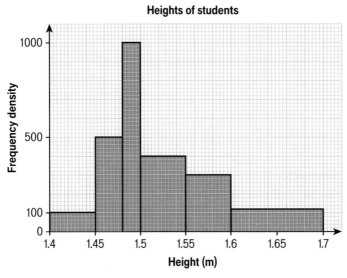

a Five students are between 1.4 metres and 1.45 metres tall.
Work out the frequency density for that class.

b Draw a frequency table for the data in the histogram.

c How many students are there in total?

d Work out an estimate for the median height.

e Work out an estimate for the mean height from your frequency table.

f Estimate how many of the students are taller than the mean height.

g Work out an estimate for the lower quartile.

14.6 Comparing and describing distributions

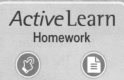

Active Learn
Homework

- Solve problems by comparing distributions.

Warm up

1 Fluency What are the measures of

a spread

b average?

2 The table gives the masses, in kilograms, of male and female giant tortoises.

Male	280	283	288	290	292	299	300	305	310
Female	260	261	263	265	269	270	271	273	274

a Work out the mean mass for each gender.

b Write down the range of masses for each gender.

Key point

The interquartile range measures the spread of the middle 50% of the data.
To describe a data set (or population) give a measure of average and a measure of spread.
To compare data sets, compare a measure of average and a measure of spread.

3 Compare the masses of the two genders of tortoise in **Q2**.

4 Future skills The heights, in centimetres, of female African and Asian elephants are shown in the table.

African	270	275	281	286	290	292	295
Asian	220	221	223	224	226	227	229

a Use the median and the interquartile range to describe the distribution of the heights of these two elephant species.

b Compare the heights of the two species.

5 Future skills The lengths, in minutes, of telephone calls to a helpline are recorded.
10, 11, 13, 15, 17, 18, 18, 19, 21, 22, 95

a Work out the mean call length.

b Work out the median call length.

c Work out the range and interquartile range.

d Reflect Which measures of average and spread best represent this data?
Explain why.

6 **Reasoning** Ten male and ten female cyclists compete in a road race.
The times, in minutes, to complete the course are recorded.
  Males: 68, 70, 75, 76, 77, 79, 81, 83, 90, 120
  Females: 71, 75, 76, 78, 83, 86, 89, 90, 91, 92

  a Explain which of the median and interquartile range, or the mean and range, should be used to describe each set of data.

  b Compare the distributions of the times for males and females.

Exam-style question

7 The table shows the lengths of delays to trains at Stratfield station.

Length of delay, x (minutes)	Frequency
$0 \leqslant x < 5$	3
$5 \leqslant x < 10$	7
$10 \leqslant x < 15$	12
$15 \leqslant x < 20$	6
$20 \leqslant x < 25$	2

  a Draw a cumulative frequency graph. **(3 marks)**

The shortest delay was 0 minutes.
The longest delay was 24 minutes.

  b On a copy of the grid, draw a box plot for the information about the delays.

Train delays at Stratfield station **(3 marks)**

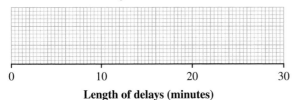

Length of delays (minutes)

Exam tip

Use the scale and grid given for your box plot.

The box plot shows the delays to trains at Westford station.

Train delays at Westford station

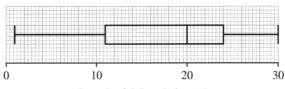

Length of delays (minutes)

  c Compare the distributions of the lengths of delays at the two stations. **(2 marks)**

8 Problem-solving This back-to-back stem-and-leaf diagram shows the average speeds, in miles per hour, of cars passing two checkpoints.

Checkpoint A

Checkpoint A		Checkpoint B
8 3 2 1	1	
5 4 2 1	2	3 4 6 8
6 5 4 2 2 1	3	1 3 4 5 6 6 7 9 9
5 4 3 2	4	1 2 2 2 3 4 6 7 8
2 1	5	

Key: (Checkpoint A) 1 | 2 represents 21 mph
(Checkpoint B) 2 | 3 represents 23 mph

a Describe the speeds at the two checkpoints.

b Compare the speeds at the two checkpoints.

9 Problem-solving The cumulative frequency graph shows the ages of male and female members of a health club.

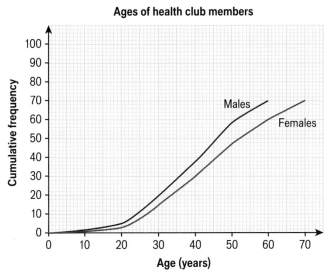

Ages of health club members

Compare the two sets of data.

10 Problem-solving / Future skills Students were timed on solving a puzzle. Then they used 'brain training' software each day for a week, and then they were timed solving a similar puzzle. The frequency table shows the results.

Time taken, t (seconds)	Frequency (before training)	Frequency (after training)
$12 \leqslant t < 15$	4	5
$15 \leqslant t < 18$	6	8
$18 \leqslant t < 21$	9	14
$21 \leqslant t < 24$	16	16
$24 \leqslant t < 27$	10	14
$27 \leqslant t < 30$	9	3
$30 \leqslant t < 33$	6	0

a Compare the distributions of the 'before' and 'after' data.

b Do the results support the software company's claim that their 'brain training' improves puzzle-solving?

14 Check up

Active Learn
Homework

Sampling

1   The president of a club wants to survey its 250 members.
    She decides to take a sample of 50 members.
    The president lists the members alphabetically and numbers them.
    She uses a random number table to select a sample.
    Use the table below to write down the numbers of the first 5 people to include in the sample.

    0489797466253111571640234021255613306...

2   Ravi captures 30 squirrels in a wood.
    He marks each squirrel with dye.
    He then lets the squirrels go.
    The next day he captures 21 squirrels.
    7 of these squirrels have been marked with dye.

    a Work out an estimate for the number of squirrels in the wood.

    b Write down any assumptions you have made.

Graphs and charts

3   This table shows the masses of 90 emperor penguins.

    a Draw a cumulative frequency table.

    b Draw a cumulative frequency graph.

Mass, m (kg)	Frequency
$20 < m \leqslant 23$	1
$23 < m \leqslant 26$	4
$26 < m \leqslant 29$	8
$29 < m \leqslant 32$	21
$32 < m \leqslant 35$	32
$35 < m \leqslant 38$	18
$38 < m \leqslant 41$	6

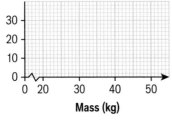

    c Use your graph to estimate the median mass of the penguins.

    d Find estimates for the lower quartile, upper quartile and interquartile range.

    e Estimate how many penguins weigh

       i   less than 36 kg

       ii more than 30 kg

    The smallest mass was 21 kg.
    The greatest mass was 40 kg.

    f Draw a box plot for the masses of the penguins.

4 The heights of 100 pine trees are given in this table.

a Draw a histogram for this data.

b Estimate how many trees are taller than 23 m.

Height, h (m)	Frequency
$0 < h \leqslant 10$	3
$10 < h \leqslant 15$	7
$15 < h \leqslant 20$	14
$20 < h \leqslant 22$	31
$22 < h \leqslant 25$	27
$25 < h \leqslant 30$	16
$30 < h \leqslant 40$	2

Comparing data

5 The box plots show the times for 15 boys and 15 girls to run 100 m.

Compare the distributions of the times for the girls and the boys.

Times to run 100 m

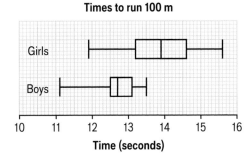

6 The stem-and-leaf diagram shows the ages of people attending a birthday party at a hotel.

a What is the median age?

b Work out the interquartile range.

c At a second party, the median age was 22 and the interquartile range was 10.
Compare the two distributions.

```
0 | 3 5 6 8 9
1 | 2 2 5 8 8 9
2 | 1 3 6 6 7 8 8 9 9
3 | 2 4 4 5 6 6 7
4 | 1 1 4 5 5 8
5 | 4 7
```

Key: 1 | 2 represents 12 years

7 **Reflect** How sure are you of your answers? Were you mostly

Just guessing 😞   Feeling doubtful 😐   Confident 🙂

What next? Use your results to decide whether to strengthen or extend your learning.

Challenge

8 The histogram shows the masses of a group of porcupines.

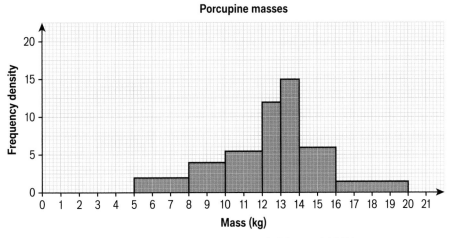

Porcupine masses

Estimate how many porcupines weigh between 8.5 kg and 16.5 kg.

14 Strengthen

Active Learn
Homework

Sampling

1 Use this random number display to select 5 people from a list of 30 people.

> 027921512108015701873373177018402124206662...

a Write the list as a sequence of two-digit numbers: 02, 79, ...

b Cross out the numbers over 30.

c Cross out any repeats.

d Write down, in order, the numbers of the first 5 people selected.

2 Use the random number table from **Q1** to select a random sample of 8 people from a list of 70 people.

> **Q2 hint**
>
> Cross out all numbers over 70.

3 Melanie captures 20 ducks from a lake.
She tags each duck with a ring on its leg.
She then lets the ducks go.
The next day she captures 16 ducks.
4 of these ducks have been tagged.

a In the sample of 16 ducks, what fraction of the ducks were tagged?

Assuming the sample is representative of the whole population,
Mel has tagged the same fraction of the population.

b Copy and complete.

In the sample: 4 tagged ducks = ☐ of the sample
In the population: 20 tagged ducks = ☐ of the population
                Total population = ☐ ducks

> **Q3b hint**
>
>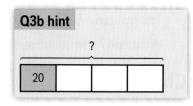

4 Sula captures 15 geese from a lake.
She tags each goose with a ring on its leg.
She then lets the geese go.
The next day she captures 12 geese.
4 of these geese have been tagged.

a Work out an estimate for the number of geese at the lake.

b Write down any assumptions you have made.

> **Q4a hint**
>
> This is the population.

> **Q4b hint**
>
> What have you assumed about the sample? See **Q3**.

Graphs and charts

1 The table shows the masses of 60 sheep.

a Copy and complete this cumulative frequency table.

Mass, m (kg)	Cumulative frequency
$70 < m \leqslant 75$	1
$70 < m \leqslant 80$	$1 + 8 = \square$
$70 < m \leqslant 85$	
$70 < m \leqslant 90$	
$70 < m \leqslant 95$	
$70 < m \leqslant 100$	

Mass, m (kg)	Frequency
$70 < m \leqslant 75$	1
$75 < m \leqslant 80$	8
$80 < m \leqslant 85$	19
$85 < m \leqslant 90$	15
$90 < m \leqslant 95$	10
$95 < m \leqslant 100$	7

Q1a hint

For the class $70 < m \leqslant 85$, add the frequencies for all the groups $\leqslant 85$.

b Copy and complete the cumulative frequency graph.

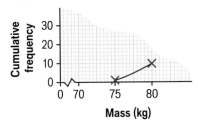

Q1b hint

Plot the top value in each class against frequency.
Draw a smooth curve through the points.

2 The cumulative frequency graph shows information about the amounts of pocket money 40 children receive.

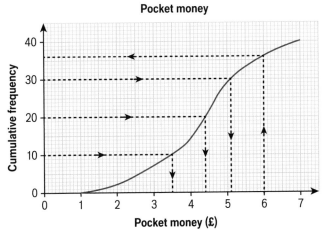

a Read off the values for the median, lower quartile and upper quartile.

b Work out the interquartile range for the pocket money.

Q2a hint

Median is halfway up cumulative frequency axis.
Lower quartile is $\frac{1}{4}$ of the way up.
Upper quartile is $\frac{3}{4}$ of the way up.

Q2b hint

Interquartile range = upper quartile − lower quartile

c Read off the number of children who received $\leqslant$ £6.

d Work out the number of children who received more than £6.

3 Draw a box plot for the data in **Q2**.

  a Copy the *x*-axis scale from **Q2**.

  b Mark on the median, lower and upper quartiles with vertical lines.

  c Complete the box with ends at the lower and upper quartiles.

  d Mark on the lowest possible and highest possible values from the graph and join with lines.

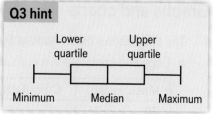

Q3 hint

Lower quartile   Upper quartile

Minimum   Median   Maximum

 4 The lengths of some caterpillars are shown in the table.

Length, l (mm)	Frequency	Class width	Frequency density
$10 < l \leqslant 15$	2	$15 - 10 = 5$	$\frac{2}{5} = 0.4$
$15 < l \leqslant 20$	8	$20 - 15 = \square$	$\frac{8}{\square} = \square$
$20 < l \leqslant 30$	15		
$30 < l \leqslant 40$	12		
$40 < l \leqslant 60$	5		

  a For the class $15 < l \leqslant 20$, work out

   i the class width

   ii frequency density $= \dfrac{\text{frequency}}{\text{class width}}$

  b Copy and complete the table.

  c Copy and complete the histogram.

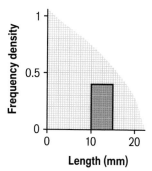

Length (mm)

Q4c hint

Plot frequency **density** on the *y*-axis.

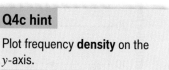

5 This histogram shows how long it took 30 students to complete a test.

  a Work out

   class width $\times$ frequency density

   for each bar, to find the frequency for each class interval.

  b Add the frequencies for the first two bars to find how many students took less than 30 minutes.

  c How many students took 30–35 minutes?

  d What fraction of the 30–35 bar is for 30–33?

  e Estimate how many students took between 30 and 33 minutes.

  f Estimate how many students took less than 33 minutes.

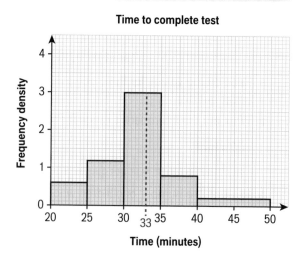

Time to complete test

Time (minutes)

Comparing data

1   The box plots show the scores achieved by some boys and girls in a test.

Scores in arithmetic test

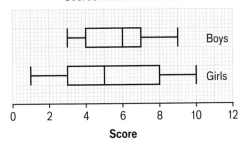

a   Copy and complete this table.

	Lower quartile	Median	Upper quartile	Interquartile range
Boys				
Girls				

b   Copy and complete.

   i   The median for the boys is _____ than the median for the girls.

   ii   The _____ have a smaller interquartile range than the _____ .

2   The stem-and-leaf diagram shows the masses, in kilograms, of female wild boars.

```
5 | 1 2 2 4 5 9
6 | 0 3 5 7 7 7 8 9
7 | 2 3 3 5 6 9
8 | 0 0 2
```

Key: 6 | 0 represents 60 kg

a   How many boars are recorded in the stem-and-leaf diagram?

b   Work out $\dfrac{n+1}{2}, \dfrac{n+1}{4}$ and $\dfrac{3(n+1)}{4}$

   n is the value you worked out in part **a**.

c   Find these data values by counting from the first one.

d   Write down the median value, the LQ and the UQ.

e   Work out the interquartile range of the masses.

f   For male wild boars:

Median	IQR
69 kg	15.5 kg

Write sentences like the ones in **Q1b** to compare the distributions of the male and female masses.

14 Extend

1 **Future skills** The speeds of 75 cheetahs are recorded in this table.

a Draw an appropriate graph for the data.

b Find an estimate for

   i the median

   ii the interquartile range

Speed, s (mph)	Frequency
$20 < s \leqslant 25$	3
$25 < s \leqslant 30$	8
$30 < s \leqslant 35$	19
$35 < s \leqslant 40$	21
$40 < s \leqslant 45$	15
$45 < s \leqslant 50$	9

2 **Future skills** Preeti and Nik own a nail bar. They have a list of 342 customers.
They want to find out their customers' views on their prices, products and opening hours.

a What is the population for this survey?

b Preeti suggests they take a random sample of 20 customers.
Nik suggests they take a random sample of 40 customers.
Which size sample would give the most representative results for the population?

c 38 customers answer the question
'How much, on average, do you spend on your nails each month?'
The mean amount for these customers is £26.50.
The nail bar sees around 200 customers each month.
Estimate the total amount 200 customers would spend in a month.
What assumptions have you made?

3 **Future skills** The lengths of calls received by two call centres are recorded and shown on this graph.

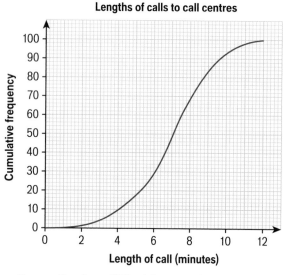

Lengths of calls to call centres

a Draw a line from 10% of the cumulative frequency across to the graph and down to the x-axis.
Copy and complete.
10% of the calls were less than ☐ minutes.

b Work out the limits between which the middle 80% of the data lies.

4 **Problem-solving** A naturalist captures 30 bats in a cave and tags them.
There are approximately 600 bats in the colony.
The naturalist returns a month later and captures 40 bats.
How many would she expect to find tagged?

 5 **Reasoning** The times, in seconds, for 15 runners to complete a hurdles race are recorded.
13.5, 14.1, 14.2, 14.3, 14.5, 14.7, 14.7, 14.9, 15.0, 15.1, 15.2, 15.5, 15.7, 15.8, 28.6
a Work out the median time taken.
b Work out the lower quartile, upper quartile and interquartile range.
c Work out $1.5 \times$ interquartile range.
d An outlier can be defined as 'a data value which lies more than $1.5 \times$ the interquartile range below the lower quartile or above the upper quartile'.
Use this definition to decide whether any of the data points are outliers.

 6 **Problem-solving** The table shows the times taken for players to complete a round of golf.

Time, t (hours)	Frequency
$3.0 < t \leqslant 3.5$	7
$3.5 < t \leqslant 3.8$	12
$3.8 < t \leqslant 4.0$	18
$4.0 < t \leqslant 4.5$	24
$4.5 < t \leqslant 6.0$	15

a Draw a histogram for this data.
b Players who took longer than 4 hours 20 minutes were given a two-shot penalty.
Estimate the number of players who were given a two-shot penalty.

7 **Reasoning** The masses of 100 camels are given in this table.

Mass, m (kg)	Frequency
$300 < m \leqslant 500$	8
$500 < m \leqslant 600$	10
$600 < m \leqslant 700$	15
$700 < m \leqslant 750$	18
$750 < m \leqslant 800$	36
$800 < m \leqslant 900$	11
$900 < m \leqslant 1100$	2

a Draw a histogram for this data.
b Work out an estimate for the mean mass.
c Work out an estimate for the median mass.
d Copy and complete.
75% of the camels have a mass less than ☐ kg (to the nearest kilogram).

8 The incomplete histogram and frequency table show the same information about the times that some vehicles spent in a car park.

Time in car park

(histogram: Frequency density on vertical axis from 0 to 3, Time (minutes) on horizontal axis from 0 to 160)

> This is typical of A level histogram questions where you will often need to match up a frequency from a frequency table with an area of a bar of a histogram. Frequency density × class width = area of bar, which might not be equal to the frequency but is proportional to it.

Time, t (minutes)	Frequency
$0 < t \leqslant 30$	33
$30 < t \leqslant 60$	45
$60 < t \leqslant 80$	
$80 < t \leqslant 120$	64
$120 < t \leqslant 150$	18

a Copy and complete the histogram and work out the missing number in the frequency table.

b 60 vehicles were in the car park for less than T minutes.
Work out an estimate for the value of T.

> **Q8a hint**
>
> Frequency density × class width = area of bar, which is proportional to frequency. Use this to work out the heights of the missing bars and the missing frequency.

> **Q8b hint**
>
> The value of T will lie somewhere in the second class. Work out the area needed (along with the 33 from the first class) to make a total of 60 vehicles.

9 A variable, y, was measured to the nearest whole number. 70 observations were taken and are shown in the table.

y	4–5	6–7	8–11	12–15
Frequency	8	16	28	18

> Questions similar to this have been set at A level very recently.

a Write down the boundaries for the 6–7 class.

A histogram was drawn, and the bar representing the 6–7 class was 1 cm wide and 4 cm high.

b Work out the width and height of the bar representing the 8–11 class.

> **Q9b hint**
>
> The area of each bar is proportional to the frequency. Look at the boundaries of the two classes and deduce the width of the 8–11 bar.

14 Test ready

Summary of key points

To revise for the test:

- Read each key point, find a question on it in the mastery lesson, and check you can work out the answer.

- If you cannot, try some other questions from the mastery lesson or ask for help.

Key points

1   A **population** is the set of items that you are interested in.
    A **census** is a survey of the whole population.
    A **sample** is a smaller number of items from the population.
    A good-sized sample is at least 10% of the population.
    In order to reduce **bias**, the sample must represent the whole population.
    In a **random** sample every item is equally likely to be chosen.   → **14.1**

2   The mean, median, mode and range are called **summary statistics**.
    The summary statistics of an unbiased sample are estimates for the summary statistics
    for the whole population. In general, the larger the sample, the better the estimate.   → **14.1**

3   To estimate the size of the population N of an animal species:

    - Capture and mark a sample size n.

    - Recapture another sample of size M. Count the number marked (m).

    $$\frac{n}{N} = \frac{m}{M}$$

    So, $N = \dfrac{n \times M}{m}$

    This is the **Petersen capture–recapture method**.
    Assumptions:

    - The population has not changed between the release and recapture times.

    - The probability of being captured is the same for all individuals.

    - Marks or tags are not lost.   → **14.1**

4   A **cumulative frequency table** shows how many data values are less than or equal
    to the **upper class boundary** of each data class.
    The upper class boundary is the highest possible value in each class.   → **14.2**

5   A **cumulative frequency graph** has data values on the x-axis and cumulative frequency
    on the y-axis.   → **14.2**

6   For a set of n data values on a cumulative frequency graph

    - the estimate for the median is the $\dfrac{n}{2}$th value

    - the estimate for the **lower quartile** (LQ) is the $\dfrac{n}{4}$th value

    - the estimate for the **upper quartile** (UQ) is the $\dfrac{3n}{4}$th value

    - the **interquartile range** (IQR) = UQ − LQ measures the spread of the middle 50%
      of the data.   → **14.2**

7 For a set of n data values

   • the lower quartile (LQ) is the $\frac{n+1}{4}$th value

   • the upper quartile (UQ) is the $\frac{3(n+1)}{4}$th value   → 14.3

8 A **box plot**, sometimes called a **box-and-whisker diagram**, displays a data set to show the median and quartiles.
   Summary statistics for a set of data are the averages, ranges and quartiles.   → 14.3

9 **Comparative box plots** are box plots for two different sets of data drawn in the same diagram.   → 14.3

10 A **histogram** is a type of frequency diagram used for grouped continuous data.
   In a histogram for unequal class intervals the area of the bar represents the frequency.
   The height of each bar is the **frequency density**.

   $$\text{Frequency density} = \frac{\text{frequency}}{\text{class width}}$$ → 14.4

11 The interquartile range measures the spread of the middle 50% of the data.
   To describe a data set (or population) give a measure of average and a measure of spread.
   To compare data sets, compare a measure of average and a measure of spread.   → 14.6

12 The median and interquartile range are not affected by extreme values or outliers.
   When there are extreme values, the median and interquartile range should be used rather than the mean and range.   → 14.6

Sample student answers

Exam-style question

The box plots show the masses of fish caught in two different lakes.

Tracey says,
   'The fish in lake A are smaller than the fish in lake B.'

Is Tracey correct? You must give a reason for your answer.   **(1 mark)**

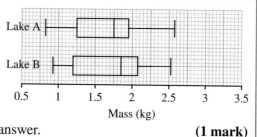

Here are three students' attempts to answer this question.

Student A

Yes, the interquartile range for A is 0.7 kg.
This is smaller than the interquartile range for B, which is 0.875 kg.

Student B

Yes, the median mass of fish in lake A is 1.75 kg.
This is lower than the median mass of fish in lake B, which is 1.85 kg.

Student C

Yes, the fish in lake A are smaller on average than the fish in lake B.

---

a Which student has written the best answer?

b What mistakes did the other two students make?

14 Unit test

Active Learn
Homework

F
H

1 A market research company wants to get the views of members of the public on new chocolate bars.

   a Give two reasons why a sample should be taken rather than a census. **(2 marks)**

   b They propose to take their sample from the school nearest to the company headquarters. Explain whether this will be a good sample. **(1 mark)**

2 Mr Jones is planning an end-of-year show. He asks a sample of 25 students which type of show they would like to watch. Each student chooses one type of show. The table shows information about his results.

There are 1200 students in the school.

Show	Number of students
talent show	4
Shakespeare play	3
modern play	5
musical	13

   a Work out how many of the 1200 students will want to watch the musical. **(2 marks)**

   b State any assumption you made. **(1 mark)**

   c Mrs Brown, the maths teacher, says that the sample of 25 students is too small to be representative.
      Suggest a better sample size and explain why it is more suitable. **(1 mark)**

3 The masses of 100 octopuses are shown in this table.

   a Draw a cumulative frequency graph. **(3 marks)**

   b Estimate the median mass. **(1 mark)**

   c Find the interquartile range. **(2 marks)**

   d Show that over 60% of the octopuses weigh more than 6.5 kg. **(2 marks)**

Mass, m (kg)	Frequency
$3 < m \leqslant 4$	4
$4 < m \leqslant 5$	11
$5 < m \leqslant 6$	15
$6 < m \leqslant 7$	18
$7 < m \leqslant 8$	23
$8 < m \leqslant 9$	17
$9 < m \leqslant 10$	12

4 The box plot shows information about the times taken by a group of men to complete an obstacle race.

Here are the times (in minutes) of the 19 women who completed the obstacle race.

   12, 15, 16, 16, 17, 18, 18, 20, 21, 22,
   22, 23, 23, 24, 24, 26, 29, 31, 32

Time to complete obstacle course

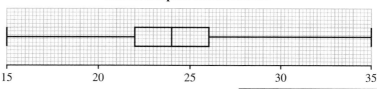

Time (minutes)

Median	
Lower quartile	
Upper quartile	
Shortest time	12
Longest time	32

   a Copy and complete the table to show information about the women's times. **(2 marks)**

   b Mia says that the women were faster at the race than the men. Is Mia correct? You must give a reason for your answer. **(1 mark)**

   c Mia says that the men's times vary more than the women's times. Is Mia correct? You must give a reason for your answer. **(1 mark)**

5 The ages of 15 people in the cast for a show are
6, 15, 23, 18, 17, 9, 11, 32, 31, 14, 45, 25, 26, 26, 15

   a Draw a box plot. **(3 marks)**

   b Outliers are defined as being data points at least $1.5 \times$ interquartile range
above the upper quartile or below the lower quartile.
Use this definition to work out if the data set contains outliers. **(1 mark)**

6 Leo is studying the red squirrels in a forest.
He captures 14 red squirrels.
Here are their masses in grams.
287, 319, 321, 315, 298, 302, 289, 312, 288, 315, 327, 290, 284, 306
He tags them before releasing them.
A week later he captures 18 red squirrels in the same area.
6 of them are tagged.

   a Estimate the number of red squirrels in the forest. **(2 marks)**

   b Estimate the median and range of masses of the squirrels in the forest. **(1 mark)**

   c Write down two assumptions you have made. **(2 marks)**

7 The lengths of 150 dolphins are recorded in this table.

   a Draw a histogram for this information. **(3 marks)**

   b Estimate the median length of the dolphins. **(3 marks)**

   c Estimate how many dolphins are over
3.25 m long. **(3 marks)**

Length, l (m)	Frequency
$1.5 < l \leqslant 2.0$	7
$2.0 < l \leqslant 2.5$	19
$2.5 < l \leqslant 2.8$	31
$2.8 < l \leqslant 3.0$	52
$3.0 < l \leqslant 3.5$	34
$3.5 < l \leqslant 5.0$	7

8 The histogram shows the lengths of
50 Barbour's seahorses.

   a Work out an estimate for the fraction
of the seahorses that have a length
between 14 cm and 16 cm. **(2 marks)**

   b Explain why your answer to part **a**
is an estimate. **(1 mark)**

Lengths of Barbour's seahorses

(TOTAL: 40 marks)

9 **Challenge** Draw a pie chart to represent the seahorse data from **Q8**.

10 **Reflect** Choose **A**, **B** or **C** to complete each statement about statistics.

In this unit, I did ...   **A** well   **B** OK   **C** not very well
I think _____ is ...   **A** easy   **B** OK   **C** hard
When I think about doing _____ , I feel ...   **A** confident   **B** OK   **C** unsure
Did you answer mostly **A**s and **B**s? Are you surprised by how you feel about _____ ?
Why?
Did you answer mostly **C**s? Find the three questions in this unit that you found the hardest.
Ask someone to explain them to you. Then complete the statements above again.

15 Equations and graphs

Prior knowledge

15.1 Solving simultaneous equations graphically

Homework

- Solve simultaneous equations graphically.

Warm up

1 Fluency a What is the equation of this graph?

b Draw the graph of $x^2 + y^2 = 25$.

Key point

You can solve a pair of simultaneous equations by plotting the graphs of both equations and finding the point(s) of intersection.

2 a Draw a coordinate grid with −10 to 10 on both axes.
Draw the graphs of

   i $2y - 4x = 8$ **ii** $y - x = 6$

b Write down the x- and y-values at the point of intersection.

c Check that your x- and y-values satisfy both equations.

3 Reasoning a Match the equations to the three lines A, B and C shown.

   i $x + y = 5$ **ii** $y - 2x = 2$ **iii** $4y + x = 8$

b Hence write down the solutions to these pairs of simultaneous equations.

   i $x + y = 5$ **ii** $x + y = 5$ **iii** $4y + x = 8$
     $y - 2x = 2$ $4y + x = 8$ $y - 2x = 2$

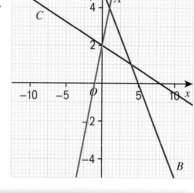

Exam-style question

4 The graphs with equations $2x + 3y = \frac{3}{5}$ and $3x - 2y = \frac{96}{18}$ have been drawn on the grid shown.

Using the graphs, find estimates for the solution of the simultaneous equations

$$2x + 3y = \frac{3}{5}$$
$$3x - 2y = \frac{96}{18}$$

(2 marks)

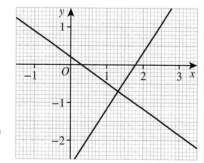

Exam tip

Estimate does not mean 'guess'. Read values from graphs as accurately as you can.

5 Solve each pair of simultaneous equations graphically.

   a $y = 2x + 4$ **b** $2y = 7x - 3$ **c** $0 = y + 2x - 5$ **d** $2x + 3y = -13$

       $y = -2x + 8$ $x + 2y = 21$ $y = 2x + 9$ $x + y = -5$

   e Show that solving the equations in part **d** algebraically gives the same solutions.

Example

Solve this pair of simultaneous equations graphically.

$$y = x^2 + x - 4 \qquad (1)$$
$$y - 2x + 2 = 0 \qquad (2)$$

(1) $y = x^2 + x - 4$

x	-3	-2	-1	0	1	2	3
y	2	-2	-4	-4	-2	2	8

> Construct a table of values and plot the graph.

(2) $y - 2x + 2 = 0$
            $y = 2x - 2$

> Rearrange the equation to make y the subject.

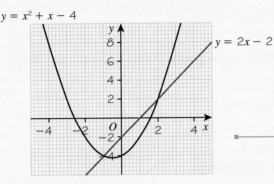

> Plot the linear graph on the same grid using the y-intercept and gradient.

The solutions are
$x = 2$, $y = 2$ and $x = -1$, $y = -4$

> Write down the pair of x- and y-values for each intersection.

6 Use a graphical method to find approximate solutions to this pair of simultaneous equations.

$$y + x^2 = 2 \qquad y + 1 = x$$

> **Q6 hint**
>
> Start by rearranging the equations, then plot the graphs.

7 **Reasoning** Solve this pair of simultaneous equations

   $3x + 2y = 5 \qquad x^2 + y = 7$

   a graphically

   b algebraically

   c **Reflect** Which method gives the more accurate solutions? Explain.

Exam-style question

8 **a** On a suitable grid draw the graph of $x^2 + y^2 = 13.69$.   **(2 marks)**

    b Hence find estimates for the solutions of the simultaneous equations

       $x^2 + y^2 = 13.69$

       $2x + y = 12$ **(3 marks)**

9 Use a graphical method to find an estimate for the solution(s) to the simultaneous equations

   $x^2 + y^2 = 9$

   $x + y = 1$

15.2 Representing inequalities graphically

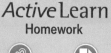

Active Learn
Homework

- Represent inequalities on graphs.
- Interpret graphs of inequalities.

Warm up

1 **Fluency** Solve the inequalities and represent the solutions using **set notation**.

   a $2x \leqslant 12$ **b** $3x - 5 > -11$ **c** $3(x - 5) \leqslant x - 7$

Key point

The points that satisfy an inequality such as $y \leqslant 3$ can be represented on a graph.

- A solid line means that the shaded area includes the points on the line.

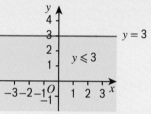

- A dotted line means that the shaded area does *not* include the points on the line.

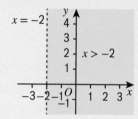

2 **a** Write down the inequality represented by each shaded region.

   i

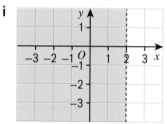

   ii

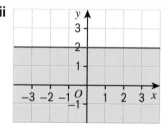

   iii

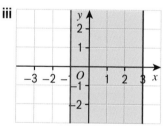

   b On a suitable coordinate grid, shade the region of points whose coordinates satisfy

   i $x \leqslant -2$ **ii** $y > 0$ **iii** $4 > x$ **iv** $-2 < x < 1$

   v $2 \leqslant y < 3.5$ **vi** $-5.5 \leqslant x \leqslant -3$ **vii** $-4 \leqslant y < -3.5$ **viii** $-1 < x \leqslant \frac{3}{4}$

3 **a** Draw a coordinate grid with -5 to 5 on both axes.

   b Draw the graph of $y = 2x + 1$ with a dotted line.

   c Copy and complete.
      At the point $(2, 3)$, $y = \square$ and $2x + 1 = \square$
      Does the point $(2, 3)$ satisfy the inequality $y < 2x + 1$?

   d Shade the region of points that satisfy $y < 2x + 1$.

> **Q3d hint**
>
> If $(2, 3)$ satisfies the inequality, then all points in that region will also satisfy the inequality.

4 Draw a coordinate grid with -6 to 6 on both axes.
Shade the region that satisfies each inequality.

   a $y > x - 3$ **b** $y \leqslant 2x - 2$ **c** $y \geqslant 3x - 3$ **d** $y < -x + 1$

Example

On a coordinate grid, shade the region that satisfies all the inequalities

$x < 5$ $\qquad y \leqslant 2x + 4$ $\qquad y \leqslant 1$ $\qquad y > -2$

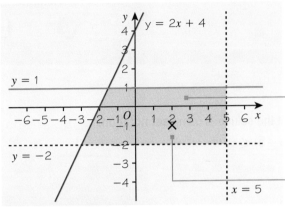

Draw dotted lines $x = 5$ and $y = -2$.
Draw solid lines $y = 2x + 4$ and $y = 1$.

Shade the required region.

Test a point. For $(2, -1)$
$y \leqslant 1$ and $y > -2$: the y-coordinate is -1
$x < 5$: the x-coordinate is 2
$2x + 4 = 8$: y-coordinate $\leqslant 8$

5 On a coordinate grid, shade the region that satisfies all these inequalities.

$y > 2$ $\qquad x + y < 7$ $\qquad y > 3x$

Label the region R. **(3 marks)**

6 **Reasoning** The diagrams show a shaded region bounded by three lines. For each diagram

i write down the equations of the lines

ii write down the three inequalities satisfied by the coordinates of the points in this region

a

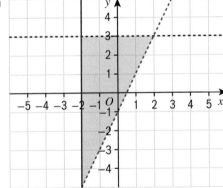

b

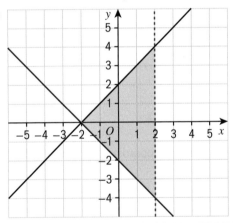

7 Write down the three inequalities that define the shaded region. **(4 marks)**

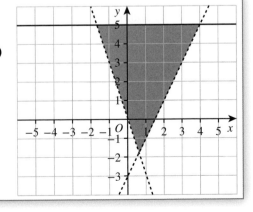

8 **Problem-solving** How many points with integer coordinates satisfy all these inequalities?
$$y > 0.5x - 3 \qquad y > -x \qquad y < 2$$

9 **a** Draw the graph of $y = x^2$ for values of x from -3 to $+3$.

b **Problem-solving** Shade the region that satisfies $y \geqslant x^2$ and $y < 7$.

10 Here is the graph of $y = 2x^2 - 2x - 4$.

a For what values of x is the graph on or below the x-axis?
Write your answer as an inequality: $\square \leqslant x \leqslant \square$

b For what values of x is $0 < 2x^2 - 2x - 4$?
Write your answer as two inequalities: $x < \square$ and $x > \square$

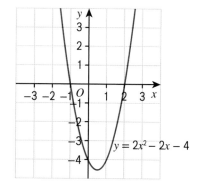

$y = 2x^2 - 2x - 4$

Key point

You can write solution sets using **set notation**.

- The inequality $x^2 - 9 \leqslant 0$ is satisfied when $-3 \leqslant x \leqslant 3$.
  This is written: $\{x : -3 \leqslant x \leqslant 3\}$.

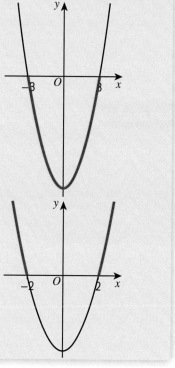

- The inequality $0 < x^2 - 4$ is satisfied when $x < -2$ or $x > 2$.
  This is written: $\{x : x < -2\} \cup \{x : x > 2\}$.
  The symbol $\cup$ means that the solution includes all the values satisfied by either inequality.

---

11 **a** Draw the graph of $y = 3x^2 + 3x - 6$, marking clearly the points where the graph intersects the x-axis.

b **Reasoning** From the graph, identify the values of x for which $0 > 3x^2 + 3x - 6$.
Write your answer

i as an inequality **ii** using set notation

12 **a** Draw the graph of $y = -x^2 + 1$.

b **Reasoning** Use your graph to find the values of x that satisfy $-x^2 + 1 \leqslant 0$.
Write your answer using set notation.

13 **Problem-solving a** Draw the graph of $y = 2x^2 + 4x - 6$.

b By drawing a suitable line on your graph, find the values of x which satisfy $0 \leqslant 2x^2 + 5x - 3$.
Write your answer using set notation.

15.3 Quadratic equations

*Active*Learn
Homework

- Find roots of equations.
- Sketch quadratic graphs.

Warm up

1 **Fluency** Solve these equations by factorising.

a $x^2 + 3x + 2 = 0$ **b** $2x^2 + 5x - 12 = 0$

2 Write in the form $a(x+p)^2 + q$ by completing the square.

a $x^2 + 2x - 5$ **b** $2x^2 + 8x + 4$

3 Here is the graph of $y = -x^2 + 2x + 6$.

a What are the coordinates of the maximum point?

b What is its line of symmetry?

4 Sketch **a** $y = x^2$ **b** $y = -x^2$

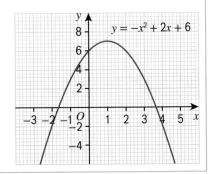

Key point

The lowest or highest point of the parabola, where the graph turns, is called the **turning point**.
The turning point is either a minimum or a maximum point.
$\smile$ is a minimum, $\frown$ is a maximum.
The x-values where the graph intersects the x-axis are the solutions, or **roots**, of the equation $y = 0$.

Exam-style question

5 Here is the graph of $y = x^2 + 4x + 3$.

a Use the graph to find the roots of the equation $x^2 + 4x + 3 = 0$. **(1 mark)**

b Write down the coordinates of the turning point. **(1 mark)**

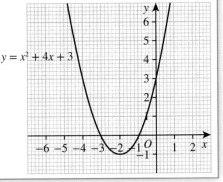

6 **a** Plot the graph of $y = x^2 + 8x + 15$.

b Use your graph to find the solutions to the equation $x^2 + 8x + 15 = 0$.

c **Reflect** Describe the position of the line of symmetry of the graph in relation to the roots of the equation.

d Write down the coordinates of the turning point.

e Write down the coordinates of the y-intercept.

f Does the graph have a maximum or a minimum point?

g **Reflect** How can you tell from the equation whether a quadratic graph will have a maximum or minimum point?

Key point

To sketch a quadratic graph:
- calculate the solutions to the equation $y = 0$ (points of intersection with the x-axis)
- find the y-intercept
- find the coordinates of the turning point and whether it is a maximum or a minimum.

7 **a** Sketch a pair of axes and label the origin O.

   b Work out the solutions of the equation $x^2 + 4x + 3 = 0$ to find its roots. Mark and label the roots on the x-axis of your sketch graph.

   c Substitute $x = 0$ into the equation $y = x^2 + 4x + 3$ to find the y-intercept.
   Mark and label the y-intercept on your sketch graph.

   d Decide whether the curve has a maximum or a minimum turning point.

   e Use symmetry to find the x-value where the line of symmetry crosses the x-axis (at the midpoint of the section of the axis between the roots). This is the x-coordinate of the turning point.

   f Substitute your value of x from part **e** into the equation $y = x^2 + 4x + 3$ to find the y-coordinate.

   g Mark and label the turning point on your sketch graph. Join the points with a smooth curve.

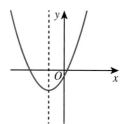

8 Follow the steps in **Q7** to sketch the graphs of

   a $y = x^2 + 8x + 15$ **b** $y = x^2 - x - 12$

9 **Reasoning** Match each equation to one of the graphs below, explaining your reasoning.

   a $y = x^2 + 5x + 6$ **b** $y = x^2 + 5x - 6$

   c $y = -x^2 + 2x + 8$ **d** $y = -2x^2 + 6x - 8$

A
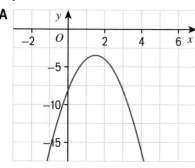

B
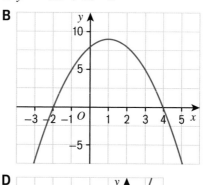

C
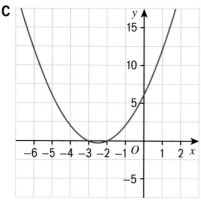

D
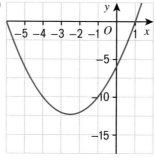

10 Here is the sketch of a curve.

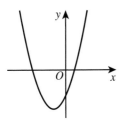

The equation of the curve is $y = x^2 + ax + b$, where a and b are integers.
The points $(1, 0)$ and $(0, -4)$ lie on the curve.
Find the coordinates of the turning point of the curve. **(4 marks)**

Key point

To find the coordinates of the turning point, write the equation in completed square form:
$y = a(x+b)^2 + c$
$(x+b)^2 \geqslant 0$, so the minimum for y is when $x+b = 0$ and $y = c$.

Example

a Does the graph of $y = x^2 + 8x + 15$ have a maximum or a minimum point?

b Find the coordinates of the turning point.

a   Minimum   ◄──────── The coefficient of x^2 is positive, so the turning point is a minimum.

b $y = x^2 + 8x + 15$
      $= (x+4)^2 - 16 + 15$ ◄──── Write the quadratic function in completed square form.
      $= (x+4)^2 - 1$ ◄──

      $x + 4 = 0$, so $x = -4$ The smallest value that y can take is -1.
                             This occurs when $(x+4)^2 = 0$.
      Minimum at $(-4, -1)$ ◄─── $(x+4)^2$ cannot be less than 0 because a square is always positive.
                             Solve the equation to find the x-coordinate.

11   **Reasoning** For each quadratic function, work out the coordinates of the turning point and state whether it is a maximum or a minimum.

    a $y = x^2 - 2x + 4$

    b $y = -x^2 - 6x - 11$

    c $y = x^2 - 10x + 23$

    d $y = 2x^2 + 12x + 13$

    e $y = 3x^2 - 12x + 13$

    f $y = -2x^2 - 4x + 2$

    g   **Reflect** What do you notice about the completed square form and the coordinates of the turning point?

Q11b hint

$y = -(x^2 + 6x + 11)$
$ = -((x+3)^2 + 2)$
$ = -(x+3)^2 - 2$
              ↑
     y-coordinate of
     the turning point

Key point

When a quadratic is written in completed square form $y = a(x+b)^2 + c$
the coordinates of the turning point are $(-b, c)$.

12 **a** Find the coordinates of the roots of $y = x^2 - 2x - 8$.

  b Where does the graph of $y = x^2 - 2x - 8$ cross the y-axis?

  c Write $x^2 - 2x - 8$ in completed square form.

  d Hence write down the coordinates of the turning point of $y = x^2 - 2x - 8$.

  e Is the turning point a maximum or a minimum?
    Explain your answer.

  f Using your answers to parts **a** to **e**, sketch the graph of $y = x^2 - 2x - 8$.

13 Use the method in **Q12** to sketch these graphs.

  a $y = x^2 - 2x - 3$ **b** $y = -x^2 - 2x + 8$ **c** $y = 2x^2 - 4x - 6$ **d** $y = 3x^2 - 3$

Exam-style question

14 Given that $x^2 - 4x + 7 = (x - a)^2 + b$ for all values of x,

  a find the value of a and the value of b.   **(2 marks)**

  b Hence write down the coordinates of the turning point on the
    graph of $y = x^2 - 4x + 7$.   **(1 mark)**

15 **a** Write down the coordinates of the turning point of the graph of $y = (x + 3)^2 - 5$.

  b Substitute $y = 0$ into the equation $y = (x + 3)^2 - 5$ and hence find the coordinates
    of the roots.
    Give your answers in surd form.

16 Find the roots of these equations given in completed square form.
  Give your answers in surd form.

  a $(x + 1)^2 - 3 = 0$ **b** $2(x - 1)^2 - 18 = 0$ **c** $3(x + 2)^2 - 4 = 0$

17 By writing the equations in completed square form, calculate the roots of these equations.
  Give your answers in surd form.

  a $x^2 + 4x - 3 = 0$ **b** $2x^2 - 8x - 1 = 0$ **c** $3x^2 + 18x - 12 = 0$

18 **Reasoning** Give three reasons why the graph shown is *not* $y = -2x^2 + 4x + 6$.

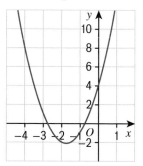

19 **Problem-solving** Find the equation of this quadratic graph.

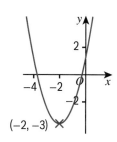

$(-2, -3)$

15.4 Using quadratic graphs

Active Learn
Homework

- Find roots of quadratic equations.
- Solve quadratic inequalities.

Warm up

1 **Fluency** Use the quadratic formula to calculate the roots of each equation. Give your answers correct to 1 decimal place.

a $x^2 - 4x - 7 = 0$ b $2x^2 - 3x - 4 = 0$ c $-3x^2 + 2x + 5 = 0$

2 Find the roots of these equations by writing the equation in completed square form. Give your answers in surd form where necessary.

a $x^2 + 6x + 5 = 0$ b $2x^2 + 4x - 1 = 0$ c $3x^2 + 6x - 1 = 0$

3 **Reasoning** a Match each equation to one of the graphs below.

  i $y = x^2 - 4x - 7$ ii $y = 3x^2 - x - 5$

b Hence estimate the solutions to the equations.

  i $x^2 - 4x - 7 = 0$ ii $3x^2 - x - 5 = 0$

c Use the quadratic formula to find the solutions of the equations in part **a** correct to 3 significant figures. Check your answers to part **b**.

A
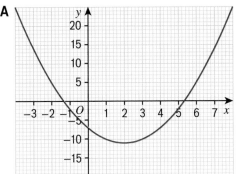

B
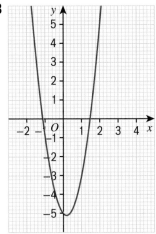

Exam-style question

4 **a** Copy and complete the table of values for
$y = 2x^2 + 4x - 2$.   **(2 marks)**

x	-4	-3	-2	-1	0	1	2
y							

Exam tip

For greater accuracy, use a ruler and draw lines on the graph to read off values.

b Plot the graph of $y = 2x^2 + 4x - 2$.   **(2 marks)**

c Use your graph to estimate the roots of the equation $y = 2x^2 + 4x - 2$.   **(3 marks)**

d Write the expression $2x^2 + 4x - 2$ in the form $a(x+b)^2 + c$.   **(2 marks)**

5 Write the equation for each quadratic graph, by writing $y = (x-a)(x-b)$ and expanding.

a

b

c

6 **Reasoning** For each graph
   i find the coordinates of the turning point   **ii** find the y-intercept   **iii** sketch the graph.
   a $y = x^2 + 2x + 2$ **b** $y = -x^2 - 4x - 7$ **c** $y = 2x^2 - 4x + 3$
   d **Reflect** Explain why these equations have no solutions when $y = 0$.

7 **Reasoning** **a** Find the turning point of the graph of the equation $y = 2(x-3)^2 + 5$.
   b Explain why you cannot find the roots of the equation $y = 2(x-3)^2 + 5$.

> **Key point**
>
> The quadratic equation $y = ax^2 + bx + c$ is said to have **no real roots** if its graph does not cross the x-axis. If its graph just touches the x-axis, the equation has **one repeated root**.

8 **Reasoning** By finding the turning point and whether it is a maximum or a minimum, decide whether each quadratic equation has
   • no roots   • two roots   • one repeated root
   a $y = x^2 + 6x + 11$ **b** $y = x^2 + 4x - 3$ **c** $y = x^2 - 6x - 12$ **d** $y = 3x^2 + 12x + 12$
   e $y = -x^2 + 2x - 5$ **f** $y = 2x^2 - 8x + 5$ **g** $y = -2x^2 - 12x - 18$ **h** $y = -3x^2 - 2x + 1$

> **Exam-style question**
>
> **9** **a** By completing the square, find the roots of the equation $x^2 - 4x - 3 = 0$, giving your answer in surd form. **(3 marks)**
>   **b** Show algebraically that $x^2 - 7x + 13 = 0$ has no real roots. **(3 marks)**

> **Exam tip**
>
> Write down all your working.

10 **a** Solve the quadratic equation $x^2 + x - 2 = 0$.
   b Sketch the graph of $y = x^2 + x - 2$.
   c Write the set of values of x that satisfy $x^2 + x - 2 < 0$.
   d Write the set of values of x that satisfy $x^2 + x - 2 \geqslant 0$.

> **Q10c hint**
>
> $\{x : \square < x < \square\}$

> **Q10d hint**
>
> $\{x : x \leqslant \square\} \cup \{x : x \geqslant \square\}$

> **Key point**
>
> To solve a quadratic inequality:
> • solve as a quadratic equation
> • sketch the graph
> • use the graph to find the sets of values that satisfy the inequality.

11 Find the set of values that satisfy each inequality.
   a $x^2 - 2x - 3 < 0$ **b** $x^2 + 3x - 10 < 0$ **c** $x^2 + 5x + 4 > 0$

12 Find the set of values that satisfy each inequality.

  a $x^2 + 7x + 10 < 0$ **b** $x^2 - 6x + 8 > 0$ **c** $x^2 - 6x + 5 < 0$

13 **Reasoning** **a** Sketch the graph of $y = x^2 - x - 6$.

  b Hence find the values of x which satisfy the inequality $6 < x^2 - x$.
  Write your answer using set notation.

14 **Reasoning** Sketch graphs to find the values of x which satisfy these inequalities.
  Write your answers using set notation.

  a $x^2 + x < 12$ **b** $2x^2 \geqslant 2x + 4$ **c** $x^2 \geqslant 9$

Example

Solve $\dfrac{5n}{n^2 + 6} \leqslant 1$.

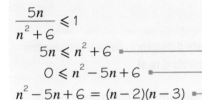

$\dfrac{5n}{n^2 + 6} \leqslant 1$ — Multiply both sides of the inequality by $n^2 + 6$. n^2 is positive, so $n^2 + 6$ is positive.

$5n \leqslant n^2 + 6$

$0 \leqslant n^2 - 5n + 6$ — Rearrange so one side of the inequality is zero.

$n^2 - 5n + 6 = (n - 2)(n - 3)$ — Factorise.

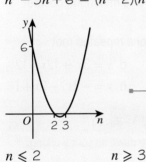

Sketch the graph.

$n \leqslant 2$ $n \geqslant 3$

$\{n : n \leqslant 2\} \cup \{n : n \geqslant 3\}$ — Write the set of values that satisfy $0 \leqslant n^2 - 5n + 6$.

15 Solve $10 < \dfrac{x^2 - 6}{3}$.

16 Solve $\dfrac{6n}{n^2 + 8} > 1$.

17 **a** Solve each side of this inequality separately.

$$3 < \frac{x^2 + 11}{5} \leqslant 15$$

  b Show the solutions to each side of the inequality on a number line.

  c Hence write the values that satisfy $3 < \dfrac{x^2 + 11}{5} \leqslant 15$ using set notation.

Exam-style question

18 Solve $7 < \dfrac{m^2 + 5}{2} < 15$.
Show all your working.   **(5 marks)**

Exam tip

When the question doesn't tell you how to write your answer, you can either write an inequality or use set notation.

15.5 Cubic equations

- Expand triple brackets.
- Find the roots of cubic equations.
- Sketch graphs of cubic equations.

  Active Learn
Homework

Warm up

1 Fluency Expand and simplify

a $(x+2)(x+3)$ **b** $(x-3)(x+4)$ **c** $(2x+1)(x-5)$ **d** $(3x-4)(x-2)$

2 Find the roots of each quadratic equation by factorising.

a $x^2-3x-10=0$ **b** $x^2-1=0$ **c** $2x^2-4x-6=0$ **d** $6x^2+10x-4=0$

3 Expand the expression $(x^2+4x+1)(x+2)$ by multiplying each term in the first bracket by each term in the second bracket. Then simplify.

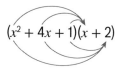

$(x^2 + 4x + 1)(x + 2)$

4 Copy and complete to expand the expression.

$$(x+2)(x+4)(x+3) = (x^2+\square x+\square)(x+3)$$
$$= x^3+\square x^2+\square x+\square$$

5 Expand these expressions.

a $(x+2)(x+5)(x+1)$ **b** $(x-3)(x+4)(x-2)$ **c** $(x+2)(x-1)(x+3)$

d $x(x+5)(x-4)$ **e** $(x+1)^2(x-1)$ **f** $(x+3)^3$

g $2x(x+3)(x-5)$ **h** $(x+5)(x-5)(2x-3)$ **i** $(x+5)(2x-3)(x-5)$

j Reflect What do you notice about your answers to parts **h** and **i**? Was one easier to expand than the other? Explain.

Exam-style question

6 Show that $(x+1)(x+2)(x-1)$ can be written in the form
$$ax^3+bx^2+cx+d$$
where a, b, c and d are positive integers.   **(3 marks)**

Key point

A **cubic** function is one whose highest power of x is x^3.
It is written in the form $y=ax^3+bx^2+cx+d$.

- When $a>0$ the function looks like

 or

- When $a<0$ the function looks like

or

The graph intersects the y-axis at the point $y=d$.
The graph's roots can be found by finding the values of x for which $y=0$.

7 Here is the graph of $y = x^3 + 3x^2 - 6x - 8$.

   a What are the roots of the equation
      $x^3 + 3x^2 - 6x - 8 = 0$?

   b Where does the graph cross the y-axis?

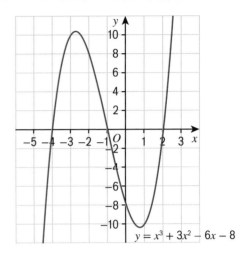

$y = x^3 + 3x^2 - 6x - 8$

Example

Sketch the graph of $y = (x-1)(x-2)(x+2)$.

When $x = 0$,
$(x-1)(x-2)(x+2) = (-1) \times (-2) \times 2 = 4$

The graph crosses the y-axis at 4.

$(x-1)(x-2)(x+2) = (x^2 - 3x + 2)(x+2)$
$\qquad\qquad\qquad\qquad = x^3 - x^2 - 4x + 4$

The coefficient of x^3 is 1, so $a = 1$

Since $a > 0$ the graph has the shape

$0 = (x-1)(x-2)(x+2)$

$x - 1 = 0 \quad$ or $\quad x - 2 = 0 \quad$ or $\quad x + 2 = 0$
$\qquad x = 1 \qquad\qquad\quad x = 2 \qquad\qquad\quad x = -2$

| The graph crosses the y-axis when $x = 0$, so the y-intercept is the product of the number terms. |
| Expand the brackets. |
| Find the roots of the equation. |
| If the product of any expressions is zero, one of them must itself be zero. |

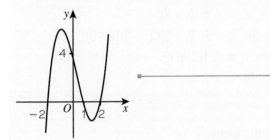

| Sketch the graph, marking on the points of intersection with the x- and y-axes. |

---

8   **a** What are the roots of the equation $y = (x+1)(x+2)(x+5)$?

     b Where does the graph of $y = (x+1)(x+2)(x+5)$ cross the y-axis?

     c Sketch the graph of $y = (x+1)(x+2)(x+5)$.

9 **Reasoning** Match each equation to one of the graphs below.

a $y = (x+1)(x+2)(x+3)$

b $y = (x+2)^2(x+5)$

c $y = (x-2)(x+2)(x+4)$

d $y = -x(x+1)(x-3)$

e $y = (1-x)(x+2)(x+4)$

f $y = -x^2(x+3)$

A

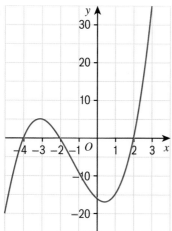

B

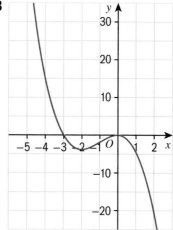

C

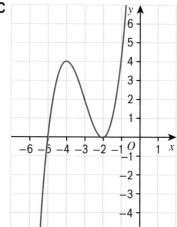

D

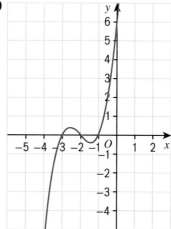

E

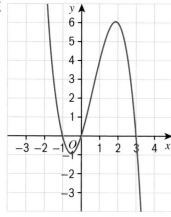

F

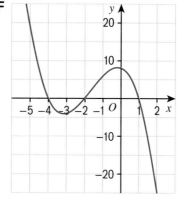

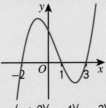
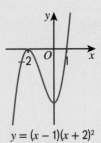
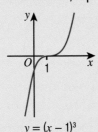
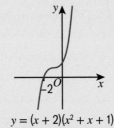
10 How many solutions does each cubic equation have?

   a $(x+1)(x-3)(x+4) = 0$ b $(x+3)^3 = 0$ c $-x(x+1)(x-3) = 0$

   d $x^2(x+4) = 0$ e $(x^2+2x+5)(x-2) = 0$ f $(10-x)(x+4)(x-1) = 0$

11 Sketch the graphs, marking clearly the points of intersection with the x- and y-axes.

   a $y = (x-3)(x+2)(x-1)$ b $y = x(x+1)(x-4)$ c $y = (-x+1)(x+3)(x-1)$

   d $y = (x+1)^2(x+3)$ e $y = (x+2)^3$ f $y = (x-1)(x^2+3x+6)$

13 **Problem-solving** The graph shown has equation $y = x^3 + ax^2 + bx + c$.

   Work out the values of a, b and c.

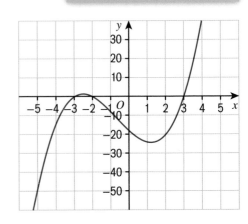

14 **Problem-solving** A graph has equation $y = -x^3 + ax^2 + bx + c$.
   It crosses the x-axis at $x = 1$, $x = 4$ and $x = -2$.
   Without drawing the graph, work out the values of a, b and c.

15.6 Using iteration to solve equations

*Active*Learn
Homework

- Solve quadratic and cubic equations using an iterative process.

Warm up

1 Fluency Solve

a $x^2 = 64$ **b** $x^3 = 125$ **c** $n^2 = 17.64$ **d** $m^3 = 29.791$

2 Make x^2 the subject of

a $y = x^2 + 4$ **b** $x^2 - 3 = x$ **c** $x^2 - 2x = 5$ **d** $x^2 + 3x = 4$

Key point

To find an accurate root of a quadratic equation you can use an **iterative** process.
Iterative means carrying out a process repeatedly.

Example

a Show that the equation $x^2 + x = 5$ has a solution between 1 and 2.

b Show that the equation $x^2 + x = 5$ can be rearranged to give $x = \sqrt{5 - x}$.

c Starting with $x_0 = 2$, use the iteration formula $x_{n+1} = \sqrt{5 - x_n}$ three times to find an estimate for a
solution of $x^2 + x = 5$.

a When $x = 1$, $y = x^2 + x = 2$
When $x = 2$, $y = x^2 + x = 6$

When $x = 1$, $y < 5$ so the
solution must be greater than 1.
When $x = 2$, $y > 5$ so the
solution must be less than 2.
Therefore the solution is
between 1 and 2.

> Substitute the given values
> into the equation.

> Explain why the solution
> must be between 1 and 2.

b $x^2 + x = 5$
$x^2 = 5 - x$
$x = \sqrt{5 - x}$

> Rewrite the iteration formula,
> substituting x_0 for x_n.

c $x_{n+1} = \sqrt{5 - x_n}$

$x_1 = \sqrt{5 - x_0}$

$x_1 = \sqrt{5 - 2} = \sqrt{3} = 1.732050808$

$x_2 = \sqrt{5 - x_1} = \sqrt{5 - \text{Ans}} = 1.807746993$

$x_3 = 1.786687719$

> 1st iteration: Use x_0 to find x_1.

> 2nd iteration: Use x_1 to find x_2.
> Use the Ans key on
> your calculator.

> 3rd iteration: You can use
> the = key to give the
> next iteration.

Unit 15 Equations and graphs   151

3 Show that

    a the equation $x^2 - x = 3$ has a solution between $x = 2$ and $x = 3$

    b the equation $2x^2 - x = 2$ has a solution between $x = 1$ and $x = 2$

    c the equation $x^2 + 2x = 1$ has a solution between $x = 0$ and $x = 1$

4 Rearrange each equation to make the highest power of x the subject.
Then take roots of each side to write an iteration formula $x = \ldots$

    a $x^2 + 3 = x$ **b** $x^2 - 2x = 7$ **c** $x^3 + x = 4$

    d $x^3 - x = 11$ **e** $x^2 - 5x - 1 = 0$ **f** $x^3 - 2x^2 + 3 = 0$

5 Starting with the value of x_0, use the iteration formula four times to find an estimate for the solution of each equation.

    a $x^2 - 2x - 4 = 0$ $x_{n+1} = \sqrt{4 + 2x_n}$ $x_0 = 3$

    b $x^2 - 5x - 4 = 0$ $x_{n+1} = \sqrt{5x_n + 4}$ $x_0 = 5.5$

    c $x^2 - x - 3 = 0$ $x_{n+1} = \dfrac{3}{x_n - 1}$ $x_0 = -1.5$

Exam-style question

6   **a** Show that the equation $x^3 - x = 3$ has a
      solution between 1 and 2.   **(2 marks)**

    b Show that the equation $x^3 - x = 3$ can be
      rearranged to give $x = \sqrt[3]{x + 3}$.   **(1 mark)**

    c Starting with $x_0 = 1$, use the iteration
      formula $x_{n+1} = \sqrt[3]{x_n + 3}$ three times to find an
      estimate for a solution of $x^3 - x = 3$.   **(3 marks)**

Exam tip

In 'Show that ...' questions, set
out your working clearly so that
someone else can follow and
understand it.

---

7 Copy and complete to rearrange each equation.

    a $x^2 - 3x + 7 = 0$ **b** $x^3 + 5x - 8 = 0$

       $x(x - 3) + 7 = 0$ $x(\Box + 5) - 8 = 0$

          $x(x - 3) = \Box$ $x(\Box + 5) = \Box$

              $x = \dfrac{\Box}{\Box - \Box}$ $x = \dfrac{\Box}{\Box + 5}$

8 Show that

    a the equation $x^2 + 4x - 2 = 0$ can be rearranged to give $x = \dfrac{2}{x + 4}$

    b the equation $x^3 - 2x + 5 = 0$ can be rearranged to give $x = \dfrac{-5}{x^2 - 2}$

9   **a** Starting with $x_0 = 0.5$, use the iteration formula $x_{n+1} = \dfrac{-1}{x_n^2 - 2}$ three times to
      find an estimate for a solution of the equation $x^3 - 2x + 1 = 0$.

    b Substitute your solution from part **a** into $x^3 - 2x + 1$, to see if it gives a value close to zero.

10 a Starting with $x_0 = 1.4$, use the iteration formula $x_{n+1} = \dfrac{7}{x_n^2 + 3}$ three times to find an estimate to 5 decimal places for the solution of the equation $x^3 + 3x - 7 = 0$.

b Substitute your solution from part **a** into $x^3 + 3x - 7$ and comment on the accuracy of the solution.

11 a Show that the equation $x^3 - 5x + 3 = 0$ has a solution between $x = 1$ and $x = 2$. **(2 marks)**

b Show that the equation $x^3 - 5x + 3 = 0$ can be rearranged to give $x = \dfrac{-3}{x^2 - 5}$. **(2 marks)**

c Starting with $x_0 = 1$, use the iteration formula

$x_{n+1} = \dfrac{-3}{x_n^2 - 5}$ three times to find an

estimate for a solution of $x^3 - 5x + 3 = 0$. **(3 marks)**

d By substituting your answer to part **c** into $x^3 - 5x + 3$, comment on the accuracy of your estimate for a solution to $x^3 - 5x + 3 = 0$. **(2 marks)**

12 Reasoning The iteration formula $x_{n+1} = \dfrac{6}{x_n^2 + 3}$ is used to estimate a solution to an equation

$x^3 + ax^2 + b = 0$, where a and b are integers.
Find the values of a and b.

13 Using $x_{n+1} = \dfrac{1}{x_n^2 - 4}$ with $x_0 = 1.5$,

a find the values of x_1, x_2 and x_3. **(3 marks)**

b Explain the relationship between the values of x_1, x_2 and x_3 and the equation $x^3 - 4x - 1 = 0$. **(2 marks)**

14 Use an iteration formula to estimate the one root of $x^3 + 2x - 6 = 0$ to 4 decimal places. The first steps have been done for you.

$x^3 = -2x + 6$
$x = \sqrt[3]{-2x + 6}$
$x_{n+1} = \sqrt[3]{-2x_n + 6}$
$x_0 = \square$
$x_1 = \sqrt[3]{-2x_0 + 6} = \square$
$x_2 = \sqrt[3]{-2x_1 + 6} = \square$

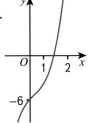

15 Use an iteration formula to estimate the negative root of the equation $x^3 - x^2 - 4x + 3 = 0$ to 5 decimal places.

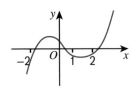

15 Check up

Active Learn
Homework

Simultaneous equations and inequalities

1 Use a graphical method to solve the simultaneous equations
$$2y - x = -8$$
$$x + y = 8$$

2 Draw a coordinate grid with x-axis from -5 to 5 and y-axis from -5 to 11.
Draw graphs to solve the simultaneous equations
$$y = x^2 - 2x - 4$$
$$y = -x - 2$$

3 Write down the three inequalities that define the shaded region of the graph.

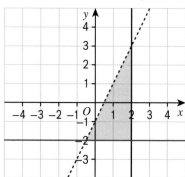

Graphs of quadratic equations

4 Here is the graph of $y = x^2 - 2x - 2$

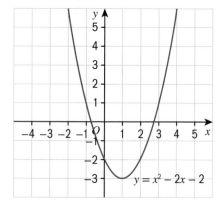

Estimate the roots of the equation $x^2 - 2x - 2 = 0$.

5 **a** Sketch the graph of $y = x^2 + x - 12$.

b Showing all your working, mark clearly the points of intersection with the y-axis, the roots of the equation $y = x^2 + x - 12$ and the turning point.

c Find the set of values that satisfy $x^2 + x - 12 < 0$.
Write your answer using set notation.

6 Find

a the y-intercept

b the turning point

of the graph of the equation $y = x^2 - 6x + 1$

7 This graph has equation $y = x^2 + bx + c$, where b and c are integers.

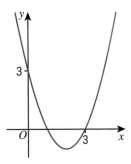

a Find the equation of the graph.

b Find the coordinates of the turning point.

8 Solve $\dfrac{x^2 + 11}{3} \geqslant 9$.

Write your answer using set notation.

Cubic equations

9 Expand and simplify $(x - 3)(x + 1)(x + 7)$.

10 Showing all your working, sketch the graph of $y = (x - 2)^2(x + 5)$, marking clearly the x- and y-intercepts.

Iteration

11 **a** Show that $x^2 + x = 1$ can be rearranged to give $x = \sqrt{1 - x}$.

b Starting with $x_0 = 0.6$, use the iteration formula $x_{n+1} = \sqrt{1 - x_n}$ three times to find an estimate for a solution to $x^2 + x = 1$.

12 **Reflect** How sure are you of your answers? Were you mostly

Just guessing 😟   Feeling doubtful 😐   Confident 🙂

What next? Use your results to decide whether to strengthen or extend your learning.

Challenge

13 A function has roots −1 and 3. The function could be quadratic *or* cubic.
List as many possible equations of the function as possible.

15 Strengthen

Simultaneous equations and inequalities

1 The diagram shows the lines $2y + x = 2$ and $x - y = -4$.

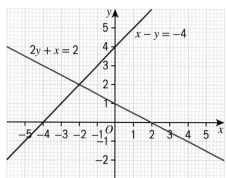

a Write down the coordinates of the point of intersection. $(\Box, \Box)$

b Write down the solution to the pair of simultaneous equations $2y + x = 2$ and $x - y = -4$

$x = \Box, y = \Box$

2 **a** Copy and complete the table of values for $y = x^2 - 3x + 1$.

x	−2	−1	0	1	2	3	4
y		5		−1			

b On suitable axes plot the graph of $y = x^2 - 3x + 1$.

c On the same axes draw the graph of $y = -x + 4$.

d Write down the coordinates of the points of intersection. $(\Box, \Box)$ and $(\Box, \Box)$

e Write down the solutions to the pair of simultaneous equations $y = x^2 - 3x + 1$ and $y = -x + 4$

$x = \Box, y = \Box$ and $x = \Box, y = \Box$

3 Solve each pair of simultaneous equations graphically.

a $y = 2x + 4$ and $y = -x + 7$

b $y = -x^2 + 3x + 4$ and $y = x + 1$

4 **a** Find the equations of the three lines that enclose the shaded region.

b Write each equation and its possible inequality signs. For example $x = -2$ $\leqslant$ $\geqslant$

> **Q4b hint**
>
> Solid line: $\leqslant$ or $\geqslant$
> Dotted line: $<$ or $>$

c Choose a point in the shaded region, for example $(-3, -2)$. Substitute the x- and y-values into each equation. $x = -2$ $\leqslant$ $\geqslant$
Choose the correct inequality sign. $-3 \Box -2$

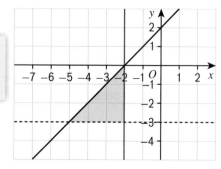

5 Write down the three inequalities that describe the shaded area.

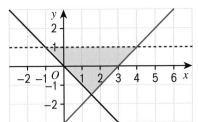

Graphs of quadratic equations

1 For each of these graphs write down

i the solutions to $y = 0$

ii the intercept with the y-axis

a

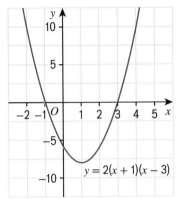

$y = 2(x + 1)(x - 3)$

Q1a hint

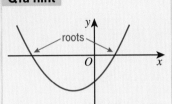

Find the x-values when $y = 0$.

Q1b hint

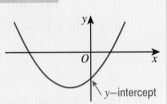

Find the y-values when $x = 0$.

b

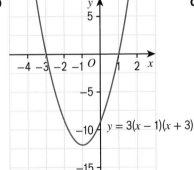

$y = 3(x - 1)(x + 3)$

c

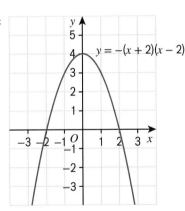

$y = -(x + 2)(x - 2)$

d

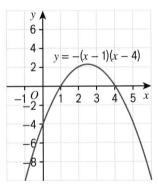

$y = -(x - 1)(x - 4)$

2 A curve has the equation $y = x^2 + 2x - 8$.

Copy and complete to find the roots and y-intercept.

a When $y = 0$

$$\underline{\hspace{3cm}} = 0$$

So $(x - \square)(x + \square) = 0$

There are two possible solutions:

$x - \square = 0$ hence $x = \square$

$x + \square = 0$ hence $x = \square$

So the roots are $x = \square$ and $x = \square$

b When $x = 0$

$y = x^2 + 2x - 8$

$y = 0^2 - \square - \square$

$y = \square$

So the intercept with the y-axis is at $y = \square$

3 Use the method in **Q2** to find the roots and y-intercept of

a $x^2 - 2x - 15 = 0$

b $-x^2 - 6x + 16 = 0$

c $2x^2 - 4x - 6 = 0$

d $3x^2 + 12x - 15 = 0$

4 Follow these steps to find the turning point of $y = x^2 + 2x - 8$.

Q4a hint

To find the missing term in the bracket, divide the coefficient of the x term by 2.

a Complete the square. $y = (x + \square)^2 - \square - 8$
$$y = (x + \square)^2 - \square$$

b At the turning point, $(x + \square) = 0$.
Write down the x-coordinate of the turning point of the graph $y = x^2 + 2x - 8$.
Find the y-coordinate of the turning point.

c Decide if the turning point is a maximum or a minimum.

5 Find the coordinates of the turning point of each graph in **Q3**.
Then decide whether the turning point is a maximum or a minimum.

6 Sketch the graphs of these quadratic equations.
a $y = -x^2 - 6x + 16$
b $y = 2x^2 - 4x - 6$
c $y = 3x^2 + 12x - 15$

Q6 hint

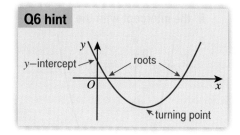

7 Here is a sketch graph of $y = x^2 + bx + c$.

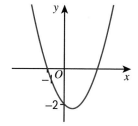

a Substitute the x- and y-values from the coordinates of the marked points into $y = x^2 + bx + c$ to give a pair of simultaneous equations.
$(-1, 0)$: $0 = \square^2 + b\square + c$
$(0, -2)$: $-2 = \square^2 + b\square + c$

b Solve to find the values of b and c.

c Substitute b and c into the equation $y = x^2 + bx + c$.

8 a Find the roots of $y = x^2 - 2x - 15$.

b Use this sketch graph of $y = x^2 - 2x - 15$ to find the values of x that satisfy
  i $x^2 - 2x - 15 < 0$ **ii** $x^2 - 2x - 15 > 0$

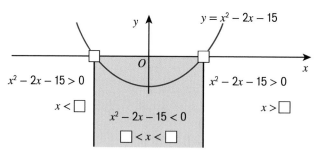

c Write each set of values using set notation.
  i $\{x : \underline{\qquad}\}$ **ii** $\{x : \underline{\qquad}\} \cup \{x : \underline{\qquad}\}$

9 a Rearrange $\dfrac{x^2 + 4}{8} \leqslant 5$ into the form $x^2 - c \leqslant 0$.

b Sketch the graph of $y = x^2 - c$.

c Find the values of x that satisfy $x^2 - c \leqslant 0$.

Cubic equations

1 Follow these steps to expand and simplify $(x-3)(x+1)(x+5)$.

   a Expand and simplify $(x-3)(x+1) = x^2 - \square x - 3$.

   b Use the grid method to expand $(x^2 - \square x - 3)(x+5)$.

   c Simplify by collecting like terms.

$\times$	x^2	$-\square x$	$-\square$
x	x^3		
$+5$			-15

2 For each of these graphs write down

   i the solutions of the equation $y = 0$ **ii** the y-intercept

  a 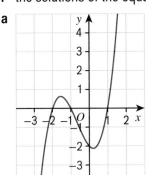   **b** 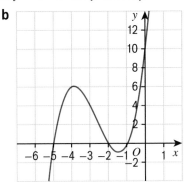   **c**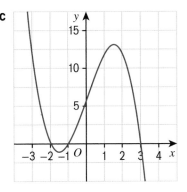

3 A curve has the equation $y = (x-3)(x+4)(x-1)$.
Copy and complete to find the roots and y-intercept.

   a When $y = 0$
     $(x-3)(x+4)(x-1) = 0$
     There are three possible solutions:
      $x - \square = 0$ hence $x = \square$
      $x + \square = 0$ hence $x = \square$
      $x - \square = 0$ hence $x = \square$
     So the roots are $x = \square$, $x = \square$ and $x = \square$

   b When $x = 0$
     $y = (x-3)(x+4)(x-1)$
     $y = (0-3)(0+4)(\square - \square)$
     $y = \square$

4 Use the method in **Q3** to find the roots and y-intercept of

   a $y = (x+3)(x-7)(x+2)$ **b** $y = (10-x)(x+2)(x+1)$

   c $y = (2x-5)(x+2)(x-1)$ **d** $y = (x-3)^2(3x-6)$

Iteration

1 **a** Copy and complete to find an iteration formula to find the roots of the equation $y = x^2 - 4x + 1$.

     $0 = x^2 - 4x + 1$

     $x^2 = 4x - \square$

     $x = \sqrt{4x - \square}$

     $x_{n+1} = \sqrt{4x_n - \square}$

  b Use the starting value $x_0 = 3.5$.
     Work out the values of x_1 to x_5. $x_1 = \sqrt{4 \times 3.5 - \square} = \square$

2 **a** Rearrange the equation $0 = x^2 - 8x + 1$ to find an iteration formula that you could use to find a root of the equation $y = x^2 - 8x + 1$.

  b Starting with $x_0 = 7.5$, find the root correct to 3 decimal places.

15 Extend

1   **a** Shade the region that satisfies the inequalities
$$y \leqslant 2x+5 \qquad x \leqslant y+1 \qquad y \leqslant -2$$
   b Calculate the area of the shaded triangle.

2   **Reasoning** Work out how many real roots each equation has. Show your working.
   a $x^2+2x+1=0$ **b** $x^2+2x+3=0$
   c $x^2-1=0$ **d** $x^2+5x+7=0$
   e $10+3x-x^2=0$ **f** $-x^2-2x-3=0$

3   **Reasoning**
   a Write down a quadratic function with
      i a maximum at $(-3, 4)$
      ii a minimum at $(4, -3)$
   b Write down a quadratic equation with roots at $x = -3$ and $x = 0$.

Exam-style question

4   A is a curve with equation $y = x^2+4x-3$.
   B is a line with equation $y = 2x+5$.
   A intersects B at the points P and Q.
   Work out the exact length of the straight line PQ.

   (6 marks)

Exam tip

Read the question carefully. What do you need to work out first?

Exam-style question

5   Show that
   $x^3-4x = x(x-2)(x+2)$ **(2 marks)**

Exam tip

'Show that' means you must write down each stage of your working clearly.

---

6   **Problem-solving**
   The general term of a sequence is $-n^2+4n+20$.
   Explain why the largest term in the sequence occurs at $n = 2$.

Q6 hint

Consider the graph of $y = -n^2+4n+20$.

Exam-style question

7   n is an integer such that $4n-3 \geqslant 13$ and $\dfrac{n^2+10}{7n} < 1$.
   Find all the possible values of n.   **(5 marks)**

---

8   **Reasoning** Use a graphical method to find approximate solutions to each pair of simultaneous equations.
   a $x^2+y^2 = 18$ and $y = x^2+3x+2$
   b $y = \dfrac{1}{x}$ and $y = 4x-1$

9 **Reasoning** How many pairs of integer coordinates satisfy the inequalities
$y \geqslant 2x^2 - 4x - 16$ and $y < -15$?

10 Find the set of values of x that satisfy

Q10b hint

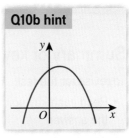

a $2x^2 - x - 3 < 0$

b $-x^2 + 3x + 10 > 0$

Exam-style question

11 Here are a rectangle and a right-angled triangle.

All measurements are in centimetres.
The area of the rectangle is greater than the area of
the triangle.
Find the set of possible values of x.   **(5 marks)**

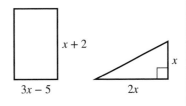

Working towards A level

12 Here is the graph of $y = ax^3 + bx^2 + cx + d$, where a, b, c and d are real constants.

Find the values of a, b, c and d.

Q12 hint

Use the given x-values to write the equation in
the form $y = (x-p)(x-q)(x-r)$ and expand the
brackets. Now look at the curve. Do your values of
the coefficients give the correct y-intercept?

In A level you might have to 'think outside the box'
to solve the problem.

13 $f(x) = x^3 + x^2 - 3x - 2$

a Show that $f(x) = 0$ has a root lying between $x = 1$ and $x = 2$.

b Show that $f(x) = 0$ can be written as $x = \sqrt{\dfrac{3x+2}{x+1}}$.

c Starting with $x_0 = 1.5$, use the iteration

formula $x_{n+1} = \sqrt{\dfrac{3x_n + 2}{x_n + 1}}$ to find a root, α,

of the equation $f(x) = 0$.

Give your answer correct to
4 decimal places.

A level questions on iterative methods often ask
you to manipulate the equation into a form where
iteration can be used, and to give your solutions to
a specified degree of accuracy.

15 Test ready

Summary of key points

To revise for the test:

- Read each key point, find a question on it in the mastery lesson, and check that you can work out the answer.

- If you cannot, try some other questions from the mastery lesson or ask for help.

Key points

1 You can solve a pair of simultaneous equations by plotting the graphs of both equations and finding the point(s) of intersection. → **15.1**

2 The points that satisfy an inequality such as $y \leqslant 3$ can be represented on a graph.

- A solid line means that the shaded area includes the points on the line.
- A dotted line means that the shaded area does *not* include the points on the line.

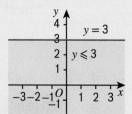 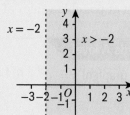

→ **15.2**

3 You can write solution sets using **set notation**.

- The inequality $x^2 - 9 \leqslant 0$ is satisfied when $-3 \leqslant x \leqslant 3$. This is written: $\{x : -3 \leqslant x \leqslant 3\}$.

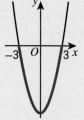

- The inequality $0 < x^2 - 4$ is satisfied when $x < -2$ or $x > 2$. This is written: $\{x : x < -2\} \cup \{x : x > 2\}$.
  The symbol $\cup$ means that the solution includes all the values satisfied by either inequality.

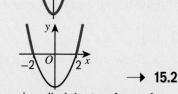

→ **15.2**

4 The lowest or highest point of the parabola, where the graph turns, is called the **turning point**. The turning point is either a minimum or a maximum point.
$\smile$ is a minimum, $\frown$ is a maximum.
The x-values where the graph intersects the x-axis are the solutions, or **roots**, of the quadratic equation $y = 0$. → **15.3**

5 To sketch a quadratic graph

- calculate the solutions to the equation '$y = 0$' (points of intersection with the x-axis)
- find the y-intercept
- find the coordinates of the turning point and whether it is a maximum or a minimum. → **15.3**

6 To find the coordinates of the turning point, write the equation in completed square form:
$y = a(x + b)^2 + c$
$(x + b)^2 \geqslant 0$, so the minimum for y is when $x + b = 0$ and $y = c$. → **15.3**

7 When a quadratic is written in completed square form $y = a(x + b)^2 + c$ the coordinates of the turning point are $(-b, c)$. → **15.3**

8 The quadratic equation $y = ax^2 + bx + c$ is said to have **no real roots** if its graph does not cross the x-axis. If its graph just touches the x-axis, the equation has **one repeated root**.　→ **15.4**

9 To solve a quadratic inequality: solve as a quadratic equation, sketch the graph, and use the graph to find the sets of values that satisfy the inequality.　→ **15.4**

10 A **cubic** function is one whose highest power of x is x^3.
It is written in the form $y = ax^3 + bx^2 + cx + d$.

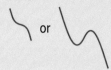

- When $a > 0$ the function looks like
- When $a < 0$ the function looks like

or　　　　or

The graph intersects the y-axis at the point $y = d$.
The graph's roots can be found by finding the values of x for which $y = 0$.　→ **15.5**

11 When the graph of a cubic function y crosses the x-axis three times, the equation $y = 0$ has three solutions. When the graph of a cubic function y crosses the x-axis once and touches the x-axis once, 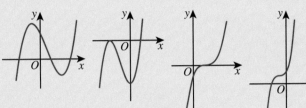 the equation $y = 0$ has three solutions but one of them is repeated. When the graph of a cubic function y crosses the x-axis once, the equation $y = 0$ can have one distinct, repeated solution or only one real solution.　→ **15.5**

12 To find an accurate root of a quadratic equation you can use an **iterative** process. Iterative means carrying out a process repeatedly.　→ **15.6**

Sample student answers

The graph of $y = x^2 + bx + c$ is shown on the grid.

Work out the value of b and the value of c.　**(4 marks)**

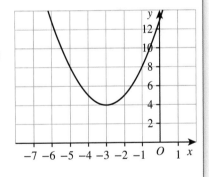

Student A	Student B	Student C
When $x = 0$, $y = 13$	The turning point is at	The turning point is at
$13 = 0^2 + 0 + c$	$(-3, 4)$	$(-3, 4)$
$c = 13$	$y = (x + 3)^2 + 4$	$y = (x + 3)(x - 4)$
$b = -3$ since the smallest	$= x^2 + 6x + 9 + 4$	$= x^2 + 3x - 4x - 12$
value of y is when $x = -3$	$= x^2 + 6x + 13$	$= x^2 - x - 12$
	$b = 6$, $c = 13$	$b = -1$, $c = -12$

Which student gives the best answer, and why?

15 Unit test

*Active*Learn
Homework

1 By drawing suitable graphs, find an estimate for the solutions to this pair of simultaneous equations.

$x^2 + y^2 = 16$

$x + 4y = 5$

(5 marks)

2 Match each equation to one of the graphs below.
Show your working.

a $y = x^2 - 4x + 4$ **(1 mark)**

b $y = (x - 3)(x + 1)$ **(1 mark)**

c $y = -x^2 + 5x + 6$ **(1 mark)**

d $y = (2 - x)^2 - 3$ **(1 mark)**

A
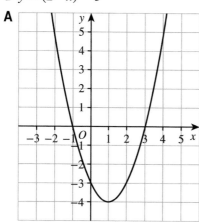

B
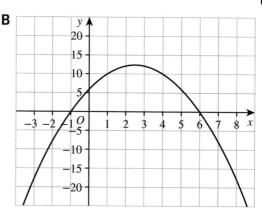

C
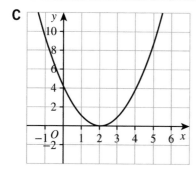

D
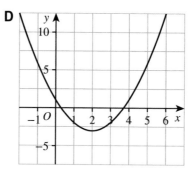

3 Draw a coordinate grid with −10 to +10 on both axes.
Shade the region that satisfies the inequalities

$y - 2x < 7$ and $y > 4$

(3 marks)

4 a Expand the expression $(x - 3)(x + 4)(x - 2)$. **(2 marks)**

b What type of graph is $y = (x - 3)(x + 4)(x - 2)$? **(1 mark)**

c Write down the solutions to the equation $(x - 3)(x + 4)(x - 2) = 0$. **(1 mark)**

d Where does the graph intersect the y-axis? **(1 mark)**

5 Sketch the graph of $y = x^2 - 4x - 1$. **(5 marks)**

6 Write down the equation of a graph with a maximum point at $(3, -4)$. **(3 marks)**

7 Calculate the exact area of the region defined by the inequalities

$x^2 + y^2 \geqslant 25$, $x \leqslant 5$ and $y \leqslant 5$ **(4 marks)**

8 A is the graph of $y = x^2 - 3x + 5$. B is the graph of $y - x = 5$.

The graphs intersect at points P and Q.

Find the length of PQ. Give your answer as a surd. **(6 marks)**

9 Find the values of x that satisfy $x^2 + x - 6 > 0$. **(5 marks)**

10 State whether each of these equations has real roots.

You must show your working.

a $x^2 - 6x + 13 = 0$ **(1 mark)**

b $2x^2 + 4x - 5 = 0$ **(2 marks)**

c $-x^2 - 4x + 9 = 0$ **(2 marks)**

11 m is an integer such that $\dfrac{m^2 + 8}{9m} \leqslant 1$ and $3m - 4 > 11$.

Find all the possible values of m. **(5 marks)**

12 a Using $x_{n+1} = \sqrt[3]{x_n - 8}$ with $x_0 = -2.2$, find the values of x_1, x_2 and x_3. **(3 marks)**

b Explain the relationship between the values of x_1, x_2 and x_3 and the equation

$x^3 - x + 8 = 0$. **(2 marks)**

(TOTAL: 55 marks)

13 Challenge Two broadband providers offer the following prices.

ONline	**Stream Speed**
No monthly cost	Monthly tarrif £20
£2 per GB of data	£1.50 per GB of data

a For each company, form an equation to calculate the monthly cost, with $y = $ total monthly cost and $x = $ GBs of data used.

b Use a graphical method to work out how many GBs of data are used if the cost for both companies is the same.

c When is Stream Speed cheaper than ONline?

14 Challenge Use a graphical method to find approximate solutions to each pair of simultaneous equations. Give all the solutions for values of x between $0°$ and $360°$.

a $y = \sin x$ and $y = 0.6$

b $y = \tan x$ and $y = 0.2$

15 Reflect In this unit, which was easiest, and which was hardest, to work with?

- Graphs for solving simultaneous equations

- Graphs of inequalities

- Graphs of quadratic functions

- Graphs of cubic functions

Copy and complete these sentences.

I find graphs _____ easiest, because _____

I find graphs _____ hardest, because _____

16 Circle theorems

16.1 Radii and chords

Prior knowledge

- Solve problems involving angles, triangles and circles.
- Understand and use facts about chords and their distance from the centre of a circle.
- Solve problems involving chords and radii.

Active Learn
Homework

Warm up

1 Fluency Work out the size of each angle marked with a letter. Give reasons for your answers.

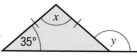

2 Prove that triangles ABD and ACD are congruent.

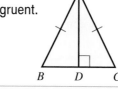

---

3 **a** Draw a circle, radius 3 cm and centre O.

    b Mark and label two points A and B on the circumference.

    c Use a ruler to draw the triangle AOB.

    d What type of triangle is AOB?

    e **Reflect** Explain why every triangle AOB, where O is the centre of the circle and A and B are points on the circumference, is isosceles.

4 **Reasoning** In each diagram, O is the centre of the circle. Work out the size of each angle marked with a letter.

    a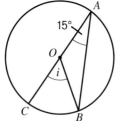
    b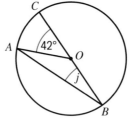
    c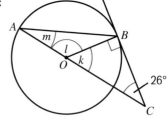

Key point

A **chord** is a straight line connecting two points on a circle.

---

5 **Reasoning** O is the centre of a circle. OA and OB are radii. OM is perpendicular to the chord AB.

    a Prove that triangles OAM and OBM are congruent.

    b Show that $AM = MB$.

    c Use your answer to part **b** to show that M is the midpoint of AB.

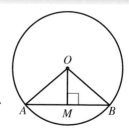

6 **a** Draw three circles.

   b In each circle draw a chord AB.

   c Draw a line from the centre O of each circle to the midpoint of AB.

   d What size is the angle between the line from the centre and the chord?

> ### Key point
>
> A **theorem** is a rule that can be proved by a chain of reasoning.
> *Theorem*: The perpendicular from the centre of a circle to a chord bisects the chord, and the line drawn from the centre of a circle to the midpoint of a chord is at right angles to the chord.

Example

O is the centre of a circle. The length of chord AB is 18 cm.
OM is perpendicular to AB.
Work out the length of AM. You must give a reason.

$AB = 18\,cm$

So $AM = \dfrac{18}{2}$ ◂──────── The perpendicular from the centre of a circle to a chord bisects the chord. So the length of AM is exactly half the length of AB.

$AM = 9\,cm$ (the perpendicular ◂── from the centre of a circle to a chord bisects the chord) └─── The question asks you to give a reason.

7 **Reasoning** O is the centre of a circle. M is a point on chord AB.
The length of chord AB is 12 cm. OM is perpendicular to AB. OM is 8 cm.

   a Draw a diagram and mark on all the information you are given in the question.

   b Work out the length of AM. Give a reason for each stage of your working.

   c Use Pythagoras' theorem to find the length of the radius of the circle.

> ### Exam-style question
>
> **8** O is the centre of a circle.
> $OA = 17$ cm and $AB = 16$ cm.
> M is the midpoint of AB.
> Work out the length of OM.
> State your reasons clearly.
>   **(3 marks)**
>
>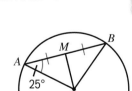
>
> ### Exam tip
>
> Mark lengths and angles on the diagram as you find them.

9 **Reasoning** O is the centre of a circle, radius 26 cm.
The distance from O to the midpoint of chord AB is 24 cm.
Work out the length of chord AB.

> ### Q9 hint
>
> Draw a diagram.

10 **Reasoning** O is the centre of a circle.
M is the midpoint of chord AB.
Angle $OAB = 25°$

   a What size is angle AMO?

   b Work out angle AOM.

   c Work out angle AOB.

   Give reasons for each stage of your working.

16.2 Tangents

- Understand and use facts about tangents at a point and from a point.
- Solve angle and length problems involving circles and tangents.

Warm up

1 **Fluency a** Find the missing angle in this triangle.

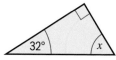

32°   x

b Find the missing length in this triangle.

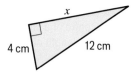

x

4 cm   12 cm

2 Use trigonometry to find the missing length in this triangle.

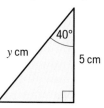

40°
y cm   5 cm

Key point

A **tangent** is a straight line that touches a circle at only one point.
Theorem: The angle between a tangent and the radius is 90°.

O

tangent

3 PQ is a chord in a circle, centre O.
RP and RQ are tangents.

a Write down the size of

   i angle OPR **ii** angle OQR

   Give a reason.

b Work out the size of

   i angle OPQ **ii** angle OQP

   Give a reason.

c Work out the size of angle QPR.

4 **Reasoning** CA and CB are two tangents to a circle with centre O.

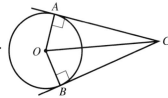

a Prove that triangles ACO and BCO are congruent.

b What can you say about the lengths AC and BC?

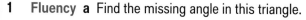
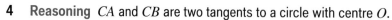
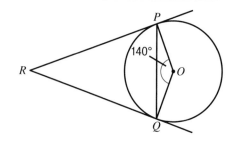

Theorem: Tangents drawn to a circle from a point outside the circle are equal in length.

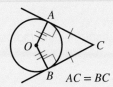

$AC = BC$

5   **Problem-solving / Reasoning** Each diagram shows a circle, centre O, and tangents from P to the circle.
    Work out the size of each angle marked with a letter.
    Give reasons for your answers.

a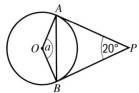

b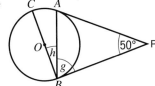

c

d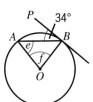

e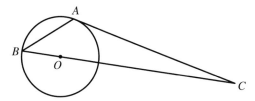

f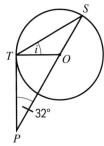

6

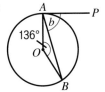

A and B are points on a circle, centre O.
AC is a tangent to the circle.
BOC is a straight line.
Angle $OBA = y°$
Find the size of angle ACO, in terms of y.
Give your answer in its simplest form.
Give reasons for each stage of your working.   **(5 marks)**

Label angle y on the diagram.
Draw in any extra lines you need.

7   **Reasoning** A and B are points on the circumference of a circle, centre O.
    TA and TB are tangents to the circle. Show that angles x and y are equal, and that angles a and b are equal.

Use congruence.

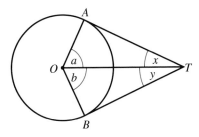

Example

A and B are points on a circle, centre O and radius 6 cm.
CA and CB are tangents.
$OC = 10$ cm
Calculate the size of

a angle COB

b angle AOB

Give your answers correct to 1 decimal place.

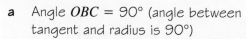

Label the angle to find as x.

a Angle $OBC = 90°$ (angle between tangent and radius is 90°)

$$\cos x = \frac{6}{10}$$

Use trigonometry.

$$x = \cos^{-1}\left(\frac{6}{10}\right) = 53.130...$$

Angle $COB = 53.1°$ (to 1 d.p.)

b Angle $AOC =$ angle COB
(triangles AOC and BOC are congruent)
Angle $AOB =$ angle $AOC +$ angle COB
$$= 53.130...° + 53.130...°$$
$$= 106.3°$$ (to 1 d.p.)

8 Each diagram shows a circle, centre O, with tangents from C.
Work out the lengths and angles marked with letters.
Give your answers correct to 1 decimal place.

a

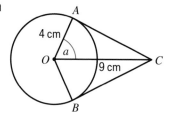

b

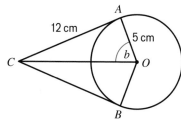

c

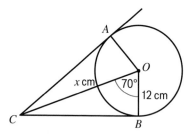

d

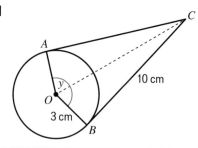

Exam-style question

9 A, B and D are points on a circle of radius 8 cm, centre O.
CB and CA are tangents to the circle.
$CO = 15$ cm
Work out the length of arc ADB.
Give your answer to 3 significant figures. **(5 marks)**

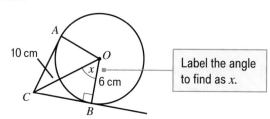

16.3 Angles in circles 1

- Understand, prove and use facts about angles subtended at the centre and the circumference of circles.
- Understand, prove and use facts about the angle in a semicircle.

Active Learn
Homework

Warm up

1 Fluency Copy each diagram and colour the arc that the marked angle stands on.

a

b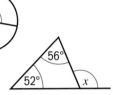

2 Work out the size of angle x in this diagram.
Give reasons for your answer.

Key point

An angle **subtended** by an arc is any angle with 'arms' that start and finish at the ends of the arc.

3 a Draw a circle, centre O and radius 3 cm.
Mark three points X, Y and Z on the circumference.

b Draw an angle subtended by the arc XY **i** with vertex at Z **ii** with vertex at O

c Measure the angles at Z and O. What do you notice?

Key point

Theorem: The angle at the centre of a circle is twice the angle at the circumference when both are subtended by the same arc.

Example

Prove that the angle at the centre of a circle is twice the angle at the circumference when both are subtended by the same arc.

$AO = OC$ (radii of same circle)
Angle ACO = angle $OAC = x$
(base angles of isosceles triangle are equal)
Similarly, angle BCO = angle $OBC = y$
Angle $AOD = 2x$ (exterior angle equals the sum of the two interior opposite angles)
Similarly, angle $BOD = 2y$
Angle $ACB = x + y$
Angle $AOB = 2x + 2y = 2(x + y) = 2$(angle ACB)

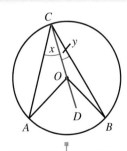

Draw the line CO and extend it to point D.
Label angle $ACO = x$ and angle $BCO = y$.

4 **Reasoning** In each diagram, O is the centre of the circle.
Work out the size of each angle marked with a letter.
Give reasons for your answers.

a

b

c

5 **a** Draw a circle with radius 5 cm.
Then draw in a diameter and label it AC.
Choose any point on the circumference and label it B.
Now draw a triangle between this point and the two ends
of the diameter by drawing in lines AB and BC.

b Measure angle B.

c Repeat for other points on the circumference. What do you notice?

6 **Reasoning** AOB is a diameter of this circle.

a What size is angle AOB?

b Prove that the angle in a semicircle is 90°.

Key point

Theorem: The angle in a semicircle is a right angle.

Angle $ABC = 90°$

---

7 **Reasoning** In each diagram, O is the centre of the circle.
Work out the size of each angle marked with a letter. Give reasons for your answers.

a

b

c

d

e

f

8 **Reasoning** Mario says that the size of angle a is 65°.
Andy says that the size of angle a is 115°.

a Show that Andy is correct.

b Explain the mistake that Mario has made.

9 In each diagram, O is the centre of the circle. Find the size of each angle marked with a letter.

a

b

c ABC is a tangent.

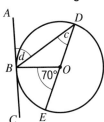

10 **Reasoning** Lucy says that the size of angle a is $15°$.
Sue says that the size of angle a is $60°$.
Who is correct? Show working to explain.

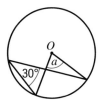

11 **Problem-solving** A, B and C are points on the circumference of a circle, centre O.
AC is the diameter of the circle.
$AB = 4$ cm and $BC = 8$ cm.
Show that the area of the circle is $20\pi\,\text{cm}^2$.

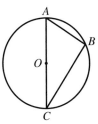

12 O is the centre of a circle, radius $24\,\text{mm}$.
A, B and C are points on the circumference.
a Find the length of the arc AC.
b Calculate the shaded area.
Give reasons for each stage of your working.

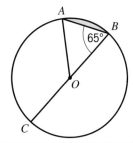

13 B, C and D are points on the
circumference of a circle, centre O.
AB and AD are tangents to the circle.
Angle $DAB = 55°$
Work out the size of angle BCD.
Give a reason for each stage in your working.

Diagram NOT
accurately drawn

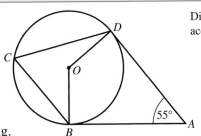

(4 marks)

Exam tip

Each reason you give must be a statement of a
mathematical rule and not just your calculations.

14 O is the centre of a circle, radius $5\,\text{cm}$.
BOD is a straight line.
Angle $BAO = 25°$
Find the length OD, correct to 1 decimal place.
Give reasons for each stage of your working.

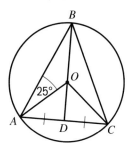

16.4 Angles in circles 2

- Understand, prove and use facts about angles subtended at the circumference of a circle.
- Understand, prove and use facts about cyclic quadrilaterals.
- Prove the alternate segment theorem.

*Active*Learn
Homework

Warm up

1 Fluency What is the shaded part of this circle called?

2 Write an expression in terms of x for angle BAC.

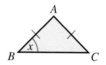

3 In each diagram, O is the centre of the circle.

a Work out the size of each angle marked with a letter.

b Work out the size of angle d in terms of x.

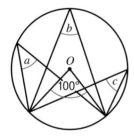

c What do you notice about all the angles at the circumference in the same segment?

Key point

Theorem: Angles subtended at the circumference by the same arc are equal. Another form of the same theorem is that angles in the same segment are equal.

4 Reasoning In each diagram, O is the centre of the circle. Work out the size of each angle marked with a letter. Give reasons for each step in your working.

a 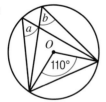 **b** 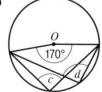 **c**  **d**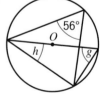

5 a Explain how you know that

  i $a = b$ **ii** $c = d$ **iii** $e = f$

b What do your answers to part **a** tell you about triangles ABE and CBD?

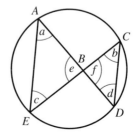

6 **Reasoning** In each diagram, O is the centre of the circle.

a Work out the sizes of angles a, b and c in each diagram.

i

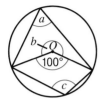

ii

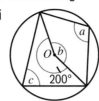

b Work out angle $a +$ angle c for each diagram. What do you notice?

7 **Reasoning** $ABCD$ is a cyclic quadrilateral. O is the centre of the circle.

a Write the obtuse angle AOC in terms of x.

b Write the reflex angle AOC in terms of y.

c Copy and complete.
Obtuse angle $AOC +$ reflex angle $AOC = 360°$
$\square \quad + \quad \square \quad = 360°$

d Factorise your expression from part **c** to show that $x + y = 180°$.

e **Reflect** How does this prove the theorem about opposite angles in a cyclic quadrilateral?

8 **Reasoning** In each diagram, O is the centre of the circle.

a Work out the size of each angle marked with a letter.
Give reasons for each step of your working.

i

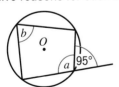

ii

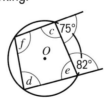

b What do you notice about the exterior angle of a cyclic quadrilateral and the opposite interior angle?

9 **Reasoning** Follow these steps to prove that an exterior angle of a cyclic quadrilateral is equal to the opposite interior angle.

a Copy the diagram.

b Copy the working and complete the reasons.
Angle $x +$ angle $y = 180°$ because _____ .
Angle $x +$ angle $z = 180°$ because _____ .
So angle $y =$ angle z.

10 **Reasoning** In each diagram, O is the centre of the circle.
Work out the size of each angle marked with a letter. Give reasons for each step in your working.

a

b

c

d

e

11 The points A, B, C and D lie on a circle.
ABE is a straight line.
$BA = AD$
$BC = CD$
Angle $BDC = 65°$
Work out the size of angle CBE.
You must give a reason for each stage of your working.

(5 marks)

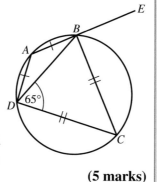

12 **Reasoning** O is the centre of a circle. AT is a tangent to the circle.

a Copy the diagram.

b Copy the working and complete the reasons.
$\angle OAT = 90°$ because the angle between the tangent and the _____ $= 90°$.
$\angle OAB = 90° - 58° = 32°$
$OA = OB$ because radii _____.
$\angle OAB = \angle OBA$ because the base angles of _____ triangle are equal.
$\angle AOB = 180° - 32° - 32° = 116°$ because angles in a _____ add up to ____ .
$\angle ACB = 116° \div 2 = 58°$ because the angle at the _____ is twice the _____.

13 **Reasoning** Repeat **Q12** but this time with angle $BAT = 72°$.

a What is the size of angle ACB?

b What do you notice about angle BAT and angle ACB in **Q12** and in part **a**?

c Repeat the steps in **Q12** but with angle $BAT = x$ to prove that angle $BAT =$ angle ACB.

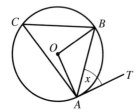

In this diagram
- angle *BAT* is in one segment of the circle
- angle *BCA* is in the **alternate segment**.

Theorem: The angle between the tangent and the chord is equal to the angle in the alternate segment. This is called the **alternate segment theorem**.

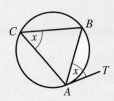

14 **Reasoning** *O* is the centre of a circle.
A, *B* and *C* are points on the circumference of the circle.
DCT is a tangent to the circle at point *C*.
Angle *BAC* = 53° and angle *ACB* = 62°.
Work out the size of angle *ACT*.
Give reasons for each step in your working.

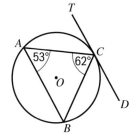

15

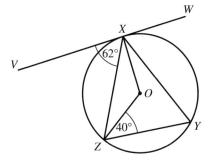

Do not assume lengths are equal just because they look equal. The diagram is not drawn accurately.

X, *Y* and *Z* are points on the circumference of a circle, centre *O*.
VXW is the tangent to the circle at *X*.
Angle *ZXV* = 62°
Angle *OZY* = 40°
Work out the size of angle *OXY*.
You must show all your working.

(3 marks)

16 *B* and *D* are two points on the circumference of a circle, centre *O*.

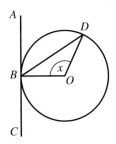

Diagram NOT accurately drawn

The straight line *ABC* is the tangent to the circle at *B*.
Angle *BOD* = *x*
Prove that angle $ABD = \frac{1}{2}x$.

(5 marks)

16.5 Applying circle theorems

ActiveLearn
Homework

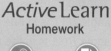

- Solve angle problems using circle theorems.
- Find the equation of the tangent to a circle at a given point.

Warm up

1 Fluency Use trigonometry to find
  a length x
  b length y, giving your answer in surd form

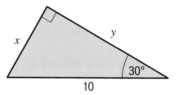

2 Find the equation of the line that is perpendicular to $y = 2x - 3$ and passes through the point $(-1, 2)$.

3 Reasoning In each diagram, O is the centre of the circle.
Work out the size of each angle marked with a letter.
Give reasons for each step in your working.

a
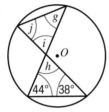

b
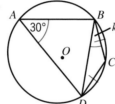

c
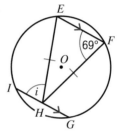

4 Reasoning In each diagram, AT is a tangent to the circle.
Work out the size of each angle marked with a letter.
Give reasons for each step in your working.

a
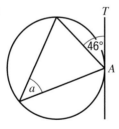

b
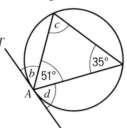

c
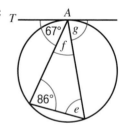

5 Reasoning Work out the size of each angle marked with a letter.
Give reasons for each step in your working.

a
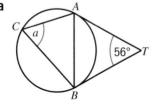

b
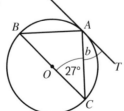

c
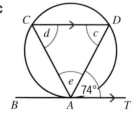

6 **Reasoning** O is the centre of the circle. DAT and BT are tangents to the circle. Angle $CAD = 50°$ and angle $ATB = 48°$. Work out the size of

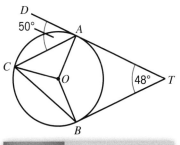

   a angle CAO **b** angle AOB **c** angle AOC

   d angle COB **e** angle CBO

Give reasons for each step in your working.

7 A, C and D are points on the circumference of a circle, centre O.

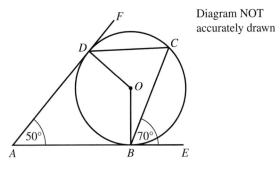

Diagram NOT accurately drawn

ABE and ADF are tangents to the circle.

Angle $DAB = 50°$ Angle $CBE = 70°$

Work out the size of angle ODC.   **(3 marks)**

Example

Find the equation of the tangent to the circle $x^2 + y^2 = 25$ at the point $A(3, -4)$.

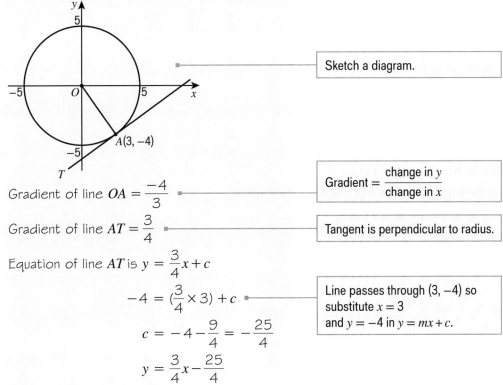

Sketch a diagram.

Gradient of line $OA = \dfrac{-4}{3}$

   $\text{Gradient} = \dfrac{\text{change in } y}{\text{change in } x}$

Gradient of line $AT = \dfrac{3}{4}$

   Tangent is perpendicular to radius.

Equation of line AT is $y = \dfrac{3}{4}x + c$

      $-4 = (\dfrac{3}{4} \times 3) + c$

   Line passes through $(3, -4)$ so substitute $x = 3$ and $y = -4$ in $y = mx + c$.

      $c = -4 - \dfrac{9}{4} = -\dfrac{25}{4}$

      $y = \dfrac{3}{4}x - \dfrac{25}{4}$

Equation of line AT is $4y = 3x - 25$

8 Problem-solving

  a Find the equation of the tangent to the circle $x^2 + y^2 = 169$ at the point $B(5, -12)$.

  b Find the equation of the tangent to the circle $x^2 + y^2 = 225$ at the point $C(9, 12)$.

  c Find the equation of the tangent to the circle $x^2 + y^2 = 100$ at the point $D(-8, 6)$.

  d Find the equation of the tangent to the circle $x^2 + y^2 = 289$ at the point $E(-8, -15)$.

9 C is the circle with equation $x^2 + y^2 = 9$.

  $P(2, \sqrt{5})$ is a point on C.

  Find an equation of the tangent to C at the point P.   **(3 marks)**

10 Reasoning The diagram shows a circle, centre O.
  AB is the tangent to the circle at point A.

  a Explain why AOB is a right-angled triangle.

  B has coordinates $(0, -6)$.

  b Use trigonometry to find the radius of the circle.
  Give your answer in surd form.

  c Hence write the equation of the circle.

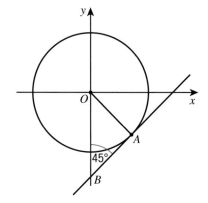

11 Reasoning The diagram shows a circle, centre O.
  AB is the tangent to the circle at point A.

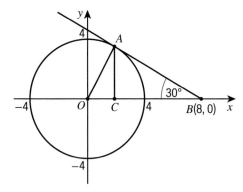

  a Write down the length of OA.

  b Find the length AB. Give your answer in surd form.

  c Use trigonometry to find the length CB.

  d Hence find the length CA.

  e Show that the coordinates of A are $(2, 2\sqrt{3})$.

16 Check up

Active Learn
Homework

Chords, radii and tangents

1 Draw a circle with radius 6 cm.
Draw and label clearly a chord, a tangent and a segment.

2 In each diagram, O is the centre of the circle.
Work out the size of each angle marked with a letter.
Give reasons for each step in your working.

a
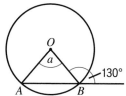

b
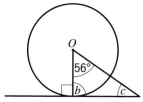

c
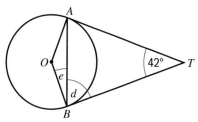

 3 O is the centre of a circle with radius 8.5 cm.
AB is a chord with length 15 cm.
Angle $OMB = 90°$
Work out the length of OM.

Circle theorems

4 In each diagram, O is the centre of the circle.
A, B, C and D are all points on the circumference.
Work out the size of each angle marked with a letter.
Give reasons for each step in your working.

a
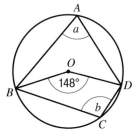

b
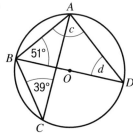

5 Work out the size of
  a angle a
  b angle b
Give reasons for each step in your working.

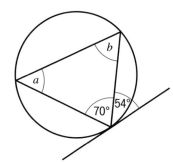

6 A, B, C and D are points on the circumference of a circle, centre O.
Angle $AOC = x$
Find the size of angle ADC in terms of x.
Give a reason for each stage of your working.

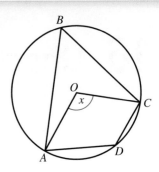

Proofs and equation of tangent to a circle at a given point

7   Find the equation of the tangent to the circle $x^2 + y^2 = 676$ at the point $A(10, -24)$.

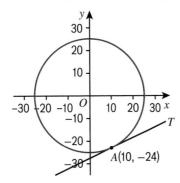

8   Prove that the angle at the centre of a circle is equal to twice the angle at the circumference when both are subtended by the same arc.

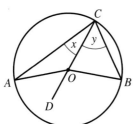

9   **Reflect** How sure are you of your answers? Were you mostly

Just guessing 😞   Feeling doubtful 😐   Confident 😊

What next? Use your results to decide whether to strengthen or extend your learning.

Challenge

10 a Copy this diagram of a circle with tangent CT parallel to the line AB.
Choose a value for x and write in the sizes of the angles in the triangle.
Repeat with different values of x.
b What type of triangle is ABC?

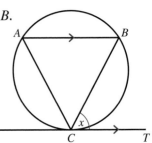

16 Strengthen

Active Learn
Homework

Chords, radii and tangents

1 Copy the diagram.
Use these words to label it.

chord   tangent   segment

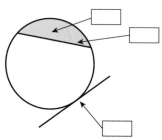

Q1 hint

A chord joins two points on the circumference. A tangent touches the circle.

2 *O* is the centre of a circle. *ABC* is a straight line.
Angle *OBC* = 160°

a Copy the diagram.
Mark the two radii with dashes to show they are equal length.

b What type of triangle is *OAB*?

c Work out the sizes of angle *x*, angle *y* and angle *z*.

3 **a** Draw a circle. Label the centre as *O*.

b Draw a radius. Label the point where it meets the circumference *A*.

c Draw a line at 90° to your radius, through point *A*.
Explain why this line is a tangent.

d Draw lines that meet your radius and the circumference at other angles. Are they tangents?

e What is the angle between a tangent and a radius?

4 *O* is the centre of a circle.
AT and *BT* are tangents.
AB is a chord. Angle *ATB* = 34°

a Copy the diagram.
Mark on any angles you know
and any lines of equal length.

b What type of triangle is *ABT*?
Work out the sizes of the two base angles in triangle *ABT*.

c What size is the angle between radius *OB* and
tangent *BT*?

d Work out the size of angle *OBA*.

Q4a hint

Tangents to a circle from
the same external point are
_____.

The angle between a radius and
a tangent is ☐°.

 5 *O* is the centre of a circle with radius 6.5 cm. Angle *OMB* = 90°

a Copy the diagram and draw the radius *OA*, labelled 6.5 cm.

b *AB* is a chord with length 12 cm. What is the length of *AM*?

c Use Pythagoras' theorem in triangle *OAM* to work out the length of *OM*.

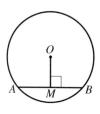

Circle theorems

1   O is the centre of a circle. P, Q and R are points on the circumference of
    the circle. Angle $QOR = 130°$.

    a Copy the diagram.

    b Colour the arc that subtends angle QOR at the centre.

    c What angle at the circumference is subtended by the same arc?

    d Use 'The angle at the centre of a circle is twice the angle at the circumference' to work out the
    size of angle QPR.

2   Work out the size of the angle marked with a letter in each of these diagrams.

    a   **b**   **c**   **d**

3   Draw a circle.
    Mark four points on the circumference and join them to make a cyclic quadrilateral.
    Measure all of the angles in your cyclic quadrilateral.
    Add the opposite angles. What do you notice?

4   P, Q, R and S are points on the
    circumference of a circle.
    Angle $QPS = 96°$ and angle $PQR = 78°$.

    a What type of quadrilateral is $PQRS$?

    b Which angle in the quadrilateral is
    opposite angle QPS?

    c Work out the size of angle QRS.

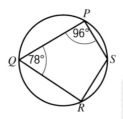

> **Q4c hint**
>
> Opposite angles in a cyclic
> quadrilateral add up to $\square$°.

5   Which of these are cyclic quadrilaterals?

    A   **B**   **C**

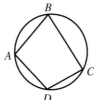
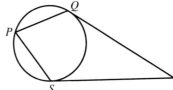
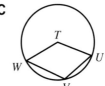

6   Draw a circle with diameter at least $5\,\text{cm}$.
    Mark two points, A and B, on the circumference.
    Draw four different triangles with base AB and
    the third vertex on the circumference.
    Measure each of the four angles at the circumference.
    What do you notice?

7   For each diagram, write down the pair(s) of equal angles.

    a   **b**   **c**

> **Q7 hint**
>
> Turning the diagram round can
> often help.

8 Each diagram shows a circle, centre O, with a tangent at point B.
Write down the angle in the alternate (unshaded) segment that is 49°.

a

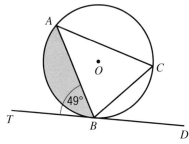

b

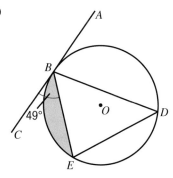

c

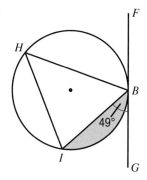

d

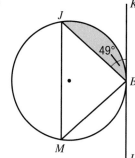

9 O is the centre of a circle.
A, B and C are points on the circumference.
TBD is a tangent to the circle at point B.
Angle $ABC = 75°$ and angle $ABT = 37°$.
Work out the sizes of angle ACB and angle CAB.
Give reasons for each step in your working.

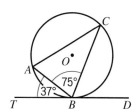

Proofs and equation of tangent to a circle at a given point

1 **Problem-solving** The diagram shows the tangent to the
circle $x^2 + y^2 = 25$ at the point $A(3, 4)$.

a What is the gradient of the radius?

b What is the gradient of the tangent AT?

> **Q1b hint**
>
> Gradient of perpendicular is $-\dfrac{1}{m}$

c Complete for the equation of the tangent: $y = \Box x + c$.

d Substitute $x = 3$ and $y = 4$ into your answer to part **c** to
work out the value of c.

e Write the equation of line AT.

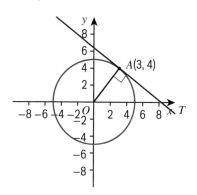

2 **Reasoning** O is the centre of a circle. AOC is a straight line.

a What size is angle AOC?

b What arc subtends angle ABC?

c Work out the size of angle ABC.

d Which circle theorem have you proved?

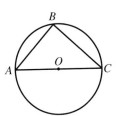

16 Extend

1 Reasoning O is the centre of a circle.
OBC is an equilateral triangle.
Angle $ABC = 130°$
Work out the sizes of angle a, angle b and angle c.
Give reasons for your answers.

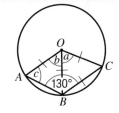

2 Reasoning O is the centre of a circle with radius 15.4 cm.
$PQ = 21.6$ cm
How far is the midpoint of PQ from the centre of the circle?
Give your answer correct to 1 decimal place.

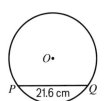

3 Reasoning O is the centre of a circle.
Angle $BAC = 3x$ and angle $ACB = 2x$.
Work out the actual size of each angle in triangle ABC.

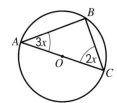

4 Reasoning O is the centre of a circle.
A, B and C are points on the circumference.
Angle $BOC = 40°$ and angle $AOB = 70°$.
Prove that AC bisects angle OCB.

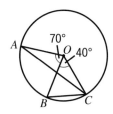

5 Reasoning O is the centre of a circle.
A, B, C and D are points on the circumference.
Angle $ABC = 114°$
Work out the size of angle COD.
Give a reason for each step of your working.

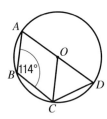

6 Problem-solving O is the centre of a circle.
A, B, C and D are points on the circumference.
Angle $BAD = 150°$
Prove that triangle OBD is equilateral.

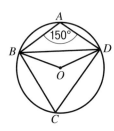

7 **Reasoning** O is the centre of a circle with radius 25 cm. AB and CD are parallel chords.
$AB = 14$ cm and $CD = 40$ cm.
MON is a straight line.
Work out the length of MN.

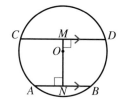

Q7 hint

First work out lengths OM and ON separately.

8 **Reasoning** CD is parallel to BT. BT is a tangent to the circle.
Prove that triangle ACD is isosceles.

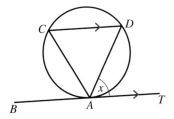

Exam-style question

9 OAC is a sector of a circle, centre O and radius 4 cm.
BA is the tangent to the circle at point A.
BC is the tangent to the circle at point C.
Angle $AOC = 60°$
Calculate the area of the shaded region.
Give your answer correct to 3 significant figures. **(5 marks)**

Exam tip

Draw any lines you need on the diagram.

10 **Problem-solving / Reasoning**
AB and BC are tangents to a circle, centre O.
Prove that $AB = BC$.

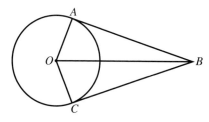

Q10 hint

Prove that triangles ABO and OBC are congruent.

11 **Problem-solving / Reasoning** Prove that the line drawn from the centre of a circle to the midpoint of a chord is perpendicular to the chord.

Q11 hint

Draw a diagram. You can use SSS to prove two triangles are congruent.

12 *AOC* and *BOD* are diameters of a circle, centre *O*.

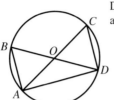

Diagram NOT accurately drawn

Prove that triangle *ABO* and triangle *CDO* are congruent. **(3 marks)**

13 The points *A*, *B*, *C* and *P* lie on a circle, centre *O*. The points *A*, *O*, *C* and *D* lie on a different circle. The two circles intersect at points *A* and *C*. *APD*, *BCD* and *AOB* are straight lines. Prove that *BD* = *AD*.

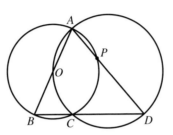

(4 marks)

14 Prove algebraically that the straight line with equation $3y + x = 10$ is a tangent to the circle with equation $x^2 + y^2 = 10$. **(5 marks)**

15 *TAP* and *TBQ* are tangents to a circle, centre *O*.
C lies on the circumference of the circle.
Angle *PAC* = 3*x*,
angle *OBC* = *x* and
angle *ATB* = *y*.
Prove that *y* = 4*x*.
Give reasons for any statements you make.

Diagram NOT accurately drawn

16 *A*, *B*, *C* and *D* are points on a circle.
PCT is a tangent to the circle at *C*.
BD is parallel to *PCT*.
DC = *DT*
Prove that *ADT* is parallel to *BC*.
Use angle *DTC* = *x* to help you.

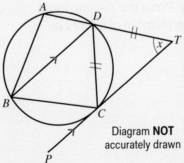

Diagram NOT accurately drawn

16 Test ready

Summary of key points

To revise for the test:

- Read each key point, find a question on it in the mastery lesson, and check you can work out the answer.

- If you cannot, try some other questions from the mastery lesson or ask for help.

Key points

1 A **theorem** is a rule that can be proved by a chain of reasoning. → **16.1**

2 A **chord** is a straight line connecting two points on a circle. → **16.1**

3 The perpendicular from the centre of a circle to a chord bisects the chord, and the line drawn from the centre of a circle to the midpoint of a chord is at right angles to the chord. → **16.1**

4 A **tangent** is a straight line that touches a circle at only one point. → **16.2**

5 The angle between a tangent and the radius is 90°. → **16.2**

6 Tangents drawn to a circle from a point outside the circle are equal in length. → **16.2**

7 An angle **subtended** by an arc is any angle with 'arms' that start and finish at the ends of the arc. → **16.3**

8 The angle at the centre of a circle is twice the angle at the circumference when both are subtended by the same arc. → **16.3**

9 The angle in a semicircle is a right angle. → **16.3**

10 Angles subtended at the circumference by the same arc are equal; or angles in the same segment are equal. → **16.4**

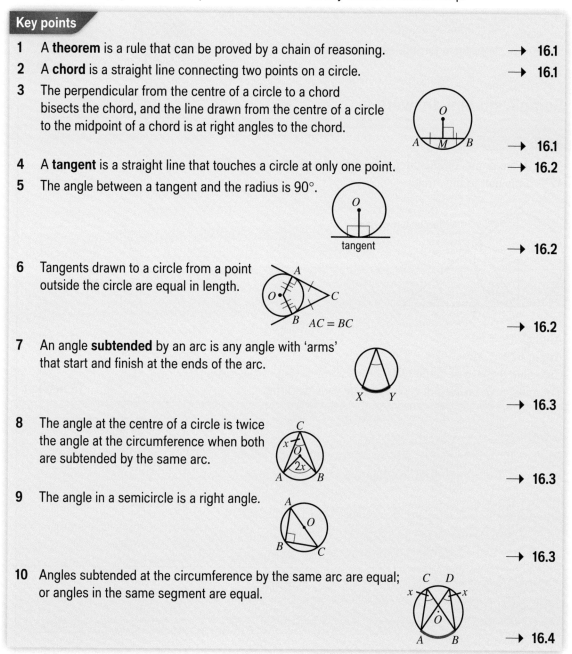

11 A **cyclic quadrilateral** is a quadrilateral with all four vertices on the circumference of a circle.

→ 16.4

12 Opposite angles of a cyclic quadrilateral add up to 180°.

→ 16.4

13 An exterior angle of a cyclic quadrilateral is equal to the opposite interior angle. → 16.4

14 The angle between the tangent and the chord is equal to the angle in the alternate segment.
This is called the **alternate segment theorem**.

→ 16.4

Sample student answers

A and B are points on the circumference of a circle, centre O.
AT is a tangent to the circle.

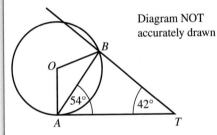

Diagram NOT accurately drawn

Angle $TAB = 54°$
Angle $BTA = 42°$
Calculate the size of angle OBT.
You must give reasons at each stage of your working.

(5 marks)

Student A

Angle $TAO = 90°$ (angle between tangent and radius $= 90°$)

Angle $OAB =$ angle $OBA = 90° - 54° = 36°$ (base angles of an isosceles triangle are equal)

Angle $ABT = 84°$ (angles in a triangle add up to 180°)

Angle $OBT = 36° + 84° = 120°$

Student B

Angle $OAB = 36°$ because $90° - 54° = 36°$

Angle $ABT = 84°$ because $180° - 54° - 42° = 84°$

Angle $OBT = 120°$ because $36° + 84° = 120°$

Which student gave the better answer, and why?

Master
p.166

Check up
p.181

Strengthen
p.183

Extend
p.186

Test ready
p.189

Unit test
p.191

16 Unit test

1 O is the centre of a circle.
ABC is a straight line.
Angle $OBC = 146°$

Work out the size of angle AOB.
Give reasons for each step in your working.

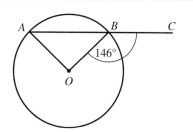

(3 marks)

2 O is the centre of a circle.
AT is a tangent and AB is a chord.
Angle $AOB = 124°$

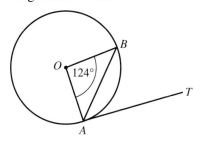

Work out the size of angle BAT.
Give reasons for each step in your working.

(3 marks)

3 O is the centre of a circle with radius 10 cm.
$OM = 6$ cm

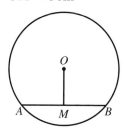

   a Write down the size of angle OMB. **(1 mark)**

   b Work out the length of chord AB. **(2 marks)**

4 O is the centre of a circle.
A, B, C and D are points on the circumference
of the circle.
Angle $BCD = 110°$

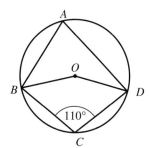

Work out the size of obtuse angle BOD.
Give reasons for each step in your working.

(3 marks)

5 O is the centre of a circle.

A, B, C and D are points on the circumference of the circle.

Angle $ADB = 19°$

Work out the size of

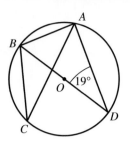

a angle ABD **(2 marks)**

b angle ACB **(1 mark)**

Give reasons for each step in your working.

6 P and Q are points on the circumference of a circle, centre O.

PR and QR are tangents to the circle.

Angle $PRQ = 42°$

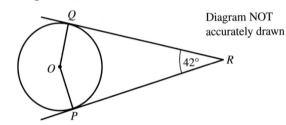

Diagram NOT accurately drawn

Find the size of angle OPQ.

Give reasons for your answer. **(4 marks)**

7 A, B, C and D are four points on the circumference of a circle.

AEC and DEB are straight lines.

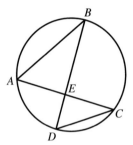

Prove that triangle ABE and triangle DEC are similar.

You must give reasons for each stage of your working. **(4 marks)**

8 A, B, C and E are points on the circumference of a circle.

DEF is a tangent to the circle.

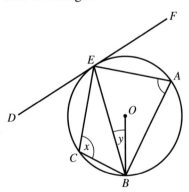

Show that $x - y = 90°$.

You must give a reason for each stage of your working. **(3 marks)**

9 A circle has equation $x^2 + y^2 = 25$.
P is the point $(-4, 3)$ on its circumference.

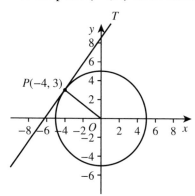

Find the equation of the tangent to the circle at P. **(5 marks)**

10 The diagram shows a circle, centre O.
AB is the tangent to the circle at the point A.
Angle $OBA = 30°$
Point B has coordinates $(12, 0)$.
Point P has coordinates $(n, 2n)$.

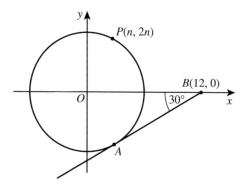

Find the value of n.
You must show all your working. **(4 marks)**

(TOTAL: 35 marks)

11 Challenge Use circle theorems to prove that an exterior angle of a cyclic quadrilateral is equal to the interior opposite angle.

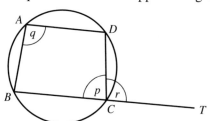

12 Reflect Look back at this unit.
Which lesson did you like most? Write a sentence to explain why.
Which lesson did you like least? Write a sentence to explain why.
Begin your sentence with: I liked lesson _____ most/least because _____

Mixed exercise 5

Exam-style question

1   Each student in a school is going to get a mug with the school name on it.
    Mrs Booth is going to order the mugs.
    Mrs Booth takes a sample of 50 students in the school.
    She asks each student to tell her the colour they would like.
    The table shows information about her results.

Mug colour	Number of students
red	13
blue	12
yellow	18
green	7

There are 1200 students in the school.

a Work out how many red mugs Mrs Booth should order.   **(2 marks)**

b Write down any assumption you made and explain how this could affect
    your answer.   **(1 mark)**

Exam-style question

2   The cumulative frequency table shows information about the heights, in centimetres,
    of 50 students.

Height, h (cm)	Cumulative frequency
$140 < h \leqslant 150$	1
$140 < h \leqslant 160$	13
$140 < h \leqslant 170$	34
$140 < h \leqslant 180$	45
$140 < h \leqslant 190$	50

a Draw a cumulative frequency graph for this information.   **(2 marks)**

b Use your graph to estimate the interquartile range.   **(2 marks)**

One of the 50 students is chosen at random.

c Use your graph to find an estimate for the probability that this person is
    between 155 cm and 175 cm tall.   **(2 marks)**

3   **Reasoning** The diagram shows a circle.
    The points A, B, C, D, E and F are all on the circle.
    Which line is the diameter of the circle?
    Give a reason for your answer.

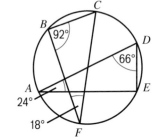

Exam-style question

4 P and Q are points on a circle with centre O.

RPS is the tangent to the circle at P.

QOR is a straight line.

Work out the size of angle QPS.

You must show your working.

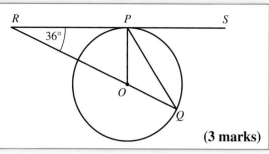

(3 marks)

Exam-style question

5 The box plot shows information about the weights of parcels processed at a post office on Monday.

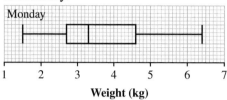

Weight (kg)

James says, 'Half of the parcels weigh less than 3 kg each.'

a Is James correct? Give a reason for your answer. **(1 mark)**

This box plot shows information about the weights of parcels processed at the post office on Tuesday.

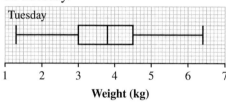

Weight (kg)

b Compare the distribution of the weights of the parcels on Monday with the distribution of the weights of the parcels on Tuesday. **(2 marks)**

6 **Reasoning** Sasha is asked to shade the region that satisfies all three of the inequalities:

$y < 3$, $y \leqslant 2x+1$, $y > x$

Sasha draws:

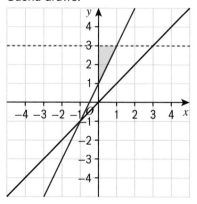

Explain why Sasha is incorrect.

7 The histogram gives information about the masses of a box of mixed tomatoes.
The histogram is incomplete.

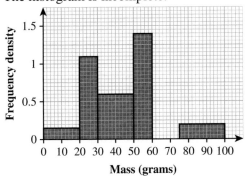

25% of the tomatoes have a mass between 60 grams and 75 grams.
None of the tomatoes have a mass greater than 100 grams.
Copy and complete the histogram. **(3 marks)**

8 **Problem-solving** The diagram shows a cyclic quadrilateral.

$p : q = 5 : 4$
Work out the size of each angle.

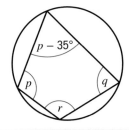

9 The histogram gives information about the length of times flights were delayed for an airline from the UK last month.

When flights are delayed for less than 3 hours customers are entitled to free food and drink.
When flights are delayed for more than 5 hours customers are entitled to a full refund.
The rest of the customers are entitled to compensation.

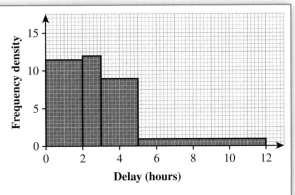

A pie chart is drawn using this information to show what customers are entitled to when their flights are delayed.

The angle of the sector for the customers entitled to compensation is $x°$.
Work out the value of x. **(4 marks)**

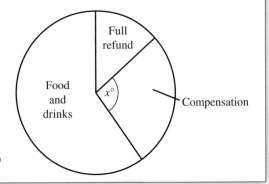

10 Problem-solving Work out the equation of the quadratic curve shown in the diagram.

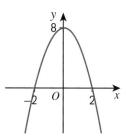

Exam-style question

11 $PQRS$ are four points on a circle.

PTR and QTS are straight lines.
Triangle PTS is an equilateral triangle.
Prove that triangle PQS is congruent to triangle SRP.

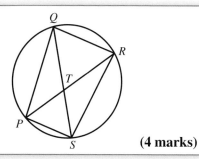

(4 marks)

12 Problem-solving Sketch the graph of $y = 2x^2 - 12x + 3$ showing the coordinates of the turning point and the exact coordinates of any intercepts with the coordinate axes.

13 Reasoning The diagram shows points P, Q and R on the circle with centre O.

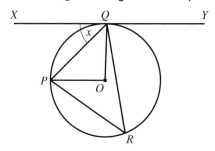

XY is the tangent to the circle at Q.
Prove angle $PRQ = x°$.

Exam-style question

14 The diagram shows a triangle.

The area of the triangle is greater than $5\,\text{cm}^2$.

a Show that $2x^2 - 19x + 44 < 0$. **(3 marks)**

b Find the range of possible values of x. **(3 marks)**

15 Problem-solving The diagram shows the circle with equation $x^2 + y^2 = 58$.

P is a point on the circle below the x-axis with x-coordinate 3.
Work out the equation of the tangent to the circle at P.

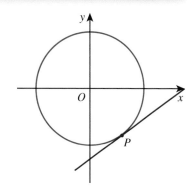

17 More algebra

Prior knowledge

17.1 Rearranging formulae

- Change the subject of a formula where the power or root of the subject appears.
- Change the subject of a formula where the subject appears twice.

Active Learn
Homework

Warm up

1 Fluency In each formula change the subject to the letter given in brackets.

a $v = u + at$ $[a]$ **b** $C = 2\pi r$ $[r]$ **c** $r = \dfrac{p}{q}$ $[p]$ **d** $r = \dfrac{p}{q}$ $[q]$

e $A = \frac{1}{2}bh$ $[h]$ **f** $t = 9s^2$ $[s]$ **g** $x = \sqrt{t}$ $[t]$ **h** $r = \sqrt{3s}$ $[s]$

2 Factorise

a $rs + r$ **b** $xy + 2y$ **c** $pq - q$ **d** $ak - 4k$

3 Make y the subject of each formula.

a $m = \dfrac{5y}{z}$ **b** $m = \dfrac{5xy}{z}$ **c** $m = \dfrac{5xy}{2z}$ **d** $m = \dfrac{5xy}{bz}$

e $n = \dfrac{x+y}{2}$ **f** $n = \dfrac{x+y}{a}$ **g** $n = \dfrac{4x+y}{a}$ **h** $n = \dfrac{4x-y}{a}$

i $n = \dfrac{x-2y}{a}$ **j** $r = x + \dfrac{y}{b}$ **k** $r = x + \dfrac{3y}{b}$ **l** $r = x - \dfrac{ky}{b}$

Key point

When the letter to be made the subject appears twice in the formula you will need to **factorise**.

Example

Make w the subject of the formula $A = wh + lh + lw$.

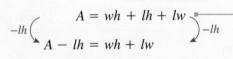

$-lh \Big(\quad A = wh + lh + lw \quad \Big) -lh$

$A - lh = wh + lw$

> w appears twice in this formula.
> Subtract lh from both sides to get the terms in w together on one side of the equals sign.

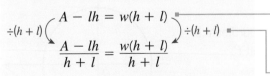

$\div(h+l) \Big(\quad A - lh = w(h + l) \quad \Big) \div(h+l)$

$\dfrac{A - lh}{h + l} = \dfrac{w(h + l)}{h + l}$

> Factorise the right-hand side, so w appears only once.

> Divide both sides by $(h+l)$.

$w = \dfrac{A - lh}{h + l}$

4 Make y the subject of each formula.

a $h = 3y + xy$ **b** $h = uy + xy$ **c** $k = y - ay$ **d** $k = by - ay$

5 Make d the subject of each formula.

a $K = 3d + nd + n$

b $L = xd + nd - x$

c $G = xd - yd + y$

d $H = ad - ac - bd$

6 **Reasoning** $5xy + 2 = w + 3xy$

a Make y the subject.

b Make x the subject.

7 **Reasoning** $H = xy + 2x + 7$

Fred rearranges the formula to make x the subject.

His answer is $x = \dfrac{H - 7 - xy}{2}$.

a What mistake has Fred made?

b Work out the correct answer.

8 Make x the subject of each formula.

a $r = \dfrac{1}{x}$

b $s = \dfrac{x+1}{x}$

c $t = \dfrac{1+7x}{x}$

d $v = \dfrac{1+bx}{x}$

e $w = \dfrac{a+bx}{x}$

f $y = \dfrac{1-x}{x}$

g $z = \dfrac{1-bx}{x}$

h $d = \dfrac{a-bx}{x}$

Q8a hint

First multiply both sides by x.

9 Make x the subject of each formula.

a $e = \dfrac{1}{x+1}$

b $f = \dfrac{x}{x+1}$

c $g = \dfrac{x}{x-1}$

d $h = \dfrac{2x}{x-1}$

e $j = \dfrac{2x}{x+1}$

f $k = \dfrac{3x}{x+1}$

g $l = \dfrac{x+1}{x-1}$

h $m = \dfrac{x+2}{x-1}$

i $n = \dfrac{x-2}{x-1}$

j $p = \dfrac{x+3}{x-1}$

k $q = \dfrac{2x+1}{x-1}$

l $r = \dfrac{2x+1}{x+1}$

Exam-style question

10 Make t the subject of the formula

$x = \dfrac{3t+2}{t-1}$ **(3 marks)**

Exam tip

Show all your working. Don't try to take shortcuts.

Key point

When the letter to be made the subject is part of a term involving a power or root, rearrange so that the whole term is on its own on one side of the equation.
Use inverse operations to eliminate the power or root.

11 Make v the subject of the formula $E = \frac{1}{2}mv^2$.

Q11 hint

First multiply both sides by 2.

12 **a** Square both sides of $H = \sqrt{x-y}$.

b Make x the subject of the formula $H = \sqrt{x-y}$.

Example

Make x the subject of the formula $P = d\sqrt{\dfrac{x}{y}}$.

$$\div d \left(\begin{array}{c} P = d\sqrt{\dfrac{x}{y}} \\ \dfrac{P}{d} = \sqrt{\dfrac{x}{y}} \end{array} \right) \div d$$ — Divide both sides by d.

$$\text{square} \left(\begin{array}{c} \dfrac{P}{d} = \sqrt{\dfrac{x}{y}} \\ \dfrac{P^2}{d^2} = \dfrac{x}{y} \end{array} \right) \text{square}$$ — Square both sides.

$$\times y \left(\begin{array}{c} \dfrac{P^2}{d^2} = \dfrac{x}{y} \\ \dfrac{yP^2}{d^2} = x \end{array} \right) \times y$$ — Multiply both sides by y.

or $\quad x = \dfrac{yP^2}{d^2}$

13 Make x the subject of each formula.

a $M = 3\sqrt{\dfrac{x}{4}}$ **b** $L = 2\sqrt{\dfrac{x}{n}}$

c $T = 2p\sqrt{\dfrac{x}{k}}$ **d** $N = \sqrt{\dfrac{5x}{t}}$

e $P = \sqrt{\dfrac{xy}{z}}$ **f** $w = \sqrt{\dfrac{1}{x}}$

g $y = 4\sqrt{\dfrac{1}{x}}$ **h** $h = 7\sqrt{\dfrac{a}{x}}$

14 Make x the subject of each formula.

a $y = x^2 + n$ **b** $y = \dfrac{x^2}{n}$ **c** $y = \dfrac{x^2 + 1}{n}$ **d** $y = \dfrac{x^2}{n} + 1$

e $z = \dfrac{1}{x^2}$ **f** $z = \dfrac{n}{x^2}$ **g** $z = \dfrac{1+n}{x^2}$ **h** $z = 1 + \dfrac{n}{x^2}$

i $k = (x+1)^2$ **j** $k = 3(x+1)^2$ **k** $k = \frac{1}{4}(x-1)^2$ **l** $k = \frac{1}{2}(x-1)^2$

15 In each formula change the subject to the letter given in brackets.

a $y = x^3$ $\quad [x]$ **b** $z = 4x^3$ $\quad [x]$

c $A = (4x)^3$ $\quad [x]$ **d** $B = \frac{1}{2}x^3$ $\quad [x]$

e $C = \frac{2}{3}x^3$ $\quad [x]$ **f** $D = \frac{8}{27}x^3$ $\quad [x]$

g $V = \frac{4}{3}\pi r^3$ $\quad [r]$ **h** $t = \sqrt[3]{y}$ $\quad [y]$

i $y = \sqrt[3]{5x}$ $\quad [x]$ **j** $n = \sqrt[3]{\dfrac{k}{2}}$ $\quad [k]$

k $z = \sqrt[3]{\dfrac{x}{y}}$ $\quad [x]$ **l** $z = \sqrt[3]{\dfrac{x}{y}}$ $\quad [y]$

17.2 Algebraic fractions

- Add and subtract algebraic fractions.
- Multiply and divide algebraic fractions.
- Change the subject of a formula involving fractions where all the variables are in the denominators.

Active Learn
Homework

Warm up

1 Fluency Work out

a $\frac{5}{7} \times \frac{2}{11}$ **b** $\frac{3}{4} \times \frac{2}{9}$ **c** $\frac{6}{7} \div \frac{5}{9}$ **d** $\frac{25}{32} \div \frac{35}{14}$

2 Fluency Work out

a $\frac{5}{12} + \frac{7}{18}$ **b** $\frac{7}{11} + \frac{2}{9}$ **c** $\frac{7}{11} - \frac{2}{9}$ **d** $\frac{9}{12} - \frac{4}{5}$

Key point

When multiplying algebraic fractions, cancel common factors in numerators and denominators before multiplying the fractions together.

---

3 Write as a single fraction in its simplest form. The first one has been started for you.

a $\frac{x}{2} \times \frac{x}{3} = \frac{x \times x}{2 \times 3} =$ **b** $\frac{5x}{3} \times \frac{x}{2}$

c $\frac{2x}{5} \times \frac{3m}{4}$ **d** $\frac{4}{9y} \times \frac{3}{5y}$

> **Q3c hint**
> First cancel any common factors.

4 Write as a single fraction in its simplest form. The first one has been started for you.

a $\frac{4x^2}{y^3} \times \frac{3y}{8x} = \frac{\overset{1}{\cancel{4}} x^2 \times 3y}{y^3 \times {}_2\cancel{8}\cancel{x}} =$ **b** $\frac{14x^3}{10y^2} \times \frac{25y^6}{21x^5}$ **c** $\frac{12a^2}{7x} \times \frac{14x^5}{16a^4}$

5 Write as a single fraction in its simplest form.

a $\frac{4}{x} \div \frac{3}{x}$ **b** $\frac{1}{x} \div \frac{1}{y}$

c $\frac{1}{x^3 n} \div \frac{1}{xn}$ **d** $x^3 y \div \frac{1}{xy}$

e $\frac{2y^3}{3x^5} \div \frac{8y^7}{15x^3}$ **f** $\frac{4x^2}{y} \div 2xy^2$

> **Q5a hint**
> Dividing by $\frac{3}{x}$ is equivalent to multiplying by $\frac{x}{3}$.

> **Q5d hint**
> Write $x^3 y$ as $\frac{x^3 y}{1}$.

6 Write as a single fraction in its simplest form.

a $\frac{x}{3} \div (x + 7)$ **b** $\frac{1}{x} \div (x + 2)$ **c** $\frac{x}{2} \div \frac{x+1}{3}$ **d** $\frac{x}{2} \div \frac{x-1}{4}$

Example

Simplify $\frac{x}{5}+\frac{x}{3}$.

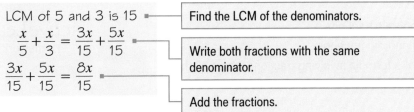

LCM of 5 and 3 is 15 — Find the LCM of the denominators.

$\frac{x}{5}+\frac{x}{3}=\frac{3x}{15}+\frac{5x}{15}$ — Write both fractions with the same denominator.

$\frac{3x}{15}+\frac{5x}{15}=\frac{8x}{15}$ — Add the fractions.

7 Simplify

a $\frac{x}{3}+\frac{x}{7}$ **b** $\frac{x}{3}-\frac{x}{12}$ **c** $\frac{x}{5}-\frac{x}{10}$ **d** $\frac{3x}{10}+\frac{x}{2}$ **e** $\frac{4x}{3}-\frac{x}{4}$ **f** $\frac{3x}{2}+\frac{5x}{3}$

8 Write down the LCM of

a $2x$ and $5x$ **b** $3x$ and $6x$

c $4x$ and $7x$ **d** $4x$ and $3x$

Q8a hint

Multiples of $2x$: $2x$, $4x$, ...
Multiples of $5x$: $5x$, $10x$, ...

9 **a** Write $\frac{1}{4x}$ and $\frac{1}{3x}$ as equivalent fractions with denominator the LCM of $4x$ and $3x$.

b Simplify $\frac{1}{4x}+\frac{1}{3x}$

10 Write as a single fraction in its simplest form.

a $\frac{1}{9x}+\frac{1}{2x}$ **b** $\frac{1}{4x}-\frac{1}{5x}$ **c** $\frac{1}{6x}+\frac{5}{9x}$ **d** $\frac{4}{5x}+\frac{1}{3x}$

11 **a** Copy and complete.   **b** Copy and complete.

$\frac{x-4}{2}=\frac{\square(x-4)}{5\times2}=\frac{\square x-\square}{10}$ $\frac{x+7}{5}=\frac{\square(x+7)}{2\times5}=\frac{\square x+\square}{10}$

c Use your answers to parts **a** and **b** to work out $\frac{x-4}{2}+\frac{x+7}{5}$.

12 Write as a single fraction in its simplest form.

a $\frac{x+2}{2}+\frac{x+1}{3}$ **b** $\frac{x+5}{2}-\frac{x-3}{7}$ **c** $\frac{x+7}{4}-\frac{2x-1}{9}$ **d** $\frac{x-1}{3}+\frac{x+3}{2}+\frac{x}{4}$

Exam-style question

13 Write $\frac{x+6}{2}+\frac{2x-3}{5}$ as a single fraction in its simplest form.   **(3 marks)**

14 Follow these steps to make x the subject of $\frac{1}{x}+\frac{1}{y}=1$

a Add $\frac{1}{x}+\frac{1}{y}$ $\frac{\square}{xy}=1$

b Multiply both sides by xy.

c Collect all terms with x on one side.

d Factorise.

e Divide to get x on its own.

15 The lens formula is $\frac{1}{f}=\frac{1}{u}+\frac{1}{v}$, where $f=$ focal length, $u=$ object distance and $v=$ image distance. Make u the subject of the formula.

17.3 Simplifying algebraic fractions

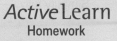

Active Learn
Homework

- Simplify algebraic fractions.

Warm up

1 **Fluency** Factorise
  a $6x + 18$ **b** $x^2 + 3x$ **c** $x^3 + 4x^2$ **d** $3x^3 - 15x$

2 Factorise
  a $x^2 - 9x + 18$ **b** $x^2 - 81$ **c** $2x^2 - 3x - 5$ **d** $5x^2 + 21x + 4$

Key point

To simplify an algebraic fraction, cancel any common factors in the numerator and denominator.

3 Simplify

  a $\dfrac{x}{xy}$

  b $\dfrac{x+6}{3(x+6)}$

  c $\dfrac{x-7}{(x-7)^2}$

  d $\dfrac{(x+2)(x-1)}{(x-1)(x-5)}$

  e $\dfrac{(x+9)(x-3)}{x(x+9)}$

  f $\dfrac{x^2(x-1)}{x(x-1)^2}$

Key point

You may need to factorise before simplifying an algebraic fraction:
- Factorise the numerator and denominator.
- Divide the numerator and denominator by any common factors.

4 **a** Factorise $x^2 - 6x$.

  b Use your answer to part **a** to simplify $\dfrac{x^2 - 6x}{x - 6}$.

5 Simplify fully

  a $\dfrac{x^2 + 8x}{x}$

  b $\dfrac{2x^2 + 5x}{x}$

  c $\dfrac{2x}{2x^2 + 6x}$

  d $\dfrac{12x^2 + 15x}{4x + 5}$

  e $\dfrac{3x + 3}{6x^2 + 6x}$

  f $\dfrac{10x - 25}{4x^2 - 10x}$

  > **Q5e hint**
  >
  > Factorise the numerator and denominator.

6 **Reasoning** Sally says, '$(x+2)$ is a factor of the numerator and the denominator of $\dfrac{x^2 + 2x}{x^2 + 2}$.'

  a Is Sally correct? Explain.

  b Can the fraction be simplified? Explain your answer.

7 For each algebraic fraction, factorise the quadratic expression. Then simplify fully.

  a $\dfrac{2(x+3)}{x^2 + 8x + 15}$

  b $\dfrac{x^2 - 4}{x - 4}$

  c $\dfrac{x^2 + 2x + 1}{x + 1}$

  d $\dfrac{x^2 - x - 6}{5(x+2)}$

Example

Simplify fully $\dfrac{x^2+5x+4}{x^2-3x-28}$.

$$\dfrac{x^2+5x+4}{x^2-3x-28} = \dfrac{(x+1)(x+4)}{(x-7)(x+4)}$$

Factorise the numerator and denominator.

$$= \dfrac{x+1}{x-7}$$

Divide the numerator and denominator by the common factor $(x+4)$.

8 Simplify fully

a $\dfrac{x^2+8x+15}{x^2+2x-15}$ **b** $\dfrac{x^2-11x+30}{x^2+x-42}$ **c** $\dfrac{x^2-25}{(x+5)^2}$ **d** $\dfrac{2x^2-5x-3}{x^2-x-6}$

Exam-style question

9 Simplify fully

$$\dfrac{2x^2+17x+21}{x^2-49}$$

(3 marks)

Exam tip

Write any factorisations clearly.

Exam-style question

10 Simplify fully

$$\dfrac{2x^2-9x+4}{3x^2-12x}$$

(3 marks)

---

11 Simplify fully

a $\dfrac{2x^2-x-3}{3x^2+x-2}$ **b** $\dfrac{5x^2+14x-3}{6x^2+23x+15}$ **c** $\dfrac{25x^2-1}{25x^2+10x+1}$ **d** $\dfrac{2x^2-23x+30}{4x^2-9}$

12 a Copy and complete.

$6-x = -(\square - \square)$

b Simplify

   i $\dfrac{6-x}{x-6}$ **ii** $\dfrac{36-x^2}{x^2-3x-18}$

13 Simplify fully

a $\dfrac{16-x^2}{x^2-4x}$ **b** $\dfrac{x^2-12x+36}{2x^2-72}$ **c** $\dfrac{12x-3x^2}{5x^2-19x-4}$ **d** $\dfrac{10x-6x^2}{6x^2-19x+15}$

14 Reasoning Show that $\dfrac{(x^2-36)(2x^2+8x)}{x^2+10x+24} = 2x(x-6)$.

15 Reasoning Show that $\dfrac{(x^2+x-12)(x^2+2x-3)(10x^2+12x)}{(9-x^2)(5x^2+26x+24)(7x-7)} = -\dfrac{2x}{7}$.

Q15 hint

Start by factorising the numerator and then the denominator.

17.4 More algebraic fractions

Active Learn
Homework

- Add and subtract more complex algebraic fractions.
- Multiply and divide more complex algebraic fractions.

Warm up

1 Fluency Write as a single fraction.
a $\dfrac{3x^2}{m^2} \times \dfrac{5m}{4x}$ **b** $\dfrac{5y}{2} \div \dfrac{2y}{15}$

2 Write as a single fraction in its simplest form.
a $\dfrac{1}{3x} - \dfrac{1}{8x}$ **b** $\dfrac{x-1}{3} + \dfrac{x+5}{4}$

3 Write as a single fraction in its simplest form.

a $x + 3 \times \dfrac{x-4}{(x+3)^2}$ **b** $\dfrac{x+2}{x-1} \times \dfrac{x-1}{x+5}$ **c** $\dfrac{x-4}{6} \times \dfrac{2}{3x-12}$

d $\dfrac{5}{x+2} \div \dfrac{15}{8x+16}$ **e** $\dfrac{2x+6}{x+7} \div \dfrac{x+3}{x-1}$ **f** $\dfrac{(x+4)^2}{x-2} \div \dfrac{x+4}{x}$

> **Q3c hint**
> First factorise $3x - 12$.

4 a Factorise **i** $x^2 - 9$ **ii** $x^2 + 5x + 6$

b Write $\dfrac{x^2 - 9}{4} \times \dfrac{8}{x^2 + 5x + 6}$ as a single fraction in its simplest form.

5 Write as a single fraction in its simplest form.

a $\dfrac{x^2 - 7x + 10}{x^2 + 4x + 3} \times \dfrac{x^2 - 9}{x^2 - x - 20}$ **b** $\dfrac{14x + 21}{2x^2 + 7x + 6} \div \dfrac{x^2 - 10x + 21}{x^2 + 9x + 14}$

Key point

The lowest common denominator of two algebraic fractions is the lowest common multiple of the two denominators.

6 Write down the LCM of
a x and $x + 2$ **b** $x + 2$ and $x + 3$ **c** $x + 1$ and $x - 1$ **d** $3x$ and $5x$

Example

Write $\dfrac{7}{x+2} - \dfrac{3}{x+3}$ as a single fraction in its simplest form.

$$\dfrac{7(x+3)}{(x+2)(x+3)} - \dfrac{3(x+2)}{(x+2)(x+3)}$$

> Convert each fraction to an equivalent fraction with the common denominator $(x+2)(x+3)$.

$$= \dfrac{7(x+3) - 3(x+2)}{(x+2)(x+3)}$$

$$= \dfrac{7x + 21 - 3x - 6}{(x+2)(x+3)}$$

> Expand and simplify.

$$= \dfrac{4x + 15}{(x+2)(x+3)}$$

7 Simplify fully

a $\dfrac{1}{x+4}+\dfrac{1}{x+5}$ **b** $\dfrac{3}{x+1}+\dfrac{4}{x-1}$ **c** $\dfrac{7}{x-5}-\dfrac{1}{x+3}$

d $\dfrac{1}{2x-3}-\dfrac{1}{2x+4}$ **e** $1+\dfrac{3}{x+2}+\dfrac{2}{x+1}$ **f** $5-\dfrac{2}{x-1}-\dfrac{1}{x+4}$

8 Simplify fully

a $\dfrac{x+3}{x-2}+\dfrac{1}{x+2}$ **b** $\dfrac{x+3}{x+1}+\dfrac{x}{x-1}$ **c** $\dfrac{x+3}{x+4}+\dfrac{x+1}{x-4}$

d $\dfrac{x+4}{x-3}-\dfrac{x+1}{x+3}$ **e** $\dfrac{x+4}{x-2}-\dfrac{x-1}{x+2}$ **f** $1+\dfrac{x+4}{x+2}-\dfrac{x-1}{x-2}$

Exam-style question

9 $2-\dfrac{x+1}{x+2}-\dfrac{x+3}{x-2}$ can be written as a single fraction in the form $\dfrac{ax+b}{x^2-4}$, where a and b are integers.

Work out the value of a and the value of b.   **(4 marks)**

Q9 hint

Simplify, then compare with $\dfrac{ax+b}{x^2-4}$.

---

10 a Factorise

    i $3x+9$ **ii** $4x+12$

b Write down the LCM of $3x+9$ and $4x+12$.

c Write $\dfrac{1}{3x+9}+\dfrac{1}{4x+12}$ as a single fraction in its simplest form.

Q10b hint

Look at the factorised form of each expression:
$a(x+y)$
$b(x+y)$
$LCM = ab(x+y)$

11 a Factorise x^2-16.

b Write $\dfrac{1}{x+4}+\dfrac{1}{x^2-16}$ as a single fraction in its simplest form.

12 Write as a single fraction in its simplest form.

a $\dfrac{1}{3x^2+8x+5}-\dfrac{1}{3x+5}$ **b** $\dfrac{1}{x^2+7x+6}-\dfrac{1}{2x+12}$

c $\dfrac{1}{x^2+6x+8}+\dfrac{3}{x^2-3x-28}$ **d** $\dfrac{4}{25-x^2}-\dfrac{3}{5-x}$

Q12b hint

Factorise x^2+7x+6 and $2x+12$.

Exam-style question

13 Show that $3+\left[(x+4)\div\dfrac{x^2-x-20}{x+1}\right]$ simplifies to $\dfrac{ax-b}{cx-d}$, where a, b, c and d are integers.   **(4 marks)**

---

14 Write $\dfrac{1}{5x}+\dfrac{1}{5(x-1)}+\dfrac{1}{10}$ as a single fraction in its simplest form.

Q14 hint

Work out the lowest common multiple of $5x$, $5(x-1)$ and 10.

15 Reasoning Show that $\dfrac{1}{x^2+5x+6}+\dfrac{1}{5x+10}=\dfrac{x+8}{A(x+3)(x+2)}$ and find the value of A.

17.5 Proof

- Prove a result using algebra.

*Active*Learn
Homework

Warm up

1 **Fluency** What type of number is **a** $2n$ **b** $2n+1$ for any integer n?

2 Which sequences contain
a only even numbers **b** only odd numbers **c** even and odd numbers?

A $n+2$ **B** $2n$ **C** $5n$ **D** $2n-1$ **E** n^2

3 Is each of these an equation or an identity?
a $2(n+3) = 2n+6$ **b** $5n-7 = 8$ **c** $\frac{1}{2}(4n+10) = 2n+5$ **d** $2(3n-5) = 4$

Key point

To show a statement is an **identity**, expand and simplify the expressions on one or both sides of the equals sign, until the two expressions are the same.

Example

Show that $(x+4)^2 - 7 \equiv x^2 + 8x + 9$.

$\text{LHS} = (x+4)^2 - 7$
$\equiv (x+4)(x+4) - 7$ ◄— Expand the brackets on the left-hand side (LHS).
$= x^2 + 8x + 16 - 7 = x^2 + 8x + 9$
$\text{RHS} = x^2 + 8x + 9$
So $\text{LHS} = \text{RHS}$ and $(x+4)^2 - 7 \equiv x^2 + 8x + 9$ ◄— Aim to show that LHS = RHS.

---

4 **Reasoning** Show that
a $(x-3)^2 + 6x \equiv x^2 + 9$ **b** $x^2 + 8x + 49 \equiv (x+7)^2 - 6x$
c $(x-5)^2 - 4 \equiv (x-3)(x-7)$ **d** $16 - (x+2)^2 \equiv (6+x)(2-x)$

5 **Reasoning a** Show that $(x-1)(x+1) \equiv x^2 - 1$
b Use your answer to part **a** to work out **i** 99×101 **ii** 199×201

6 **Reasoning** The blue card is a rectangle of length $x+5$ and width $x+2$.
a Write an expression for the area of the blue card.
A rectangle of length $x+1$ and width x is cut out and removed.
b Write an expression for the area of the rectangle cut out.
c Show that the area of the remaining card is $6x+10$.

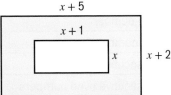

Q6c hint
Subtract your expression from part **b** from your expression from part **a**.

Exam-style question

7

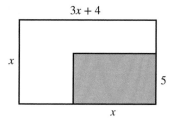

3x + 4

x

5

x

The diagram shows a large rectangle of length $(3x+4)$ cm and width x cm. A smaller rectangle of length x cm and width 5 cm is cut out and removed.
The area of the shape that is left is $70\,\text{cm}^2$.
Show that $3x^2 - x - 70 = 0$. **(3 marks)**

Exam tip

'Show that' means you need to write down every stage of your working so someone else can follow it.

8 **Reasoning** Show that $\dfrac{1}{x^2-x} - \dfrac{1}{x^2+3x} = \dfrac{A}{x(x-1)(x+3)}$ and find the value of A.

Key point

A **proof** is a logical argument for a mathematical statement.
To prove a statement is true, you must show that it will be true in *all* cases.
To prove a statement is not true you can find a **counter-example** – an example that does not fit the statement.

9 Give a counter-example to prove that these statements are *not* true.

  a All prime numbers are odd.

  b The cube of a number is always greater than its square.

  c The difference between two numbers is always less than their sum.

  d The difference between two square numbers is always odd.

Q9c hint

Try some negative numbers.

10 **Reasoning**

  a Explain why any odd number can be written as $2n+1$ or $2n-1$.

  b Let $2n$ be any even number.
    Let $2m+1$ be any odd number.
    Prove that the sum of any odd number and any even number is always odd.

Key point

For an algebraic proof, use n to represent any integer.

Even number	$2n$
Odd number	$2n+1$ or $2n-1$
Consecutive numbers	$n, n+1, n+2, \ldots$
Consecutive even numbers	$2n, 2n+2, 2n+4, \ldots$
Consecutive odd numbers	$2n+1, 2n+3, 2n+5, \ldots$

11 Reasoning

a The nth even number is $2n$. Explain why the next even number is $2n+2$.

b Prove algebraically that the product of two consecutive even numbers is always a multiple of 4.

Exam-style question

12 Prove algebraically that the product of any two odd numbers is odd. **(3 marks)**

Exam tip

Multiplying pairs of odd numbers does not **prove** the statement. You need to use algebra.

13 Reasoning Given that $2(x-a) = x+5$, where a is an integer, show that x must be an odd number.

14 Reasoning

a Work out

i $\frac{1}{5}-\frac{1}{6}$ **ii** $\frac{1}{3}-\frac{1}{4}$ **iii** $\frac{1}{7}-\frac{1}{8}$

b Use your answers to part **a** to write down the answer to $\frac{1}{9}-\frac{1}{10}$.

c Explain how you can quickly calculate $\frac{1}{99}-\frac{1}{100}$.

d i Simplify $\dfrac{1}{x}-\dfrac{1}{x+1}$.

ii Explain how this proves your answer from part **c**.

15 Reasoning Prove that the square of an even number is always a multiple of 4.

Exam-style question

16 Prove that the product of two consecutive odd numbers is always 1 less than a multiple of 4. **(4 marks)**

17 Reasoning

a Write an expression for the product of three consecutive integers, $n-1$, n and $n+1$.

b Hence show that $n^3 - n$ is always a multiple of 2.

Exam-style question

18 Prove algebraically that the sum of any two consecutive integers is equal to the difference between their squares. **(4 marks)**

Exam-style question

19 n is an integer greater than 2.
Prove algebraically that
$$n^2 - 3 - (n-3)^2$$
is always a multiple of 6. **(4 marks)**

Exam tip

Use all the information given in the question in your proof.

17.6 Surds

- Simplify expressions involving surds.
- Expand expressions involving surds.
- Rationalise the denominator of a fraction.

Active Learn
Homework

Warm up

1 Fluency Work out

a $\sqrt{5} \times \sqrt{5}$ **b** $\sqrt{5} \times \sqrt{3}$ **c** $7\sqrt{3} - 4\sqrt{3}$ **d** $3\sqrt{2} + 5\sqrt{2}$

2 Copy and complete.

a $\sqrt{6} = \sqrt{2} \times \sqrt{\square}$ **b** $\sqrt{\square} = \sqrt{5} \times \sqrt{6}$

c $\sqrt{\dfrac{\square}{\square}} = \dfrac{\sqrt{5}}{\sqrt{7}}$ **d** $\sqrt{50} = \sqrt{\square} \times \sqrt{2} = \square\sqrt{2}$

e $\sqrt{18} = \square\sqrt{2}$ **f** $\sqrt{48} = \square\sqrt{3}$

> **Q2 hint**
>
> Use $\sqrt{m} \times \sqrt{n} = \sqrt{mn}$
> and $\dfrac{\sqrt{m}}{\sqrt{n}} = \sqrt{\dfrac{m}{n}}$

3 Rationalise the denominators. Simplify your answers if possible.

a $\dfrac{1}{\sqrt{10}}$ **b** $\dfrac{3}{\sqrt{15}}$ **c** $\dfrac{8}{\sqrt{32}}$ **d** $\dfrac{2}{\sqrt{14}}$

4 Simplify

a $\sqrt{45}$ **b** $\sqrt{20}$ **c** $3\sqrt{45} + 7\sqrt{20}$ **d** $2\sqrt{75} - 3\sqrt{27}$

e $\sqrt{200} + 3\sqrt{32}$ **f** $5\sqrt{18} - \sqrt{128} + 4\sqrt{8}$ **g** $2\sqrt{98} + 5\sqrt{8}$ **h** $3\sqrt{28} - 5\sqrt{63}$

5 Factorise these expressions. The first one has been started for you.

a $\sqrt{12} + 2 = 2\sqrt{\square} + 2 = 2(\square + \square)$ **b** $9 + \sqrt{54}$

c $18 - \sqrt{45}$ **d** $\sqrt{75} - \sqrt{50}$

6 Expand and simplify

a $\sqrt{5}(4 + \sqrt{5})$ **b** $(\sqrt{7} + 1)(4 + \sqrt{7})$

c $(6 - \sqrt{2})(4 + \sqrt{2})$ **d** $(2 - \sqrt{2})^2$

e $(4 - \sqrt{10})^2$ **f** $(7 + \sqrt{3})^2$

> **Q6d hint**
>
> $(2 - \sqrt{2})^2 = (2 - \sqrt{2})(2 - \sqrt{2})$
> Your answer should be in the
> form $a + b\sqrt{2}$.

Exam-style question

7 Expand $(5 - \sqrt{5})^2$.
Write your answer in the form $a + b\sqrt{c}$,
where a, b and c are integers.   **(2 marks)**

> **Exam tip**
>
> Make sure your answer is in the
> form $a + b\sqrt{c}$.

Exam-style question

8 $\sqrt{5}(\sqrt{50} - \sqrt{8})$ can be written in the form $a\sqrt{10}$,
where a is an integer. Find the value of a.   **(3 marks)**

> **Exam tip**
>
> Your final answer should be
> an integer – not the simplified
> expression.

9 **a** Work out the area of each shape. Write your answers in the form $a + b\sqrt{2}$.

i

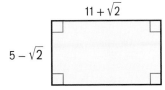

$11 + \sqrt{2}$

$5 - \sqrt{2}$

ii

$2 + \sqrt{8}$

b **Reasoning** Would the perimeter of each shape be rational or irrational? Explain.

10 Rationalise the denominators. The first one has been started for you.

a $\dfrac{3 + \sqrt{2}}{\sqrt{2}} = \dfrac{(3 + \sqrt{2})\sqrt{2}}{\sqrt{2} \times \sqrt{2}} =$

b $\dfrac{6 - \sqrt{3}}{\sqrt{3}}$

c $\dfrac{19 - \sqrt{7}}{\sqrt{7}}$

d $\dfrac{5 + \sqrt{5}}{\sqrt{5}}$

Exam-style question

11 Given that

$\dfrac{8 + \sqrt{18}}{\sqrt{2}} = a + b\sqrt{2}$, where a and b are integers,

find the value of a and the value of b. **(3 marks)**

Exam tip

Make sure you multiply both parts of the expression in the numerator by the surd.

12 **Reasoning**

a Expand and simplify $(3 + \sqrt{5})(3 - \sqrt{5})$.

b Is your answer rational or irrational?

c Multiply the numerator and the denominator of $\dfrac{1}{7 + \sqrt{2}}$ by $7 - \sqrt{2}$.

Is the new denominator rational or irrational?

Key point

To rationalise the fraction $\dfrac{1}{a\sqrt{b}}$, multiply by $\dfrac{\sqrt{b}}{\sqrt{b}}$.

To rationalise the fraction $\dfrac{1}{a \mp \sqrt{b}}$, multiply by $\dfrac{a \pm \sqrt{b}}{a \pm \sqrt{b}}$.

13 Rationalise the denominators.

a $\dfrac{1}{1 + \sqrt{2}}$

b $\dfrac{1}{5 - \sqrt{3}}$

c $\dfrac{7}{4 - \sqrt{5}}$

d $\dfrac{4}{1 + \sqrt{6}}$

e $\dfrac{\sqrt{5}}{1 - \sqrt{5}}$

f $\dfrac{6 + \sqrt{2}}{8 - \sqrt{2}}$

Q13a hint

Multiply the numerator and denominator by $(1 - \sqrt{2})$.

14 **a** Solve $x^2 - 6x + 1 = 0$ by using the quadratic formula.

b Solve the equation $x^2 + 10x + 13 = 0$ by completing the square.

c Solve the equation $x^2 - 16x + 8 = 0$.

Write all your answers in simplified surd form.

Exam-style question

15 Show that $\dfrac{(2 + \sqrt{8})^2}{\sqrt{8} - 2}$ can be written in the form $a(b + c\sqrt{2})$,

where a, b and c are integers. **(4 marks)**

17.7 Solving algebraic fraction equations

Active Learn
Homework

- Solve equations that involve algebraic fractions.

Warm up

1 **Fluency** Find the LCM of **a** x and 4   **b** $4x$ and x **c** $x+3$ and $x+2$

2 Write as a single fraction in its simplest form.

a $\dfrac{x}{4} - \dfrac{x}{5}$ **b** $\dfrac{7}{2x} - \dfrac{3}{2x}$ **c** $\dfrac{8}{x-6} + \dfrac{2}{x-6}$ **d** $\dfrac{5}{x+2} - \dfrac{3}{x+3}$

3 Solve

a $x^2 + 6x + 8 = 0$ **b** $2x^2 - 13x + 11 = 0$ **c** $5x^2 - 25x + 20 = 0$ **d** $3x^2 + 8x - 17 = 0$

Give your answers correct to 2 decimal places where appropriate.

4 **a** Write the two fractions on the left-hand side (LHS) of the equation $\dfrac{2x+1}{5} + \dfrac{x+3}{2} = 1$ with a common denominator.

b Add the fractions.

c Multiply both sides of the equation by the denominator of the LHS.

d Solve the equation. Write the solution as a fraction.

5 Solve **a** $\dfrac{3x-1}{2} + \dfrac{x+5}{3} = 4$ **b** $\dfrac{2x-5}{3} - \dfrac{x+1}{5} = \dfrac{3}{2}$

Exam-style question

6 Solve $\dfrac{3x-1}{3} - \dfrac{2x+3}{4} = \dfrac{x+1}{6}$.   **(4 marks)**

Key point

To solve an equation involving algebraic fractions, first write one side as a fraction in its simplest form.

---

7 Solve these equations. Give each answer as a simplified fraction.

a $\dfrac{3}{x} + \dfrac{2}{x} = 4$ **b** $\dfrac{6}{x-1} - \dfrac{2}{x-1} = 7$ **c** $\dfrac{3}{x+5} - \dfrac{7}{x+5} = 8$

Key point

To solve a quadratic equation, rearrange it into the form $ax^2 + bx + c = 0$.

---

8 Solve these equations.

a $\dfrac{4}{x} = \dfrac{3x-7}{5}$ **b** $\dfrac{2x+1}{3} = \dfrac{2}{x}$

c $\dfrac{5x-3}{2} = \dfrac{7}{x}$ **d** $\dfrac{10}{x} = \dfrac{2x+3}{2}$

Q8a hint

First multiply both sides by the LCM ($5x$) and simplify. Then rearrange to a quadratic equation with 0 on one side. Solve by factorising.

Solve $\dfrac{3}{2x-1}+\dfrac{4}{x+2}=2$.

$\dfrac{3(x+2)}{(2x-1)(x+2)}+\dfrac{4(2x-1)}{(x+2)(2x-1)}=2$ — Rewrite the LHS using the common denominator $(2x-1)(x+2)$.

$\dfrac{3(x+2)+4(2x-1)}{(2x-1)(x+2)}=2$ — Add the fractions.

$\dfrac{3x+6+8x-4}{(2x-1)(x+2)}=2$ — Expand the brackets in the numerator then simplify.

$\dfrac{11x+2}{(2x-1)(x+2)}=2$ — Multiply both sides by $(2x-1)(x+2)$.

$11x+2=2(2x-1)(x+2)$ — Multiply out the brackets and simplify the right-hand side.

$11x+2=4x^2+6x-4$

$4x^2-5x-6=0$ — Rearrange into the form $ax^2+bx+c=0$.

$(4x+3)(x-2)=0$,
so either $4x+3=0$ or $x-2=0$ — Solve by factorising.
The solutions are $x=-\frac{3}{4}$ and $x=2$.

9 **Reasoning**

   a Show that the equation $\dfrac{x+1}{2x}+\dfrac{x-3}{3x}=5$ can be rearranged to give $25x^2+3x=0$.

   b Hence, solve the equation $\dfrac{x+1}{2x}+\dfrac{x-3}{3x}=5$.

10 **Reasoning**

   a Show that the equation $\dfrac{x}{2x-3}+\dfrac{4}{x+1}=1$ can be rearranged to give $x^2-10x+9=0$.

   b Hence, solve the equation $\dfrac{x}{2x-3}+\dfrac{4}{x+1}=1$.

11 Solve

   a $\dfrac{1}{x-1}+\dfrac{1}{5-x}=1$ **b** $\dfrac{5}{x+2}+\dfrac{3}{x-2}=1$ **c** $\dfrac{4}{x}-\dfrac{3}{2x-1}=1$

   d $\dfrac{3}{x+1}-\dfrac{2}{x+3}=1$ **e** $\dfrac{4}{x}-\dfrac{3}{2x+3}=1$ **f** $\dfrac{4}{x-3}=2-\dfrac{3}{x+2}$

12 Solve these equations.
   Give your answers correct to 2 decimal places.

   a $\dfrac{4x-1}{2-x}=\dfrac{x}{3}$ **b** $\dfrac{1}{x-1}+\dfrac{1}{x+2}=5$ **c** $\dfrac{4}{x}-\dfrac{2}{1-x}=1$

   d $\dfrac{2}{x-5}+\dfrac{1}{x+1}=3$ **e** $x+\dfrac{3}{x}=5$ **f** $x-\dfrac{7}{x}=4$

17.8 Functions

- Use function notation.
- Find composite functions.
- Find inverse functions.

*Active*Learn
Homework

Warm up

1 Fluency Find the value of y when $x = 2$.

a $y = 5x - 3$ **b** $y = 7x + 8$ **c** $y = 2x^3$ **d** $y = 5x^2$

2 a $H = 4x$ and $x = 3t$. Substitute $x = 3t$ into $H = 4x$ to write H in terms of t.

   b $P = \dfrac{x}{3}$ and $x = \dfrac{1}{2}y$. Write P in terms of y.   **c** $y = x^2$ and $x = h + 3$. Write y in terms of h.

Key point

A **function** is a rule for working out values of y for given values of x.
For example, $y = 3x$ and $y = x^2$ are functions.
The notation $f(x)$ is read as 'f of x'. f is the function. $f(x) = 3x$ means the function of x is $3x$.

3 $f(x) = \dfrac{10}{x}$. Work out

a $f(5)$ **b** $f(-2)$ **c** $f\left(\frac{1}{2}\right)$ **d** $f(-20)$

> **Q3a hint**
> Substitute $x = 5$ into $\dfrac{10}{x}$.

4 $g(x) = 2x^3$. Work out

a $g(3)$ **b** $g(-1)$ **c** $g\left(\frac{1}{2}\right)$ **d** $g(-5)$

5 $f(x) = x + x^3$ and $g(x) = 3x^2$. Work out

a $f(1) + g(1)$ **b** $f(4) - g(2)$ **c** $f(2) \times g(4)$

d $\dfrac{g(5)}{f(3)}$ **e** $2g(10)$ **f** $3f(-1) - g(3)$

> **Q5e hint**
> First work out g(10) and then multiply the answer by 2.

6 $g(x) = 5x - 3$. Work out the value of a when

a $g(a) = 12$ **b** $g(a) = 0$ **c** $g(a) = -7$

> **Q6a hint**
> $g(a) = 5a - 3 = 12$
> Solve for a.

7 $f(x) = x^2 - 8$. Work out the values of a when

a $f(a) = 17$ **b** $f(a) = -4$ **c** $f(a) = 0$ **d** $f(a) = 12$

Write your answers as surds where appropriate.

8 $f(x) = x(x + 3)$ and $g(x) = (x - 1)(x + 5)$. Work out the values of a when

a $f(a) = 0$ **b** $g(a) = 0$ **c** $f(a) = -2$ **d** $g(a) = -8$

9 $f(x) = 5x - 4$. Write an expression in x for

a $f(x) + 5$ **b** $f(x) - 9$ **c** $2f(x)$ **d** $7f(x)$

e $f(-x)$ **f** $f(2x)$ **g** $f(4x)$

> **Q9f hint**
> Replace x in f(x) with $2x$.

h Reflect Is $2f(x)$ equal to $f(2x)$? Use your answers to parts **c** and **f** to explain.

10 f(x) = $6 - 2x$ and g(x) = $x^2 + 7$.

   a Work out gf(2). First work out f(2) and then substitute your answer into g(x).

   b Work out **i** gf(7)   **ii** fg(4)   **iii** fg(5)   **iv** fg(−1)

11 f and g are functions such that

$$f(x) = \frac{3}{x^2} \quad \text{and} \quad g(x) = 2x^3$$

   a Find f(−4).   **(1 mark)**

   b Find fg(1).   **(2 marks)**

12 **a** f(x) = $4x - 3$, g(x) = $10 - x$ and h(x) = $x^2 + 7$.

     Write an expression in x for **i** gf(x)   **ii** fg(x)   **iii** fh(x)   **iv** hf(x)

   b Show that gh(x) = $-x^2 + 3$.

Example

Find the inverse function of f(x) = $5x - 1$.

$$x \longrightarrow \boxed{\times 5} \longrightarrow \boxed{-1} \longrightarrow 5x - 1$$

Write the function as a function machine.

$$\frac{x+1}{5} \longleftarrow \boxed{\div 5} \longleftarrow \boxed{+1} \longleftarrow x$$

Reverse the function machine to find the inverse function. Start with x as the input.

The inverse function of f(x) = $5x - 1$ is f^{-1}(x) = $\dfrac{x+1}{5}$

13 Find the inverse of each function.

   a f(x) = $4x + 9$ **b** g(x) = $\dfrac{x}{3} - 4$ **c** h(x) = $2(x + 6)$ **d** k(x) = $7(x - 4)$

14 **Reasoning** f(x) = $4(x - 1)$ and g(x) = $4(x + 1)$

   a Find **i** f^{-1}(x)   **ii** g^{-1}(x)   **iii** f^{-1}(3)   **iv** g^{-1}(2)

   b Work out f^{-1}(x) + g^{-1}(x).

   c Given that f^{-1}(a) + g^{-1}(a) = 1, work out the value of a.

15 The functions f and g are such that   f(x) = $2x + 1$ g(x) = $ax + b$

   where a and b are constants.

   g(4) = 17 and f^{-1}(15) = g(2)

   Find the value of a and the value of b.   **(5 marks)**

17 Check up

Active Learn
Homework

Formulae and functions

1 In each formula, change the subject to the letter given in brackets.

a $5xy + 3x = 9 - 2y$ $[y]$

b $T = 2p\sqrt{\dfrac{x}{k}}$ $[k]$

c $x = \dfrac{3+t}{t}$ $[t]$

2 f is a function such that $f(x) = 3x + 4$. Find **a** $f(-2)$ **b** $f^{-1}(x)$

3 $f(x) = 9 - 2x$ and $g(x) = x^2 + 4x$. Work out **a** $f(2) + g(3)$ **b** $fg(5)$

4 $f(x) = 4x^2 - 7$. Find the values of a when $f(a) = 0$.

Proof

5 Show that $25 - (x+1)^2 \equiv (6+x)(4-x)$.

6 Prove that this statement is not true: 'The sum of two cubed numbers is always odd.'

Surds

7 Simplify   **a** $\sqrt{200} + 2\sqrt{50}$ **b** $(4 - \sqrt{7})^2$

8 Rationalise the denominator in **a** $\dfrac{3 - \sqrt{2}}{\sqrt{5}}$ **b** $\dfrac{3}{2 - \sqrt{3}}$

Algebraic fractions

9 Simplify fully   **a** $\dfrac{x^2 - 4}{3x + 6}$ **b** $\dfrac{x^2 + 4x - 32}{x^2 + 9x + 8}$

10 Write as a single fraction in its simplest form.

a $\dfrac{16x^2}{15y} \div \dfrac{4y^4}{5x}$

b $\dfrac{x^2 + 9x - 10}{x^2 + 5x + 4} \div \dfrac{4x - 4}{3x + 12}$

c $\dfrac{3}{x+4} + \dfrac{1}{x-5}$

d $\dfrac{4}{x^2 - 7x + 6} - \dfrac{2}{x-1}$

11 Solve the equation $\dfrac{2}{x+1} - \dfrac{1}{x+2} = 1$.

12 **Reflect** How sure are you of your answers? Were you mostly

Just guessing ☹   Feeling doubtful 😐   Confident ☺

What next? Use your results to decide whether to strengthen or extend your learning.

Challenge

13 **a** Prove that the sum of two consecutive odd numbers is a multiple of 4.

b Prove that the sum of three consecutive even numbers is a multiple of 6.

17 Strengthen

Active Learn
Homework

Formulae and functions

1 **a** Copy and complete.

  i $y = \sqrt{3}$, so $y^2 = \square$

  ii $y = \sqrt{x}$, so $y^2 = \square$

  iii $y = \sqrt{\dfrac{x}{k}}$, so $y^2 = \square$

  iv $y = 3n\sqrt{\dfrac{x}{k}}$, so $\left(\dfrac{y}{3n}\right)^2 = \square$

b Use your answer to part **a iv** to make x the subject of the formula $y = 3n\sqrt{\dfrac{x}{k}}$

> **Q1b hint**
>
> Your answer will be $x = ...$

2 Copy and complete to make x the subject.

$$3xy + 2x = 1 - y$$

$$\div(\square + \square)\left(\begin{array}{c} x(\square + \square) = 1 - y \\ \\ x = \dfrac{1-y}{\square + \square} \end{array}\right)\div(\square + \square)$$

3 Here are all the steps to make y the subject of $x = \dfrac{7+y}{y}$.

a Rewrite the formula so there is no fraction.

b Get all the terms containing y on the left-hand side and all other terms on the right-hand side.

c Factorise so that y appears only once.

d Get y on its own on the left-hand side.

Match each step to one of these rearrangements.

$y(x-1) = 7$	$xy = 7 + y$	$y = \dfrac{7}{x-1}$	$xy - y = 7$

4 **a** $y = 5x - 9$. Work out the value of y when $x = 2$.

b $f(x) = 5x - 9$. Work out $f(2)$.

c Work out

  i $f(5)$ **ii** $f(-3)$

  iii $f(0)$ **iv** $f(10)$

> **Q4b hint**
>
> $f(2)$ means substitute $x = 2$ into $5x - 9$.

5 $p(x) = 9x - 4$

a Write an expression for $p(a)$.

b Find the value of a when $p(a) = 0$.

6 $f(x) = 3x$ and $g(x) = x^2 - 1$. Work out

a $f(5)$ **b** $g(6)$ **c** $f(5) + g(6)$

d $fg(6)$ **e** $gf(5)$ **f** $fg(1)$

> **Q6d hint**
>
> Substitute your value for $g(6)$ into $f(x)$.

7 Here is a function machine for $f(x) = 3x - 5$.
To find the inverse function, reverse the machine using inverse operations.

$x \longrightarrow \boxed{\times 3} \longrightarrow \boxed{-5} \longrightarrow f(x)$

$f^{-1}(x) \longleftarrow \boxed{\div 3} \longleftarrow \boxed{+5} \longleftarrow x$

a Copy and complete the inverse function.

$f^{-1}(x) = \dfrac{x + \square}{\square}$

b Use the method in part **a** to find $f^{-1}(x)$ for each function.

i $f(x) = 2x - 9$ **ii** $f(x) = 3(x - 5)$ **iii** $f(x) = \dfrac{x + 4}{2}$ **iv** $f(x) = \dfrac{2(x + 1)}{5}$

Proof

1 **a** Expand $(x - 4)^2$.

b Expand and simplify $(x - 4)^2 - 9$.

c Expand $(x - 7)(x - 1)$.

d Use your answers to parts **b** and **c** to show that
$(x - 4)^2 - 9 \equiv (x - 7)(x - 1)$.

Q1d hint

$\equiv$ means 'is identical to'.

2 Show that $(x - 1)^2 - 16 \equiv (x - 5)(x + 3)$.

3 **a** List the first five cube numbers.

b Give a counter-example to prove that this statement is *not* true:
'The difference between two cube numbers is always odd'.

Surds

1 Copy and complete.

a $\sqrt{3} \times \sqrt{3} = \square$ **b** $\sqrt{7} \times \sqrt{\square} = 7$ **c** $2\sqrt{2} \times \sqrt{2} = \square$

d $\sqrt{5}(6 - \sqrt{5}) = \sqrt{5} \times 6 - \sqrt{5} \times \sqrt{5} = 6\sqrt{5} - \square$

e $\sqrt{180} + \sqrt{45} = \sqrt{\square}\sqrt{5} + \sqrt{\square}\sqrt{5} =$

Q1d hint

$\sqrt{5} \times 6 = 6 \times \sqrt{5} = 6\sqrt{5}$
Always write the whole number before the surd.

2 Rationalise the denominators.

a $\dfrac{12}{\sqrt{3}} = \dfrac{12}{\sqrt{3}} \times \dfrac{\sqrt{3}}{\sqrt{3}} = \dfrac{\square\sqrt{3}}{\square} = \square\sqrt{3}$

b $\dfrac{(4 + \sqrt{11})}{\sqrt{11}} = \dfrac{(4 + \sqrt{11})}{\sqrt{11}} \times \dfrac{\sqrt{11}}{\sqrt{11}} = \dfrac{4 \times \sqrt{11} + \square}{\sqrt{11} \times \square} = \dfrac{\square}{\square}$

c $\dfrac{8 - \sqrt{5}}{\sqrt{5}}$

3 Expand and simplify. Some have been started for you.

a $(4 - \sqrt{7})(2 + \sqrt{7}) = 8 + 4\sqrt{7} - \square\sqrt{7} - \square$

b $(5 - \sqrt{2})^2 = (5 - \sqrt{2})(5 - \sqrt{2}) =$

c $(3 - \sqrt{5})(3 + \sqrt{5}) = 9 + 3\sqrt{5} - \square\sqrt{5} - \square$

d $(2 + \sqrt{11})(2 - \sqrt{11})$ **e** $(4 - \sqrt{7})(4 + \sqrt{7})$

f What would you multiply these expressions by to get an integer answer?

    i $6 + \sqrt{8}$ **ii** $3 - \sqrt{11}$

Q3 hint

Multiply each term in the second bracket by each term in the first bracket.
FOIL: Firsts, Outers, Inners, Lasts.

4 Rationalise the denominators.

a $\dfrac{8}{5 - \sqrt{2}} = \dfrac{8(5 + \sqrt{2})}{(5 - \sqrt{2})(5 + \sqrt{2})} =$ **b** $\dfrac{7}{2 + \sqrt{3}}$ **c** $\dfrac{6}{7 - \sqrt{10}}$

Algebraic fractions

1 Simplify

a $\dfrac{(x+10)(x-8)}{(x+7)(x+10)}$

b $\dfrac{x}{x-2} \times \dfrac{x+5}{x}$

c $\dfrac{x+4}{x-3} \times \dfrac{x-3}{x+8} \times \dfrac{x-2}{x-4}$

d $\dfrac{x+5}{18} \times \dfrac{10}{x-1} \times \dfrac{x+1}{x+5}$

2 a Copy and complete.

i $\dfrac{15}{20} = \dfrac{\square}{\square}$ 　　**ii** $\dfrac{9}{6} = \dfrac{\square}{\square}$ 　　**iii** $\dfrac{x}{x^3} = \dfrac{\square}{\square}$ 　　**iv** $\dfrac{y^6}{y^2} = \dfrac{\square}{\square}$

b Use your answers from part **a** to fully simplify $\dfrac{15y^6}{6x^3} \times \dfrac{9x}{20y^2}$.

3 Write each of these as a single fraction in its simplest form.

a $\dfrac{4x^5}{15y^2} \times \dfrac{20y}{12x^3}$ 　　**b** $\dfrac{12y^2}{21x^2} \div \dfrac{9y^5}{14x^3}$

> **Q3b hint**
>
> Multiply the first fraction by the reciprocal of the second fraction, $\dfrac{14x^3}{9y^5}$.

4 a Factorise

i $2x+10$ 　　**ii** x^2-25

b Use your answers from part **a** to fully simplify $\dfrac{x^2-25}{2x+10}$.

> **Q4a ii hint**
>
> Use the difference of two squares.
> $x^2-25 = (x+\square)(x-\square)$

5 Simplify fully

a $\dfrac{8x+32}{x^2+12x+32} = \dfrac{8(\square+\square)}{(x+\square)(x+\square)} = \dfrac{\square}{\square}$

b $\dfrac{x^2+6x-16}{x^2-11x+18}$

6 a Factorise

i $3x+9$ 　　**ii** $x^2+9x+18$ 　　**iii** $x^2+8x+15$ 　　**iv** $2x+10$

b Use your answers to part **a** to write as a single fraction

i $\dfrac{3x+9}{x^2+9x+18} \times \dfrac{x^2+8x+15}{2x+10}$

ii $\dfrac{2x+10}{x^2+8x+15} \div \dfrac{3x+9}{x^2+9x+18}$

7 a Write down the LCM of $x+1$ and $x+2$.

b Copy and complete.

i $\dfrac{3}{x+1} = \dfrac{3(\square+\square)}{(x+1)(x+2)}$ 　　**ii** $\dfrac{5}{x+2} = \dfrac{5(\square+\square)}{(x+1)(x+2)}$

c Use your answers to part **b** to add $\dfrac{3}{x+1} + \dfrac{5}{x+2}$.

d Use your answer to part **c** to solve $\dfrac{3}{x+1} + \dfrac{5}{x+2} = 1$. Give your answer in surd form.

8 Write $\dfrac{3}{x^2-5x+4} - \dfrac{2}{x-4}$ as a single fraction in its simplest form.

> **Q8 hint**
>
> First factorise x^2-5x+4.

17 Extend

1 **Reasoning** Both Jack and Ruth make y the subject of the formula $1 - 2y = x$.

Jack's answer is $y = \dfrac{x-1}{-2}$ Ruth's answer is $y = \dfrac{1-x}{2}$

a Show that both answers are correct.

b Explain why Ruth's answer might be considered a better answer.

c Make x the subject of the formula $\dfrac{P - 2x^2}{3} = d$.

2 **Problem-solving** The total resistance of a set of resistors in a parallel circuit is given by the formula $\dfrac{1}{R} = \dfrac{1}{R_1} + \dfrac{1}{R_2}$. Make R_2 the subject of the formula.

3 **Reasoning** $\dfrac{1}{u} = \dfrac{1}{v} + \dfrac{1}{w} - \dfrac{1}{x}$

a Write down an expression for $\dfrac{1}{x}$.   **b** Show that $x = \dfrac{uvw}{uw + uv - vw}$.

4 Solve   **a** $\dfrac{4}{2x-3} = \dfrac{x}{5}$ **b** $\dfrac{4}{2-x} - \dfrac{1}{x-3} = 5$

Exam-style question

5 Find the exact solutions of $x + \dfrac{5}{x} = 12$.

                             (3 marks)

> **Exam tip**
>
> 'Find the exact solutions' means that you should not use a calculator. You should give your answers using simplified surds.

Exam-style question

6 $f(x) = 2\cos x°$

a Find f(35). Give your answer correct to 3 significant figures.   **(1 mark)**

b $g(x) = 3x - 1$
    Find fg(15). Give your answer correct to 3 significant figures.   **(2 marks)**

Exam-style question

7 The functions f and g are such that $f(x) = 3 - 4x$, $g(x) = 3 + 4x$.

a Find f(6).   **(1 mark)**

b Find gf(x).   **(2 marks)**

c Find   **i** $f^{-1}(x)$ **ii** $g^{-1}(x)$ **(2 marks)**

d Show that $f^{-1}(x) + g^{-1}(x) = 0$, for all values of x.   **(2 marks)**

Exam-style question

8 $f(x) = x + 7$, $g(x) = x^2 + 6$

a Work out   **i** fg(x)   **ii** gf(x)   **(4 marks)**

b Solve $fg(x) = gf(x)$.   **(2 marks)**

---

9 Reasoning $f(x) = \dfrac{x-7}{2}$ and $g(x) = 2x+7$

   a Work out   **i** $fg(x)$ **ii** $gf(x)$

Q9b hint

Functions f and g are inverses of each other if $fg(x) = gf(x) = x$.

   b Are $f(x)$ and $g(x)$ inverse functions? Explain your answer.

   c Check whether $f(x) = \frac{1}{4}x - 1$ and $g(x) = 4(x+1)$ are inverse functions.

10 Reasoning Show that $\dfrac{49 - x^2}{x^2 - 49} = -1$.

11 Reasoning Show that $(3n+1)^2 - (3n-1)^2$ is a multiple of 12, for all positive integer values of n.

12 Reasoning Show that $\dfrac{1}{5x^2 - 13x - 6} - \dfrac{1}{x^2 - 9} = \dfrac{Ax+B}{(x-3)(x+3)(5x+2)}$ and find the value of A and the value of B.

13 Problem-solving Newton's Law of Universal Gravitation can be used to calculate the force (F) between two objects.

$F = \dfrac{G m_1 m_2}{r^2}$, where G is the gravitational constant $(6.67 \times 10^{-11}\,\text{N}\,\text{m}^2\,\text{kg}^{-2})$, m_1 and m_2 are the masses of the two objects (kg) and r is the distance between them (m).
The gravitational force between the Earth and the Sun is $3.52 \times 10^{22}\,\text{N}$.
The mass of the Sun is $1.99 \times 10^{30}\,\text{kg}$ and the mass of the Earth is $5.97 \times 10^{24}\,\text{kg}$.
Work out the distance between the Sun and the Earth.

Exam-style question

14 S is a geometric sequence.

   a Given that $(\sqrt{x} - 2)$, 1 and $(\sqrt{x} + 2)$ are the first three terms of S, find the value of x.
    You must show all your working.   **(3 marks)**

   b Show that the 4th term of S is $9 + 4\sqrt{5}$.   **(2 marks)**

Exam-style question

15 The functions f and g are such that $f(x) = 2x + 5$ and $g(x) = x^2 - 1$.
Given that $3fg(x) = gf(x)$, show that $2x^2 - 20x - 15 = 0$.   **(5 marks)**

Working towards A level

16 A number such as 243 is divisible by 9 because the sum of its digits $(2+4+3)$ is 9.
Prove that any three-digit number 'abc' in which $a+b+c = 9$ will be divisible by 9.

Q16 hint

First write 'abc' as $100a + 10b + c$.
How can you re-write this expression introducing multiples of 9?

This question requires some 'outside the box' thinking and tests advanced problem-solving skills – a feature of many A level questions.

17 Given that $5x^2 - 8ax + 6b \equiv 5(x+2b)^2 + 2a - 3$,
work out the two possible pairs of values of a and b.

Q17 hint

Expand the brackets and equate the coefficients of the x^2, x and constant terms. Then solve two simultaneous equations, one linear and one quadratic. Don't necessarily expect the solutions to be integers.

This question combines several higher-level algebra skills – using the identity symbol, comparing coefficients, solving simultaneous equations when one is linear and one quadratic – typical of A level thinking.

17 Test ready

Summary of key points

To revise for the test:

- Read each key point, find a question on it in the mastery lesson, and check you can work out the answer.

- If you cannot, try some other questions from the mastery lesson or ask for help.

Key points

1 When the letter to be made the subject appears twice in the formula you will need to **factorise**. → **17.1**

2 When the letter to be made the subject is part of a term involving a power or root, rearrange so that the whole term is on its own on one side of the equation.
Use inverse operations to eliminate the power or root. → **17.1**

3 When multiplying algebraic fractions, cancel common factors in numerators and denominators before multiplying the fractions together. → **17.2**

4 The lowest common denominator of two fractions is the lowest common multiple of the two denominators. → **17.2**

5 To simplify an algebraic fraction, cancel any common factors in the numerator and denominator. → **17.3**

6 You may need to factorise before simplifying an algebraic fraction:
- Factorise the numerator and denominator.
- Divide the numerator and denominator by any common factors. → **17.3**

7 You may need to factorise the numerator and/or denominator before you multiply or divide algebraic fractions. → **17.4**

8 The lowest common denominator of two algebraic fractions is the lowest common multiple of the two denominators. → **17.4**

9 To show a statement is an **identity**, expand and simplify the expressions on one or both sides of the equals sign, until the two expressions are the same. → **17.5**

10 A **proof** is a logical argument for a mathematical statement.
To prove that a statement is true, you must show that it will be true in *all* cases. → **17.5**

11 To prove that a statement is not true you can find a **counter-example** – an example that does not fit the statement. → **17.5**

12 For an algebraic proof, use n to represent any integer.

Even number	$2n$
Odd number	$2n+1$ or $2n-1$
Consecutive numbers	$n, n+1, n+2, ...$
Consecutive even numbers	$2n, 2n+2, 2n+4, ...$
Consecutive odd numbers	$2n+1, 2n+3, 2n+5, ...$

→ **17.5**

13 To rationalise the fraction $\dfrac{1}{a\sqrt{b}}$, multiply by $\dfrac{\sqrt{b}}{\sqrt{b}}$. → **17.6**

14 To rationalise the fraction $\dfrac{1}{a\mp\sqrt{b}}$, multiply by $\dfrac{a\pm\sqrt{b}}{a\pm\sqrt{b}}$. → **17.6**

15 To solve an equation involving algebraic fractions, first write one side as a fraction in its simplest form. → **17.7**

16 To solve a quadratic equation, rearrange it into the form $ax^2+bx+c=0$. → **17.7**

17 A **function** is a rule for working out values of y for given values of x.
For example, $y=3x$ and $y=x^2$ are functions. → **17.8**

18 The notation $f(x)$ is read as 'f of x'. f is the function.
$f(x)=3x$ means the function of x is $3x$. → **17.8**

19 fg is a **composite function**. To work out fg(x), first work out g(x) and then substitute your answer into f(x). → **17.8**

20 The **inverse function** reverses the effect of the original function.
$f^{-1}(x)$ is the inverse function of $f(x)$. → **17.8**

Sample student answers

Exam-style question

1 Make y the subject of the formula $x = \dfrac{2-7y}{y-5}$. **(4 marks)**

$xy - 5 = 2 - 7y$

$xy + 7y = 2 + 5$

$y(x + 7) = 7$

$y = \dfrac{7}{x + 7}$

---

a Explain what common mistake the student has made right at the very start of the answer.

b Suggest a way to avoid making this mistake.

Exam-style question

2 Rationalise the denominator of $\dfrac{11}{2+\sqrt{5}}$. **(3 marks)**

$\dfrac{11}{2+\sqrt{5}} = \dfrac{11(2-\sqrt{5})}{(2+\sqrt{5})(2-\sqrt{5})}$

$= \dfrac{22 - 11\sqrt{5}}{4 + 2\sqrt{5} - 2\sqrt{5} + 5}$

$= \dfrac{22 - 11\sqrt{5}}{9}$

---

a Find the student's mistake.

b Work out the correct answer.

17 Unit test

*Active*Learn
Homework

1 Find the inverse of the function $f(x) = 5(x+4)$. **(2 marks)**

2 Show that $(x+4)^2 - (2x+7) \equiv (x+3)^2$ for all values of x. **(2 marks)**

3 Make x the subject of the formula $y = \frac{1}{2}(x+3)^2$. **(2 marks)**

4 Simplify fully $\dfrac{x-4}{3(x-4)^2}$. **(1 mark)**

5 Make x the subject of the formula $y = \dfrac{3(w-2x)}{x}$. **(3 marks)**

6 Expand and simplify

 a $(3+\sqrt{2})(4-\sqrt{2})$ **b** $(3+\sqrt{5})^2$ (part **a**) **(2 marks)**
(part **b**) **(2 marks)**

7 Write as a single fraction in its simplest form

 a $\dfrac{9x^3}{8y} \times \dfrac{4y^2}{15x^5}$ **b** $\dfrac{9}{4x} - \dfrac{2}{5x}$ (part **a**) **(3 marks)**
(part **b**) **(2 marks)**

 c $\dfrac{x^2+x-30}{x^2+10x+24} \div \dfrac{x^2-12x+35}{x^2+3x-4}$ **d** $\dfrac{5x^2-4x-12}{4x-8} \times \dfrac{5x+5}{5x^2+11x+6}$ (part **c**) **(3 marks)**
(part **d**) **(2 marks)**

8 Show that $2 - \dfrac{x-1}{x+2} - \dfrac{x-4}{x-2}$ can be written in the form $\dfrac{ax-b}{x^2-c}$,

 where a, b and c are integers. **(4 marks)**

9 Given that $\dfrac{8}{1-\sqrt{3}} = a(b+\sqrt{3})$, where a and b are integers,

 find the value of a and the value of b. **(3 marks)**

10 Solve

 a $\dfrac{5x-1}{2} = \dfrac{3}{x}$ **b** $\dfrac{5}{x-1} + \dfrac{7}{x-1} = x$ (part **a**) **(3 marks)**
(part **b**) **(4 marks)**

11 The functions f and g are such that $f(x) = x^2 - 9$ and $g(x) = 2x+1$.

 a Find $f(4) + g(2)$. **(2 marks)**

 b Find the value of a when $g(a) = 0$. **(1 mark)**

 c Find $g^{-1}(x)$. **(2 marks)**

12 Given that n can be any integer such that $n > 1$, prove that $n^2 + n$ is never an odd number. **(2 marks)**

(TOTAL: 45 marks)

13 Challenge

 a Show that $\dfrac{1}{1+\frac{1}{x}} = \dfrac{x}{x+1}$. **b** Work out the exact value of $\dfrac{1}{1+\frac{1}{9}}$.

 c A sequence has general term $\left(1+\frac{1}{n}\right)^{-1}$

  What happens to the values of the terms in the sequence as n gets very large?

14 Reflect Which lesson in this unit made you think the hardest? Explain why.

18 Vectors and geometric proof

Prior knowledge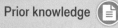

18.1 Vectors and vector notation

* Understand and use vector notation.
* Work out the magnitude of a vector.

*Active*Learn
Homework

Warm up

1 **Fluency** How far does the translation $\begin{pmatrix} 3 \\ 5 \end{pmatrix}$ move an object in

a the x-direction **b** the y-direction?

2 Work out the value of x in each right-angled triangle. Give your answer as a surd in its simplest form.

Key point

A **vector** is a quantity that has magnitude (size) and direction. Examples of vectors are force (5 N acting vertically upwards) and velocity (15 km/h due north). In the diagram, **a** is a vector. **Displacement** is change in position. A displacement can be written as $\begin{pmatrix} 2 \\ 3 \end{pmatrix}$, where 2 is the x-component and 3 is the y-component.

3 Write these vectors as column vectors.

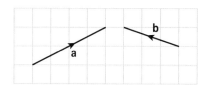

Q3 hint

Use the arrow to decide the start of the line. Move across then up or down to the end of the line.

4 On squared paper draw and label these vectors.

$\mathbf{a} = \begin{pmatrix} 1 \\ 3 \end{pmatrix}$ $\mathbf{b} = \begin{pmatrix} 3 \\ -2 \end{pmatrix}$ $\mathbf{c} = \begin{pmatrix} -4 \\ -3 \end{pmatrix}$ $\mathbf{d} = \begin{pmatrix} -5 \\ 4 \end{pmatrix}$ $\mathbf{e} = \begin{pmatrix} 0 \\ 5 \end{pmatrix}$

Key point

The displacement vector from A to B is written $\overrightarrow{AB}$.
Vectors are written as **bold** lower case letters: **a**, **b**, **c**. When handwriting, <u>underline</u> the letter: <u>a</u>, <u>b</u>, <u>c</u>

Example

Point A has coordinates (2, 3) and point B has coordinates (5, 1).

Write $\overrightarrow{AB}$ as a column vector.

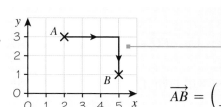

First mark the points A and B on a grid.
To move from A to B go 3 to the right and 2 down.

$\overrightarrow{AB} = \begin{pmatrix} 3 \\ -2 \end{pmatrix}$

5 The point A is $(1, 2)$, the point B is $(3, 4)$ and the point C is $(5, -1)$. Write as column vectors

a $\overrightarrow{AB}$ **b** $\overrightarrow{BC}$ **c** $\overrightarrow{AC}$

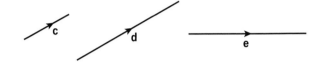

6 **Problem-solving** Point C has coordinates $(4, 3)$ and point D is such that $\overrightarrow{CD} = \begin{pmatrix} -1 \\ 2 \end{pmatrix}$. Find the coordinates of D.

Q6 hint

From C go 1 to the left and 2 up to find point D.

7 **Problem-solving** $\overrightarrow{EF} = \begin{pmatrix} 3 \\ 4 \end{pmatrix}$. F is the point $(2, 3)$. Work out the coordinates of E.

Q5 hint

Mark the points on a grid.

Key point

Equal vectors have the same magnitude and the same direction.

8 **a** Which of these vectors are equal?

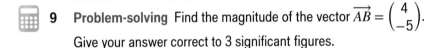

b **Reflect** Are parallel vectors always equal? Explain.

Key point

The **magnitude** of the vector $\begin{pmatrix} x \\ y \end{pmatrix}$ is its length, i.e. $\sqrt{x^2 + y^2}$.

$|\mathbf{a}|$ means the magnitude of vector $\mathbf{a}$. $|\overrightarrow{OA}|$ means the magnitude of vector $\overrightarrow{OA}$.

9 **Problem-solving** Find the magnitude of the vector $\overrightarrow{AB} = \begin{pmatrix} 4 \\ -5 \end{pmatrix}$.

Give your answer correct to 3 significant figures.

10 **Problem-solving** Work out the magnitude of each vector.
Where necessary, leave your answer as a surd.

a $\mathbf{u} = \begin{pmatrix} 6 \\ 8 \end{pmatrix}$ **b** $\mathbf{v} = \begin{pmatrix} -5 \\ 12 \end{pmatrix}$ **c** $\mathbf{w} = \begin{pmatrix} -1 \\ -3 \end{pmatrix}$ **d** $\overrightarrow{AB} = \begin{pmatrix} 8 \\ 15 \end{pmatrix}$ **e** $\overrightarrow{CD} = \begin{pmatrix} 4 \\ -6 \end{pmatrix}$

11 **Problem-solving** A is the point $(3, 4)$ and B is the point $(-3, 0)$.

a Write $\overrightarrow{AB}$ as a column vector.

b Find the length of vector $\overrightarrow{AB}$. Give your answer as a surd in its simplest form.

12 **Reasoning** In triangle ABC, $\overrightarrow{AB} = \begin{pmatrix} 20 \\ -15 \end{pmatrix}$ and $\overrightarrow{AC} = \begin{pmatrix} -7 \\ 24 \end{pmatrix}$.

a Work out the length of the side AB of the triangle.

b Show that triangle ABC is isosceles.

Q12 hint

Sketch A, B and C.

Exam-style question

13 The magnitude of $\overrightarrow{PQ}$ is 15. $\overrightarrow{PQ} = \begin{pmatrix} x \\ 9 \end{pmatrix}$

P is the point $(-4, -1)$.
Work out the coordinates of point Q.   **(3 marks)**

Exam tip

Sketch the vector and mark on the information you know.

18.2 Vector arithmetic

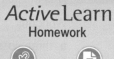
- Calculate using vectors and represent the solutions graphically.
- Identify when vectors are parallel.
- Calculate the resultant of two vectors.

Warm up

1 Fluency The components of the column vector $\begin{pmatrix} 3 \\ -5 \end{pmatrix}$ are 3 units ____ and ☐ units ____ .

2 On squared paper draw and label these vectors.

$\mathbf{a} = \begin{pmatrix} 3 \\ -2 \end{pmatrix}$ $\mathbf{b} = \begin{pmatrix} -3 \\ 2 \end{pmatrix}$

3 The point A is (4, 5) and the point B is (−1, 7).
Write $\overrightarrow{AB}$ as a column vector.

4 Reasoning The points A, B, C and D are the vertices of a quadrilateral.

A has coordinates (1, 1), $\overrightarrow{AB} = \begin{pmatrix} 1 \\ 2 \end{pmatrix}$, $\overrightarrow{BC} = \begin{pmatrix} 4 \\ 1 \end{pmatrix}$ and $\overrightarrow{CD} = \begin{pmatrix} 3 \\ -1 \end{pmatrix}$.

a Draw quadrilateral $ABCD$ on a coordinate grid.

b Write $\overrightarrow{AD}$ as a column vector.

c What type of quadrilateral is $ABCD$?

d Reflect What do you notice about $\overrightarrow{BC}$ and $\overrightarrow{AD}$?

> **Q4d hint**
> $\overrightarrow{AD} = \Box \overrightarrow{BC}$

5 Reasoning The points A, B, C and D are the vertices of a parallelogram.

A has coordinates (1, 1), $\overrightarrow{AB} = \begin{pmatrix} 2 \\ 3 \end{pmatrix}$ and $\overrightarrow{AD} = \begin{pmatrix} 4 \\ -1 \end{pmatrix}$.

a Draw parallelogram $ABCD$ on a coordinate grid.

b Write as a column vector   **i** $\overrightarrow{CB}$ **ii** $\overrightarrow{BC}$

c Reflect What do you notice about your answers to part **b**?

d What do you notice about   **i** $\overrightarrow{AB}$ and $\overrightarrow{DC}$ **ii** $\overrightarrow{AD}$ and $\overrightarrow{CB}$?

Key point

If $\overrightarrow{AB} = \overrightarrow{CD}$ then the line segments AB and CD
are equal in length and are parallel.
$\overrightarrow{AB}$ and $\overrightarrow{CD}$ have the same direction.
$\overrightarrow{BA}$ and $\overrightarrow{DC}$ have the opposite direction.

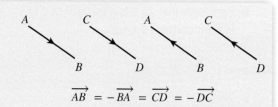

$\overrightarrow{AB} = -\overrightarrow{BA} = \overrightarrow{CD} = -\overrightarrow{DC}$

6 Reasoning In quadrilateral $PQRS$, $\overrightarrow{PQ} = \begin{pmatrix} 3 \\ 4 \end{pmatrix}$, $\overrightarrow{QR} = \begin{pmatrix} 5 \\ 0 \end{pmatrix}$,

$\overrightarrow{RS} = \begin{pmatrix} -3 \\ -4 \end{pmatrix}$ and $\overrightarrow{SP} = \begin{pmatrix} -5 \\ 0 \end{pmatrix}$.

What type of quadrilateral is $PQRS$?

> **Q6 hint**
> Look at the vectors for opposite sides.

7 Problem-solving The points G, H, I and J are the vertices of a parallelogram.

H is the point (1, 3), $\overrightarrow{GH} = \begin{pmatrix} 2 \\ 5 \end{pmatrix}$ and $\overrightarrow{JG} = \begin{pmatrix} -5 \\ -2 \end{pmatrix}$.

Work out

a three possibilities for $\overrightarrow{HI}$

b the coordinates of the point J

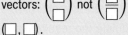

Q7 hint

Always use the correct notation when answering questions.
Write vectors like $\overrightarrow{HI}$ as column vectors: $\begin{pmatrix} \square \\ \square \end{pmatrix}$ not $\begin{pmatrix} \square \\ \overline{\square} \end{pmatrix}$ or $(\square, \square)$.
Write coordinates as $(\square, \square)$.

Key point

$2\mathbf{a}$ is twice as long as $\mathbf{a}$ and in the same direction.
$-\mathbf{a}$ is the same length as $\mathbf{a}$ but in the opposite direction.

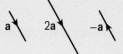

8 On squared paper draw vectors to represent

a $2\mathbf{a}$	**b** $-\mathbf{a}$
c $-2\mathbf{a}$	**d** $-\mathbf{b}$
e $3\mathbf{b}$	**f** $-2\mathbf{b}$

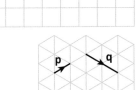

9 The vector $\mathbf{p}$ and the vector $\mathbf{q}$ are shown on an isometric grid. Draw these vectors on an isometric grid.

a $2\mathbf{p}$	**b** $\frac{1}{2}\mathbf{q}$
c $-\mathbf{p}$	**d** $-\mathbf{q}$

Key point

When a vector $\mathbf{a}$ is multiplied by a scalar k then the vector $k\mathbf{a}$ is parallel to $\mathbf{a}$ and is equal to k times $\mathbf{a}$.
A **scalar** is a number, e.g. $3, 2, \frac{1}{2}, -1, \dots$

10 Reasoning $\overrightarrow{AB} = \mathbf{a}$ and $\overrightarrow{CD} = \mathbf{a} + \mathbf{b}$.
Here are six other vectors expressed in terms of $\mathbf{a}$ and/or $\mathbf{b}$.

$\overrightarrow{EF} = -2\mathbf{a}$ $\overrightarrow{GH} = \frac{1}{2}\mathbf{a}$ $\overrightarrow{IJ} = -\mathbf{a} - \mathbf{b}$

$\overrightarrow{KL} = \frac{1}{2}\mathbf{a} + \frac{1}{2}\mathbf{b}$ $\overrightarrow{MN} = 3(\mathbf{a} + \mathbf{b})$ $\overrightarrow{OP} = -\frac{2}{3}\mathbf{a}$

a Which of these vectors are parallel to

i $\overrightarrow{AB}$ **ii** $\overrightarrow{CD}$?

b Sam says that $2\mathbf{a} + \mathbf{b}$ is parallel to $\overrightarrow{CD}$.
Is Sam correct? Explain.

Q10a hint

Look for vectors of the form
i $k \times \mathbf{a}$
ii $k \times \mathbf{a} + k \times \mathbf{b}$ or $k(\mathbf{a} + \mathbf{b})$

11 $\mathbf{m} = \begin{pmatrix} -6 \\ 4 \end{pmatrix}$. Work out

a $2\mathbf{m}$

b $\frac{1}{2}\mathbf{m}$

c $-\mathbf{m}$

d $-2\mathbf{m}$

Q11a hint

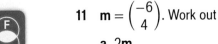

$2\mathbf{m} = \begin{pmatrix} 2 \times -6 \\ 2 \times 4 \end{pmatrix} = \begin{pmatrix} \square \\ \square \end{pmatrix}$

12 Reasoning $\overrightarrow{AB} = \begin{pmatrix} 2 \\ 1 \end{pmatrix}$

Write down the column vector for

a $2\overrightarrow{AB}$

b $3\overrightarrow{AB}$

c $-4\overrightarrow{AB}$

d $\frac{1}{2}\overrightarrow{AB}$

13 Reasoning $\overrightarrow{AB} = \begin{pmatrix} 1 \\ 2 \end{pmatrix}$ and $\overrightarrow{BC} = \begin{pmatrix} 3 \\ 1 \end{pmatrix}$.

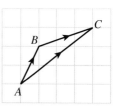

a Write down the vector $\overrightarrow{AC}$.

b **Reflect** Explain how you can find $\overrightarrow{AC}$ from $\overrightarrow{AB}$ and $\overrightarrow{BC}$.

> ### Key point
>
> The two-stage journey from A to B and then from B to C has the same starting point and the same finishing point as the single journey from A to C.
> So A to B followed by B to C is equivalent to A to C.
> $$\overrightarrow{AB} + \overrightarrow{BC} = \overrightarrow{AC}$$
> **Triangle law for vector addition:**
> Let $\overrightarrow{AB} = \mathbf{a}$, $\overrightarrow{BC} = \mathbf{b}$ and $\overrightarrow{AC} = \mathbf{c}$.
> Then $\mathbf{a} + \mathbf{b} = \mathbf{c}$ forms a triangle.
>
>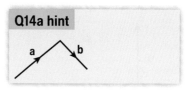

14 The vector **a** and the vector **b** are shown on the grid.

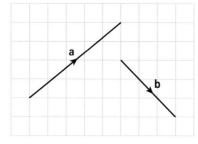

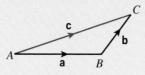
Q14a hint

a Draw vector **a** and vector **b**, moving vector **b** to the end of vector **a** so that the lines and arrows follow on.

b Draw and label the vector $\mathbf{a} + \mathbf{b}$ to complete the triangle.

c Copy and complete this vector addition.

$$\begin{array}{ccccc} \mathbf{a} & + & \mathbf{b} & = & \mathbf{a+b} \end{array}$$
$$\begin{pmatrix} 5 \\ 4 \end{pmatrix} + \begin{pmatrix} 3 \\ -3 \end{pmatrix} = \begin{pmatrix} \square \\ \square \end{pmatrix}$$

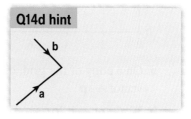
Q14d hint

d Draw and label the vector $\mathbf{a} - \mathbf{b}$.

e Check your diagram by working out

$$\mathbf{a} - \mathbf{b} = \begin{pmatrix} 5 \\ 4 \end{pmatrix} - \begin{pmatrix} 3 \\ -3 \end{pmatrix} = \begin{pmatrix} \square \\ \square \end{pmatrix}$$

15 a $\overrightarrow{AB} = \begin{pmatrix} 2 \\ 5 \end{pmatrix}$ and $\overrightarrow{BC} = \begin{pmatrix} 7 \\ -3 \end{pmatrix}$.

Find $\overrightarrow{AC}$.

b $\mathbf{a} = \begin{pmatrix} -7 \\ 2 \end{pmatrix}$ and $\mathbf{b} = \begin{pmatrix} 8 \\ -3 \end{pmatrix}$.

Find

i $\mathbf{a} + \mathbf{b}$

ii $\mathbf{b} + \mathbf{a}$

c Reflect What do you notice about your answers to parts **b i** and **ii**? Explain.

Key point

$\mathbf{a} + \mathbf{b} = \mathbf{b} + \mathbf{a}$

---

16 $\mathbf{p} = \begin{pmatrix} 1 \\ 4 \end{pmatrix}$, $\mathbf{q} = \begin{pmatrix} -2 \\ 7 \end{pmatrix}$ and $\mathbf{r} = \begin{pmatrix} 0 \\ -3 \end{pmatrix}$.

Write down the column vector for

a $-\mathbf{p}$

b $\mathbf{q} + \mathbf{p}$

c $\mathbf{p} + \mathbf{q} + \mathbf{r}$

d $\mathbf{p} - \mathbf{q}$

e $\mathbf{q} - \mathbf{r}$

f $\mathbf{q} - \mathbf{p} - \mathbf{r}$

Q16d hint

$\mathbf{p} - \mathbf{q} = \mathbf{p} + (-\mathbf{q})$

$-\mathbf{q} = \begin{pmatrix} \square \\ \square \end{pmatrix}$

Exam-style question

17 The vector **p** and the vector **q** are shown on the grid.

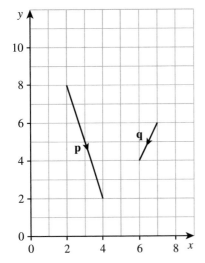

a On a copy of the grid, draw and label vector $-\frac{1}{2}\mathbf{p}$. **(1 mark)**

b Work out $\mathbf{p} + 2\mathbf{q}$ as a column vector. **(2 marks)**

Exam tip

When drawing a vector, you must include an arrow to show its direction.

18.3 More vector arithmetic

*Active*Learn
Homework

- Solve problems using vectors.
- Use the resultant of two vectors to solve vector problems.

Warm up

1 Fluency In this parallelogram, which line segments are

  a parallel

  b equal?

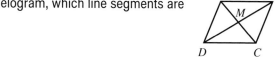

2 $\overrightarrow{AB}$ is the column vector $\begin{pmatrix} 4 \\ -3 \end{pmatrix}$ and $\overrightarrow{BC}$ is the column vector $\begin{pmatrix} 2 \\ 4 \end{pmatrix}$.

  a Find the column vector $\overrightarrow{AC}$. Draw a diagram to show your answer.

  b Work out the magnitude of $\overrightarrow{AC}$.

3 $\mathbf{p} = \begin{pmatrix} -2 \\ 6 \end{pmatrix}$

Work out   **a** $3\mathbf{p}$ **b** $-\mathbf{p}$

4 Which of these vectors are parallel to $\mathbf{a} - \mathbf{b}$?

  A $2\mathbf{a} - \mathbf{b}$ **B** $2\mathbf{a} - 2\mathbf{b}$ **C** $\mathbf{a} - 2\mathbf{b}$ **D** $-\mathbf{a} - \mathbf{b}$ **E** $-\mathbf{a} + \mathbf{b}$ **F** $\mathbf{a} + \mathbf{b}$

  Explain why.

5 $\mathbf{p} = \begin{pmatrix} -2 \\ -4 \end{pmatrix}$ and $\mathbf{q} = \begin{pmatrix} 7 \\ -5 \end{pmatrix}$.

  Write as a column vector

  a $3\mathbf{p} + \mathbf{q}$ **b** $-\mathbf{p} + 2\mathbf{q}$ **c** $2\mathbf{p} - 3\mathbf{q}$

6 $\mathbf{a} = \begin{pmatrix} -1 \\ 3 \end{pmatrix}$, $\mathbf{b} = \begin{pmatrix} -2 \\ 4 \end{pmatrix}$ and $\mathbf{a} + \mathbf{c} = \mathbf{b}$

  Calculate $\mathbf{c}$.

> **Q6 hint**
> Let $\mathbf{c} = \begin{pmatrix} x \\ y \end{pmatrix}$
> $\begin{pmatrix} -1 \\ 3 \end{pmatrix} + \begin{pmatrix} x \\ y \end{pmatrix} = \begin{pmatrix} -2 \\ 4 \end{pmatrix}$

7 $2\begin{pmatrix} x \\ y \end{pmatrix} + \begin{pmatrix} 3 \\ -1 \end{pmatrix} = \begin{pmatrix} 4 \\ -2 \end{pmatrix}$

  Find $\begin{pmatrix} x \\ y \end{pmatrix}$.

> **Q7 hint**
> $2x + 3 = \square$
> $2y - 1 = \square$

8 $\mathbf{e} = \begin{pmatrix} 5 \\ 1 \end{pmatrix}$ and $\mathbf{f} = \begin{pmatrix} -1 \\ 4 \end{pmatrix}$.

  Calculate $\mathbf{g}$ given that $2\mathbf{e} - \mathbf{g} = \mathbf{f}$.

9 Problem-solving $\mathbf{m} = \begin{pmatrix} 5 \\ 9 \end{pmatrix}$ and $\mathbf{n} = \begin{pmatrix} 3 \\ 3 \end{pmatrix}$.

  Work out the magnitude of

  a $\mathbf{m}$ **b** $2\mathbf{n}$ **c** $\mathbf{m} + \mathbf{n}$ **d** $\mathbf{m} - \mathbf{n}$ **e** $\mathbf{m} - 2\mathbf{n}$

  Give your answers as surds in their simplest form.

10 **Reasoning** $\mathbf{p} = \begin{pmatrix} -2 \\ -7 \end{pmatrix}$ and $\mathbf{r} = \begin{pmatrix} 6 \\ 8 \end{pmatrix}$.

$\mathbf{p} + \mathbf{q} = \mathbf{r}$

Work out the magnitude of $\mathbf{q}$.

11 **Reasoning** $\overrightarrow{AB} = \begin{pmatrix} 2 \\ 3 \end{pmatrix}$, $\overrightarrow{BC} = \begin{pmatrix} 1 \\ -4 \end{pmatrix}$ and $\overrightarrow{CD} = \begin{pmatrix} -4 \\ 5 \end{pmatrix}$.

a Find the column vector for $\overrightarrow{AD}$.
Draw a diagram to show this.

b Show that $\overrightarrow{AC} = \overrightarrow{DB}$.

12 **Problem-solving** In the quadrilateral $ABCD$, $\overrightarrow{AB} = \mathbf{a}$, $\overrightarrow{BC} = \mathbf{b}$ and $\overrightarrow{CD} = \mathbf{c}$.

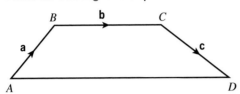

Find in terms of $\mathbf{a}$ and/or $\mathbf{b}$ and/or $\mathbf{c}$

a $\overrightarrow{BA}$ **b** $\overrightarrow{AC}$ **c** $\overrightarrow{AD}$

d **Reasoning** Alfie says, 'You can work out $\overrightarrow{BD}$ using either $\overrightarrow{BC} + \overrightarrow{CD}$ or $\overrightarrow{BA} + \overrightarrow{AD}$.'
Show that Alfie is correct.

> **Q12b hint**
>
> Start at A.
> Move along $\overrightarrow{AB}$.
> Move along $\overrightarrow{BC}$.
> Finish at C.
> $\overrightarrow{AC} = \overrightarrow{AB} + \overrightarrow{BC} = \square + \square$

13 **Reasoning** $\overrightarrow{OA} = \mathbf{a}$
M is the midpoint of OA.

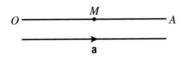

a Write down $\overrightarrow{OM}$ in terms of $\mathbf{a}$.

$\overrightarrow{OA} = \begin{pmatrix} 6 \\ 4 \end{pmatrix}$

b Express as a column vector

  i $\overrightarrow{AO}$ **ii** $\overrightarrow{OM}$

> **Q13a hint**
>
> $\overrightarrow{OM} = \dfrac{\square}{\square} \overrightarrow{OA}$

14 **Problem-solving** In the diagram, $\overrightarrow{BA} = \mathbf{a}$ and $\overrightarrow{AP} = \mathbf{b}$.

P is the midpoint of AC.
Write down in terms of $\mathbf{a}$ and/or $\mathbf{b}$

a $\overrightarrow{AC}$ **b** $\overrightarrow{BP}$ **c** $\overrightarrow{BC}$

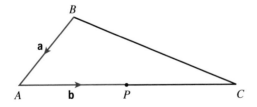

15 Reasoning $PQRS$ is a parallelogram.

$\overrightarrow{PQ} = \mathbf{a}$ and $\overrightarrow{PS} = \mathbf{b}$.

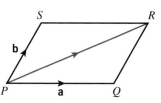

a Explain why $\overrightarrow{SR} = \mathbf{a}$.

b Find

   i $\overrightarrow{QR}$ **ii** $\overrightarrow{PR}$

Key point

In parallelogram $PQRS$ where $\overrightarrow{PQ}$ is $\mathbf{a}$ and $\overrightarrow{PS}$ is $\mathbf{b}$, the diagonal $\overrightarrow{PR}$ of the parallelogram is $\mathbf{a} + \mathbf{b}$. This is called the **parallelogram law for vector addition**.

When $\mathbf{c} = \mathbf{a} + \mathbf{b}$, the vector $\mathbf{c}$ is called the **resultant vector** of the two vectors $\mathbf{a}$ and $\mathbf{b}$.

Exam-style question

16 $CDEF$ is a parallelogram.
CD is parallel to FE.
CF is parallel to DE.
$\overrightarrow{CD} = \mathbf{p}$ and $\overrightarrow{CF} = \mathbf{q}$

Exam tip

Make sure you show that $\mathbf{p}$ and $\mathbf{q}$ are vectors by underlining: p̲ and q̲.

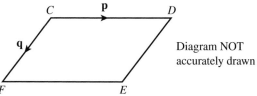

Diagram NOT accurately drawn

a Express, in terms of $\mathbf{p}$ and/or $\mathbf{q}$

   i $\overrightarrow{CE}$ **ii** $\overrightarrow{DF}$ **(2 marks)**

CE and DF are diagonals of parallelogram $CDEF$.
CE and DF intersect at X.

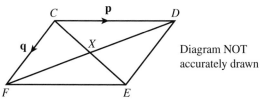

Diagram NOT accurately drawn

b Express $\overrightarrow{CX}$ in terms of $\mathbf{p}$ and/or $\mathbf{q}$.   **(1 mark)**

---

17 Reasoning $\overrightarrow{PQ} = \mathbf{a}$ and $\overrightarrow{PR} = \mathbf{b}$.

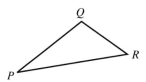

a Write $\overrightarrow{QR}$ in terms of $\mathbf{a}$ and/or $\mathbf{b}$.

b Where is the point S such that $\overrightarrow{PS} = \frac{1}{2}\mathbf{b}$?

18 Reasoning *ABCDEF* is a regular hexagon.

$\overrightarrow{AB} = \mathbf{n}$

a Explain why $\overrightarrow{ED} = \mathbf{n}$.

$\overrightarrow{BC} = \mathbf{m}$ and $\overrightarrow{CD} = \mathbf{p}$.

b Find   **i** $\overrightarrow{FE}$ **ii** $\overrightarrow{AF}$

c Find   **i** $\overrightarrow{AC}$ **ii** $\overrightarrow{AD}$

d What is $\overrightarrow{FD}$?

19 Reasoning *ABCD* is a square.
M is the midpoint of *AB*.

$\overrightarrow{DC} = \mathbf{r}$ and $\overrightarrow{DA} = \mathbf{s}$.

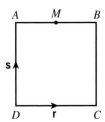

Write in terms of **r** and/or **s**

a $\overrightarrow{AB}$ **b** $\overrightarrow{BC}$

c $\overrightarrow{AM}$ **d** $\overrightarrow{DM}$

20 Reasoning Here are five vectors.

$\overrightarrow{AB} = 4\mathbf{a} - 2\mathbf{b}$ $\overrightarrow{CD} = 8\mathbf{a} + 12\mathbf{b}$ $\overrightarrow{EF} = 8\mathbf{a} - 4\mathbf{b}$ $\overrightarrow{GH} = -2\mathbf{a} + \mathbf{b}$ $\overrightarrow{IJ} = 12\mathbf{a} - 6\mathbf{b}$

a Three of these vectors are parallel. Which three?

b Add like vectors to simplify

  i $4\mathbf{p} + 3\mathbf{q} - \mathbf{p} - 6\mathbf{q}$ **ii** $2(2\mathbf{a} - 3\mathbf{b}) + \frac{1}{2}(4\mathbf{a} - \mathbf{b})$

21 Reasoning In triangle *PQR*, $\overrightarrow{PQ} = \mathbf{a}$ and $\overrightarrow{PR} = \mathbf{b}$.
M is the midpoint of *QR*.

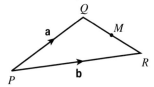

Write in terms of **a** and/or **b**

a $\overrightarrow{QR}$ **b** $\overrightarrow{QM}$ **c** $\overrightarrow{PM}$

d Point *S* is added between points *Q* and *R* to make the parallelogram *PQSR*.
Complete the parallelogram *PQSR*.
Show how you can use the parallelogram law to find $\overrightarrow{PM}$.

> **Q21c hint**
>
> Make use of the vectors you already know.
> $\overrightarrow{PM} = \overrightarrow{P\Box} + \overrightarrow{\Box M}$
> Simplify.

18.4 Parallel vectors and collinear points

- Express points as position vectors.
- Prove that lines are parallel.
- Prove that points are collinear.

Active Learn
Homework

Warm up

1 **Fluency** Here are five vectors. Three of them are parallel. Which three?

$$\begin{pmatrix} -2 \\ 4 \end{pmatrix} \quad \begin{pmatrix} 6 \\ -12 \end{pmatrix} \quad \begin{pmatrix} -1 \\ -2 \end{pmatrix} \quad \begin{pmatrix} 4 \\ -6 \end{pmatrix} \quad \begin{pmatrix} 1 \\ -2 \end{pmatrix}$$

2 P is the point $(0, 4)$. $\overrightarrow{PQ} = \begin{pmatrix} 2 \\ 1 \end{pmatrix}$. Find the coordinates of Q.

3 Work out

a $\begin{pmatrix} 2 \\ -1 \end{pmatrix} + \begin{pmatrix} -3 \\ 4 \end{pmatrix}$
b $\begin{pmatrix} 5 \\ 2 \end{pmatrix} - \begin{pmatrix} 3 \\ 1 \end{pmatrix}$
c $\begin{pmatrix} -2 \\ 3 \end{pmatrix} + \begin{pmatrix} 2 \\ -3 \end{pmatrix}$

4 Simplify $3\mathbf{c} - \frac{1}{2}\mathbf{d} + \mathbf{c} + \mathbf{d}$.

5 In triangle ABC, $\overrightarrow{AB} = \mathbf{a}$ and $\overrightarrow{AC} = \mathbf{b}$.
P is the midpoint of AB.
Q is the midpoint of AC.
Write in terms of $\mathbf{a}$ and/or $\mathbf{b}$

a $\overrightarrow{BC}$
b $\overrightarrow{AP}$
c $\overrightarrow{AQ}$
d $\overrightarrow{PQ}$

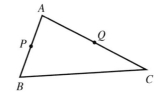

6 **Reasoning** In **Q5** what do your answers to parts **a** and **d** show about the lines BC and PQ?

7 **Problem-solving** $GHIJ$ is a quadrilateral.
$$\overrightarrow{GH} = \frac{1}{2}\mathbf{a} \qquad \overrightarrow{HI} = \frac{1}{2}\mathbf{a} - \mathbf{b} \qquad \overrightarrow{JI} = \mathbf{b}$$

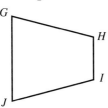

a Express $\overrightarrow{GJ}$ in terms of $\mathbf{a}$ and/or $\mathbf{b}$. Give your answer in its simplest form.

b **Reasoning** What does your answer to part **a** show about $\overrightarrow{GJ}$ and $\overrightarrow{HI}$?
Explain how you know.

8 O is the origin $(0, 0)$.
A has coordinates $(1, 5)$ and B has coordinates $(2, 4)$.
Find as column vectors

a $\overrightarrow{OA}$
b $\overrightarrow{AO}$
c $\overrightarrow{OB}$
d $\overrightarrow{AB}$

Q8d hint
Use the triangle law:
$$\overrightarrow{AB} = \overrightarrow{A\square} + \overrightarrow{\square B}$$

9 $\overrightarrow{OA} = \mathbf{a}$ and $\overrightarrow{OB} = \mathbf{b}$.
Express $\overrightarrow{AB}$ in terms of $\mathbf{a}$ and/or $\mathbf{b}$.

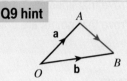
10 $\overrightarrow{OX} = 2\mathbf{a} + \mathbf{b}$ and $\overrightarrow{OY} = 3\mathbf{a} + 4\mathbf{b}$.
Express the vector $\overrightarrow{XY}$ in terms of $\mathbf{a}$ and/or $\mathbf{b}$.
Give your answer in its simplest form.

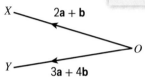

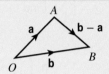

11 The points A, B, C and D have coordinates
$(1, 3)$, $(2, 7)$, $(-6, -10)$ and $(-1, 10)$ respectively.
O is the origin.

  a Work out as column vectors

    i $\overrightarrow{AB}$ **ii** $\overrightarrow{CD}$

  b Are $\overrightarrow{AB}$ and $\overrightarrow{CD}$ parallel? Explain how you know.

  c Copy and complete to state the relationship between $\overrightarrow{CD}$
    and $\overrightarrow{AB}$.
    $\overrightarrow{CD} = \square \overrightarrow{AB}$
    This means that $\overrightarrow{CD}$ is $\square$ times the length of $\overrightarrow{AB}$.

12 Reasoning The points P, Q, R and S have coordinates $(-2, 5)$, $(3, 1)$, $(-6, -9)$ and $(14, -25)$ respectively. O is the origin.

  a Work out as a column vector

    i $\overrightarrow{PQ}$ **ii** $\overrightarrow{RS}$

  b What do these results show about the lines PQ and RS?

Exam-style question

13 P is the point $(7, 5)$ and Q is the point $(-3, 1)$.

  a Find $\overrightarrow{PQ}$ as a column vector.   **(1 mark)**

  R is the point such that $\overrightarrow{QR} = \begin{pmatrix} 4 \\ 7 \end{pmatrix}$.

  b Write down the coordinates of the point R.   **(2 marks)**

  X is the midpoint of PQ. O is the origin.

  c Find $\overrightarrow{OX}$ as a column vector.   **(2 marks)**

Exam tip

In this type of vector question it can be helpful to draw a sketch.

14 Reasoning The point A has coordinates $(1, 3)$, the point B has coordinates $(4, 5)$ and the point C has coordinates $(-2, -4)$.

a Write $\overrightarrow{AB}$ as a column vector.

b $\overrightarrow{CD} = 6\overrightarrow{AB}$. Find $\overrightarrow{CD}$.

c Find the coordinates of the point D.

Q14c hint

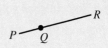

$\overrightarrow{CD} = \overrightarrow{OD} - \overrightarrow{OC}$

$\begin{pmatrix} \square \\ \square \end{pmatrix} = \begin{pmatrix} x \\ y \end{pmatrix} - \begin{pmatrix} \square \\ \square \end{pmatrix}$

15 Problem-solving $\mathbf{a} = \begin{pmatrix} 4 \\ 1 \end{pmatrix}$ and $\mathbf{b} = \begin{pmatrix} -2 \\ 3 \end{pmatrix}$.

Find a vector $\mathbf{c}$ such that $\mathbf{a} + \mathbf{c}$ is parallel to $\mathbf{a} - \mathbf{b}$.

16 Reasoning $OABC$ is a quadrilateral in which $\overrightarrow{OA} = \mathbf{a}$, $\overrightarrow{OB} = \mathbf{a} + 2\mathbf{b}$ and $\overrightarrow{OC} = 2\mathbf{b}$.

a Find $\overrightarrow{AB}$ in terms of $\mathbf{a}$ and/or $\mathbf{b}$.
What does this tell you about $\overrightarrow{AB}$ and $\overrightarrow{OC}$?

Q16 hint

Sketch a quadrilateral $OABC$.

b Find $\overrightarrow{BC}$ in terms of $\mathbf{a}$ and/or $\mathbf{b}$.
What does this tell you about $\overrightarrow{OA}$ and $\overrightarrow{BC}$?

c What type of quadrilateral is $OABC$?

Key point

$\overrightarrow{PQ} = k\overrightarrow{QR}$ shows that the lines PQ and QR are parallel. Also they both pass through point Q, so PQ and QR are part of the same straight line. P, Q and R are said to be **collinear** (they all lie on the same straight line).

17 The points A, B and C have coordinates $(2, 13)$, $(5, 22)$ and $(11, 40)$ respectively.

a Find as column vectors $\begin{pmatrix} 3 & 12 \end{pmatrix} \begin{pmatrix} 6 & 21 \end{pmatrix}$ $12, 39$

   i $\overrightarrow{AB}$ ii $\overrightarrow{AC}$

b Are the lines AB and AC parallel? Explain.

c Do both lines pass through a single point? If yes, which point?

d What do your answers to parts **b** and **c** tell you about the points A, B and C?

18 Problem-solving The point P has coordinates $(1, 3)$, the point Q has coordinates $(4, 6)$ and the point R has coordinates $(10, 12)$.
Show that points P, Q and R are collinear.

Q18 hint

First find column vectors $\overrightarrow{PQ}$ and $\overrightarrow{QR}$. Then state whether they are parallel and pass through a single point.

19 Problem-solving OAB is a triangle.

$\overrightarrow{OA} = \mathbf{a}$ $\overrightarrow{OB} = \mathbf{a} + 2\mathbf{b}$ $\overrightarrow{AC} = 2\mathbf{a} + 3\mathbf{b}$

X is the midpoint of AB.

a Find in terms of $\mathbf{a}$ and/or $\mathbf{b}$

   i $\overrightarrow{AB}$ ii $\overrightarrow{AX}$

   iii $\overrightarrow{OX}$ iv $\overrightarrow{OC}$

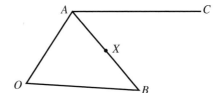

b Use your answers to parts **a iii** and **a iv** to show that O, X and C are collinear.

c **Reasoning** Antony says, 'You can also use $\overrightarrow{OX}$ and $\overrightarrow{XC}$ to show that O, X and C are collinear.'
Show that Antony is correct.

18.5 Solving geometric problems

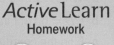
- Solve geometric problems in two dimensions using vector methods, including where vectors are divided in a given ratio.
- Apply vector methods for simple geometric proofs.

Warm up

1 Fluency What fraction of AC is AB when $AB : BC = 3 : 2$?

2 $ABCD$ is a parallelogram.
Match each displacement vector to one of the vectors below.

a $\overrightarrow{AC}$ **b** $\overrightarrow{AD}$ **c** $\overrightarrow{BD}$ **d** $\overrightarrow{CD}$ **e** $\overrightarrow{DA}$

| $\mathbf{q}$ | $-\mathbf{q}$ | $\mathbf{p}+\mathbf{q}$ | $-\mathbf{p}$ | $\mathbf{q}-\mathbf{p}$ |

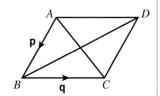

3 $\overrightarrow{PQ} = 3\mathbf{a} - 2\mathbf{b}$ and $\overrightarrow{PR} = 9\mathbf{a} - 6\mathbf{b}$.
What does this tell you about

a PQ and PR **b** the points P, Q and R?

Key point

Some vector problems involve ratios.
For example, the point X lies on AB such that $AX : XB = 1 : 3$.
$AX = \frac{1}{4}AB$
$AB = 4AX$

4 Reasoning In triangle ABO, $\overrightarrow{OA} = \mathbf{a}$ and $\overrightarrow{OB} = \mathbf{b}$.
The point X divides AB in the ratio $1 : 2$.

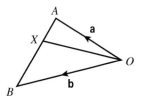

Express in terms of $\mathbf{a}$ and/or $\mathbf{b}$

a $\overrightarrow{AB}$ **b** $\overrightarrow{AX}$ **c** $\overrightarrow{OX}$

Q4b hint

$AX : XB = 1 : 2$
$AX = \dfrac{\square}{\square}AB$

Q4c hint

$\overrightarrow{OX} = \overrightarrow{OA} + \overrightarrow{\square}$

5 Problem-solving $OACB$ is a parallelogram
with $\overrightarrow{OA} = \mathbf{a}$ and $\overrightarrow{OB} = \mathbf{b}$
E is the point on AC such that $AE : EC = 1 : 3$.
F is the point on BC such that $CF : FB = 3 : 1$.

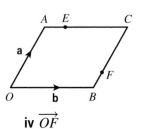

a Find in terms of $\mathbf{a}$ and/or $\mathbf{b}$

   i $\overrightarrow{AB}$ **ii** $\overrightarrow{AE}$ **iii** $\overrightarrow{OE}$ **iv** $\overrightarrow{OF}$

b Reasoning Show that EF is parallel to AB.

6 **Reasoning** XYZ is a straight line.

$\overrightarrow{OX} = \mathbf{a} - 2\mathbf{b}$ and $\overrightarrow{OY} = \mathbf{a} + 6\mathbf{b}$.

$XY : YZ = 2 : 3$

Express in terms of $\mathbf{a}$ and/or $\mathbf{b}$

a $\overrightarrow{XY}$ **b** $\overrightarrow{XZ}$ **c** $\overrightarrow{OZ}$

Give your answers in their simplest form.

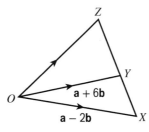

Q6b hint

$XY : YZ = 2 : 3$

$XY = \dfrac{\square}{\square} XZ$

so $\quad XZ = \dfrac{\square}{\square} XY$

Exam-style question

7 $PQRS$ is a parallelogram.

$\overrightarrow{PQ} = \mathbf{p} - 2\mathbf{q} \qquad \overrightarrow{PS} = 2\mathbf{p}$

M is the midpoint of the straight line QS.

a Express $\overrightarrow{MS}$ in terms of $\mathbf{p}$ and/or $\mathbf{q}$.   **(2 marks)**

PST is a straight line so that $PS : ST = n : 1$.

Given that $\overrightarrow{MT} = \mathbf{p} + \mathbf{q}$

b find the value of n.   **(2 marks)**

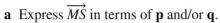

Q7b hint

$\overrightarrow{MT} = \overrightarrow{MS} + \overrightarrow{ST}$

Exam tip

When finding a missing number in a ratio, begin by writing down what you know. For example,

$PS : ST$

$2\mathbf{p} : \square$

$n : 1$

---

8 **Reasoning** $OABC$ is a quadrilateral in which $\overrightarrow{OA} = \mathbf{a}$, $\overrightarrow{OB} = \mathbf{a} + 2\mathbf{b}$ and $\overrightarrow{OC} = 4\mathbf{b}$.

D is the point such that $\overrightarrow{BD} = \overrightarrow{OC}$, and X is the midpoint of BC.

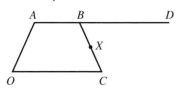

a Find in terms of $\mathbf{a}$ and/or $\mathbf{b}$

   i $\overrightarrow{OD}$ **ii** $\overrightarrow{OX}$

b Explain what your answers to parts **a i** and **a ii** mean.

Q8a i hint

$\overrightarrow{OD} = \overrightarrow{OB} + \square$

Q8a ii hint

First find $\overrightarrow{CB}$.

---

9 **Reasoning** The diagram shows a regular hexagon $ABCDEF$ with centre O.

$\overrightarrow{OA} = \mathbf{a}$ and $\overrightarrow{OB} = \mathbf{b}$

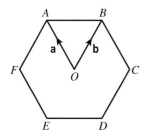

a Write down all other displacement vectors in the diagram that are equal to

   i $\overrightarrow{OA}$ **ii** $\overrightarrow{OB}$

b Express $\overrightarrow{AB}$ in terms of $\mathbf{a}$ and/or $\mathbf{b}$.

c Which other displacement vector in the diagram is equal to $\overrightarrow{AB}$?

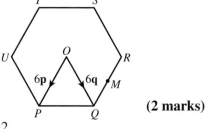

10 The diagram shows a regular hexagon
    PQRSTU with centre *O*.
    $\overrightarrow{OP} = 6\mathbf{p}$ $\overrightarrow{OQ} = 6\mathbf{q}$
    M is the midpoint of *QR*.
    a Express $\overrightarrow{TM}$ in terms of **p** and/or **q**.   **(2 marks)**
    N is the point on *PQ* extended such that $PQ : QN = 3 : 2$.
    b Prove that *T*, *M* and *N* lie on the same straight line.   **(3 marks)**

11 **Problem-solving** In triangle *OMN*, $\overrightarrow{OM} = \mathbf{m}$ and $\overrightarrow{ON} = \mathbf{n}$.
    The point *P* is the midpoint of *MN* and *Q* is the point such
    that $\overrightarrow{OQ} = \frac{3}{2}\overrightarrow{OP}$.

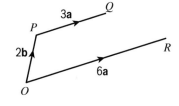

    a Find in terms of **m** and/or **n**
        i $\overrightarrow{OP}$ **ii** $\overrightarrow{OQ}$ **iii** $\overrightarrow{MQ}$
    The point *R* is such that $\overrightarrow{OR} = 3\overrightarrow{ON}$.
    b Find in terms of **m** and/or **n** the vector $\overrightarrow{MR}$.
    c **Reasoning** Explain why *MQR* is a straight line and give the value of $\dfrac{MR}{MQ}$.

12 **Problem-solving** In the diagram,
    $\overrightarrow{OR} = 6\mathbf{a}$, $\overrightarrow{OP} = 2\mathbf{b}$ and $\overrightarrow{PQ} = 3\mathbf{a}$.
    The point *M* is on *PQ* such that $\overrightarrow{PM} = 2\mathbf{a}$.
    The point *N* is on *OR* such that $\overrightarrow{ON} = \frac{1}{3}\overrightarrow{OR}$.
    The midpoint of *MN* is the point *S*.

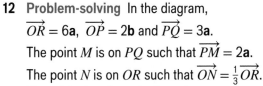

    a Find in terms of **a** and/or **b** the vector $\overrightarrow{NM}$.
    b Find in terms of **a** and/or **b** the vector $\overrightarrow{OS}$.
    c *T* is the point such that $\overrightarrow{QT} = \mathbf{a}$. Find in terms of **a** and/or **b** the vector $\overrightarrow{OT}$.
    d Show that *S* lies on the line *OT*.
    e When $\mathbf{a} = \begin{pmatrix} 8 \\ 2 \end{pmatrix}$ and $\mathbf{b} = \begin{pmatrix} 3 \\ 15 \end{pmatrix}$ find the length of *QR*.

13 **Problem-solving** The diagram shows triangle *OPQ*.
    M is the midpoint of *OP* and *N* is the midpoint of *OQ*.
    $\overrightarrow{OM} = \mathbf{p}$ and $\overrightarrow{ON} = \mathbf{q}$.
    a Find in terms of **p** and/or **q**
        i $\overrightarrow{PO}$ **ii** $\overrightarrow{QO}$ **iii** $\overrightarrow{PN}$ **iv** $\overrightarrow{QM}$

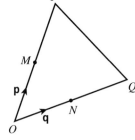

    X lies on *PN* such that $PX : XN = 2 : 1$.
    Y lies on *QM* such that $QY = \frac{2}{3}QM$.
    b Find in terms of **p** and/or **q**
        i $\overrightarrow{OX}$ **ii** $\overrightarrow{OY}$
    c **Reasoning** Explain what your answer to part **b** means.

18 Check up

Vector notation

1 The diagram shows two vectors **a** and **b**.
Write **a** and **b** as column vectors.

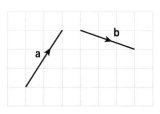

2 **a** A is the point $(4, 3)$ and B is the point $(7, -2)$.
Write $\overrightarrow{AB}$ as a column vector.

   b C is the point $(1, 4)$ and $\overrightarrow{CD} = \begin{pmatrix} 2 \\ 3 \end{pmatrix}$.
   Find the coordinates of D.

3 Find the magnitude of the vector $\begin{pmatrix} -3 \\ 5 \end{pmatrix}$.
Give your answer in surd form.

Vector arithmetic

4 The diagram shows two vectors **p** and **q**.
On squared paper draw vectors to represent

   a 2**q**   **b** **p** + **q**   **c** **p** − **q**

5 $\overrightarrow{AB} = \begin{pmatrix} 3 \\ -1 \end{pmatrix}$ and $\overrightarrow{BC} = \begin{pmatrix} -5 \\ 4 \end{pmatrix}$. Find $\overrightarrow{AC}$.

6 $\mathbf{m} = \begin{pmatrix} 4 \\ -2 \end{pmatrix}$ and $\mathbf{n} = \begin{pmatrix} -1 \\ 6 \end{pmatrix}$. Find

   a **m** + **n**   **b** **n** − **m**   **c** 3**m**

7 $\mathbf{p} = \begin{pmatrix} 5 \\ -1 \end{pmatrix}$ and $\mathbf{q} = \begin{pmatrix} 9 \\ -3 \end{pmatrix}$.
p + 2**r** = **q**
Find **r** as a column vector.

Geometric problems

8 $PQRS$ is a square.
$\overrightarrow{PQ} = \mathbf{a}$ and $\overrightarrow{QR} = \mathbf{b}$.
PR and QS are diagonals of $PQRS$.
PR and QS intersect at X.
Express in terms of **a** and/or **b**

   a $\overrightarrow{SR}$ **b** $\overrightarrow{PR}$ **c** $\overrightarrow{PX}$ **d** $\overrightarrow{SX}$

9 Which of these vectors are parallel to
   a **s** − **t**   **b** **s** + **t**?

| 3**s** + 3**t** | 2**s** − **t** | 3**s** − 3**t** | $\frac{1}{2}$**s** − $\frac{1}{2}$**t** |

10 P is the point $(4, -3)$ and Q is the point $(-2, 7)$.

   a Write down the position vector, **p**, of the point P.

   b Write down the position vector, **q**, of the point Q.

   c Work out $\overrightarrow{PQ}$.

11 The points A, B and C have coordinates $(2, 13)$, $(5, 22)$ and $(11, 40)$ respectively.

   a Find as column vectors

     i $\overrightarrow{AB}$

     ii $\overrightarrow{AC}$

   b What do these results show about the points A, B and C?

12 $OABC$ is a parallelogram.
X is the midpoint of AB and Y is the midpoint of BC.
$\overrightarrow{OA} = \mathbf{a}$ and $\overrightarrow{OC} = \mathbf{c}$.

   a Find in terms of **a** and/or **c**

     i $\overrightarrow{XB}$

     ii $\overrightarrow{XY}$

   b Show that AC is parallel to XY.

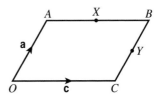

13 DEF is a straight line.
$\overrightarrow{OD} = \mathbf{a} + 3\mathbf{b}$ and $OE = 4\mathbf{a}$.
$DE : EF = 3 : 2$
Express in terms of **a** and/or **b**

   a $\overrightarrow{DE}$

   b $\overrightarrow{DF}$

   c $\overrightarrow{OF}$

14 **Reflect** How sure are you of your answers? Were you mostly

   Just guessing 😞   Feeling doubtful 😐   Confident 🙂

   What next? Use your results to decide whether to strengthen or extend your learning.

Challenge

15 P is the point $(2, 2)$ and R is the point $(6, 4)$.
$PQRS$ is a parallelogram. PR is a diagonal.

   a Choose a position for Q, such that the magnitude of vector $\overrightarrow{PQ}$ is 5, the x-coordinate is a positive integer and the y-coordinate is a negative integer.

   b For any one of the positions of Q in part **a**, write the vectors $\overrightarrow{PQ}$, $\overrightarrow{QR}$, $\overrightarrow{RS}$ and $\overrightarrow{SQ}$ as column vectors.

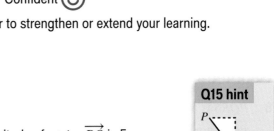

Q15 hint

18 Strengthen

Vector notation

1 Write down the column vector that describes each translation.

a A to B

b B to A

c A to C

d A to D

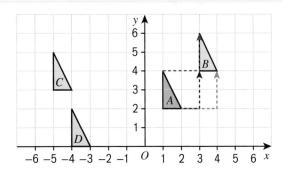

2 $\overrightarrow{OA}$ is the vector that translates O to A.
$\overrightarrow{OB}$ is the vector that translates O to B.

a $\overrightarrow{OA}$ and $\overrightarrow{OB}$ start at the same position on the grid. Write down where they start.

b Write as column vectors

i $\overrightarrow{OA}$ **ii** $\overrightarrow{OB}$

> **Q2b hint**
>
> From the start:
>
> Count across $\longrightarrow$ $\begin{pmatrix} \square \\ \end{pmatrix}$
> Count up $\longrightarrow$ $\begin{pmatrix} \\ \square \end{pmatrix}$

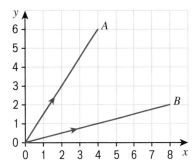

3 P is the point $(2, 3)$.

a Plot point P on a grid.

$\overrightarrow{PQ} = \begin{pmatrix} 1 \\ 5 \end{pmatrix}$

b Find the coordinates of Q.

> **Q3b hint**
>
> From P move 1 unit across and 5 units up to find the position of Q.

4 A is the point $(4, 2)$ and B is the point $(7, 1)$.

a Plot points A and B on a grid.

b Write $\overrightarrow{AB}$ as a column vector.

5 **a** Copy and complete the diagram to show the vector $\begin{pmatrix} 2 \\ -5 \end{pmatrix}$.

b Use Pythagoras' theorem to work out the magnitude of the vector $\begin{pmatrix} 2 \\ -5 \end{pmatrix}$
(the length of the hypotenuse of the triangle).
Give your answer in surd form.

c Use the method in parts **a** and **b** to find the magnitude of these vectors, giving your answers in surd form.

i $\begin{pmatrix} 3 \\ -4 \end{pmatrix}$ **ii** $\begin{pmatrix} -9 \\ 5 \end{pmatrix}$ **iii** $\begin{pmatrix} 9 \\ 7 \end{pmatrix}$ **iv** $\begin{pmatrix} -3 \\ -5 \end{pmatrix}$

Vector arithmetic

1 The diagram shows the vector **m**.

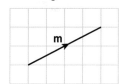

On squared paper draw vectors to represent

a **2m**   b **3m**

c $\frac{1}{2}$**m**   d **−m**

Q1a hint

Draw vector **m** twice, end to end.

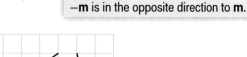

Q1d hint

−m is in the opposite direction to **m**.

2 The diagram shows the vectors **b** and **c**.

a Which diagram below shows

   i **b+c**   ii **b−c**?

Explain how you know.

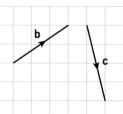

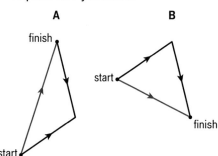

b Copy the diagrams on squared paper and label the vectors **b** and **c**.

3 $\mathbf{a} = \begin{pmatrix} 3 \\ 4 \end{pmatrix}$ and $\mathbf{b} = \begin{pmatrix} 1 \\ 2 \end{pmatrix}$.

a On squared paper draw the vectors

   i **a**   ii **b**   iii **a+b**   iv **a−b**   v **2a**   vi **2a+b**

b Use your answers to part **a** to write as column vectors

   i **a+b**   ii **a−b**   iii **2a+b**

c Work these out. The first one is started for you.

   i $\mathbf{a} + \mathbf{b} = \begin{pmatrix} 3 \\ 4 \end{pmatrix} + \begin{pmatrix} 1 \\ 2 \end{pmatrix} = \begin{pmatrix} 3+1 \\ 4+2 \end{pmatrix} =$

   ii $\mathbf{a} - \mathbf{b} = \begin{pmatrix} 3 \\ 4 \end{pmatrix} - \begin{pmatrix} 1 \\ 2 \end{pmatrix} =$ iii $2\mathbf{a} + \mathbf{b} = 2\begin{pmatrix} 3 \\ 4 \end{pmatrix} + \begin{pmatrix} 1 \\ 2 \end{pmatrix} =$

d What do you notice about your answers to parts **b** and **c**?

4 $\overrightarrow{AB} = \begin{pmatrix} 2 \\ 4 \end{pmatrix}$, $\overrightarrow{BC} = \begin{pmatrix} 4 \\ -1 \end{pmatrix}$, $\overrightarrow{CD} = \begin{pmatrix} -5 \\ 3 \end{pmatrix}$ and $\overrightarrow{BE} = \begin{pmatrix} -2 \\ -1 \end{pmatrix}$.

a Work out

   i $\overrightarrow{AB} + \overrightarrow{BC}$ ii $\overrightarrow{BC} + \overrightarrow{CD}$ iii $\overrightarrow{AB} + \overrightarrow{BE}$

b Which of your answers to part **a** gives

   i $\overrightarrow{AC}$ ii $\overrightarrow{AE}$ iii $\overrightarrow{BD}$?

Q4a i hint

$\overrightarrow{AB} + \overrightarrow{BC} = \begin{pmatrix} 2 \\ 4 \end{pmatrix} + \begin{pmatrix} 4 \\ -1 \end{pmatrix}$

Q4b i hint

Geometric problems

1 a On squared paper draw the vectors

   i $\mathbf{a} = \begin{pmatrix} 1 \\ -2 \end{pmatrix}$ **ii** $\mathbf{b} = \begin{pmatrix} 3 \\ -6 \end{pmatrix}$ **iii** $\mathbf{c} = \begin{pmatrix} -1 \\ 2 \end{pmatrix}$ **iv** $\mathbf{d} = \begin{pmatrix} 2 \\ -4 \end{pmatrix}$ **v** $\mathbf{e} = \begin{pmatrix} -1 \\ -2 \end{pmatrix}$

  b What do you notice about vectors

    i **a** and **b**   **ii** **c** and **a**   **iii** **d** and **a**?

Q1b i hint

$$\begin{pmatrix} 3 \\ -6 \end{pmatrix} = \begin{pmatrix} \square \times 1 \\ \square \times -2 \end{pmatrix} = \square \begin{pmatrix} 1 \\ -2 \end{pmatrix}$$

2 Which of these vectors are parallel to $\mathbf{a} + \mathbf{b}$?

$3\mathbf{a} + 3\mathbf{b}$	$\mathbf{a} - \mathbf{b}$	$2\mathbf{a} + \mathbf{b}$
$\mathbf{a} + 2\mathbf{b}$	$2\mathbf{a} + 2\mathbf{b}$	$\frac{1}{2}\mathbf{a} - \frac{1}{2}\mathbf{b}$

Q2 hint

Can you multiply **a** and **b** by the same number to give the vector?

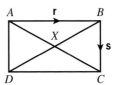

3 $ABCD$ is a trapezium.

$\overrightarrow{AB} = \mathbf{a}$ and $\overrightarrow{AD} = \mathbf{b}$

  a CD is parallel to AB.
    What does this tell you about
    the vectors $\overrightarrow{AB}$ and $\overrightarrow{CD}$?

  b $CD = 2AB$
    Copy and complete.
    $\overrightarrow{CD} = \square\overrightarrow{AB} = \square$

  c Copy and complete.
    $\overrightarrow{BC} = \overrightarrow{BA} + \overrightarrow{\square} + \overrightarrow{\square}$
    $\phantom{\overrightarrow{BC}} = -\mathbf{a} + \square + \square$
    Simplify your answer by collecting like vectors.

Q3c hint

Trace around the diagram.

4 $ABCD$ is a rectangle. $\overrightarrow{AB} = \mathbf{r}$ and $\overrightarrow{BC} = \mathbf{s}$.
  AC and BD are diagonals of $ABCD$.
  AC and BD intersect at X.

  a Which side is parallel and equal in length to

    i AB **ii** BC?

  b Draw the rectangle.
    Mark the vector (with an arrow and a letter) on

    i side CD **ii** side AD

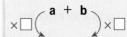

  c Express in terms of **r** and/or **s**

    i $\overrightarrow{AC}$ **ii** $\overrightarrow{BD}$

  d What is the relationship between the vectors

    i $\overrightarrow{AX}$ and $\overrightarrow{AC}$ **ii** $\overrightarrow{BX}$ and $\overrightarrow{BD}$?

  e Express in terms of **r** and/or **s**

    i $\overrightarrow{AX}$ **ii** $\overrightarrow{BX}$

Q4c i hint

Place your finger at A. Which vectors that you know do you move along to get to C?

Q4c ii hint

Look carefully at the direction of the arrows.

5 $\vec{AB} = \begin{pmatrix} 4 \\ 1 \end{pmatrix}$, $\vec{BC} = \begin{pmatrix} -3 \\ 2 \end{pmatrix}$ and $\vec{CD} = \begin{pmatrix} 2 \\ -5 \end{pmatrix}$.

   a On squared paper draw a diagram to show $ABCD$.

   b Find the column vector for $\vec{AD}$.

   c Show that $\vec{AC} = \vec{DB}$.
   What does this tell you about the vectors $\vec{AC}$ and $\vec{DB}$?

Q5a hint

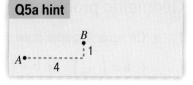

6 **Problem-solving** A, B and C are collinear.

   a Write down the column vectors

      i $\vec{AB}$ **ii** $\vec{BC}$

   b How do these vectors show that AB and BC are collinear?

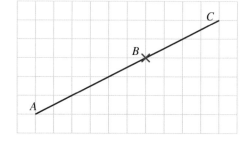

7 **a** Plot points $O(0, 0)$, $P(3, -2)$ and $Q(5, -1)$ on a grid.

   b Join with a straight line

      i O and P **ii** O and Q

   c Draw an arrow to show the direction of

      i $\vec{OP}$ **ii** $\vec{OQ}$

   d Write down the position vector, **p**, of the point P.

   e Write down the position vector, **q**, of the point Q.

   f Work out $\vec{PQ}$.

Q7c i hint

$\vec{OP}$ is the vector that translates O to P.

Q7d, e, f hint

Give your answers as column vectors $\begin{pmatrix} \square \\ \square \end{pmatrix}$.

8 A is the point $(4, 1)$, B is the point $(8, 4)$ and C is the point $(20, 13)$.

   a Find $\vec{AB}$ and $\vec{BC}$.

   b Show that AB and BC are parallel.

   c Copy and complete.
   AB and BC are _____ and both pass through the point ____.
   So ABC is a _____ line and A, B, C are collinear.

9 OAB is a triangle. $\vec{OA} = $ **a** and $\vec{OB} = $ **b**.

   a C is the midpoint of OA. Write $\vec{OC}$ in terms of **a**.

   b Find $\vec{BA}$ in terms of **a** and/or **b**.

   c D is the midpoint of BA. Find $\vec{BD}$ in terms of **a** and/or **b**.

   d Find $\vec{OD}$ in terms of **a** and/or **b**.

   e Find $\vec{CD}$ in terms of **a** and/or **b**.

   f Show that CD is parallel to OB.

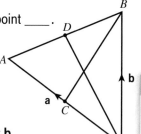

Q9c hint

$BD = \dfrac{\square}{\square} BA$

Q9f hint

CD and OB are parallel if their vectors are multiples of each other.

10 **Reasoning** $\vec{OA} = $ **a**, $\vec{OB} = $ **b** and $AP : PB = 3 : 2$.

   a Copy and complete. $AP = \dfrac{\square}{\square} AB$ $BP = \dfrac{\square}{\square} BA$

   b Express in terms of **a** and/or **b**.

      i $\vec{AB}$ **ii** $\vec{AP}$ **iii** $\vec{BA}$ **iv** $\vec{PB}$ **v** $\vec{OP}$

Q10a hint

      3   2

A P B

18 Extend

1 Problem-solving *JKLM* is a parallelogram.
The diagonals of the parallelogram intersect at *O*.
$\overrightarrow{OJ} = \mathbf{j}$ and $\overrightarrow{OK} = \mathbf{k}$.

a Write $\overrightarrow{KJ}$ in terms of **j** and/or **k**.

X is the point such that $\overrightarrow{OX} = 2\mathbf{j} - \mathbf{k}$.

b i Write down an expression, in terms of **j** and/or **k**, for $\overrightarrow{JX}$.

ii Explain why *J*, *K* and *X* lie on the same straight line.

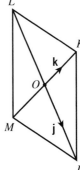

2 *ABD* is a triangle. *N* is a point on *AD*.
$\overrightarrow{AB} = \mathbf{a}$, $\overrightarrow{AN} = 2\mathbf{b}$ and $\overrightarrow{ND} = \mathbf{b}$.

a Find the vector $\overrightarrow{BD}$ in terms of **a** and/or **b**.

B is the midpoint of *AC* and *M* is the midpoint of *BD*.

b Show that *NMC* is a straight line.

> **Q2c hint**
>
> Show that *N*, *M* and *C* are collinear.

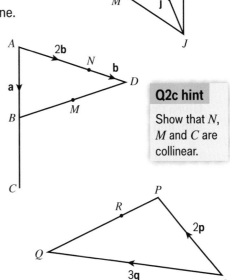

3 Problem-solving *OPQ* is a triangle.
$\overrightarrow{OP} = 2\mathbf{p}$ and $\overrightarrow{OQ} = 3\mathbf{q}$.
R is the point on *PQ* such that $PR : RQ = 2 : 3$.
Show that $\overrightarrow{OR}$ is parallel to the vector $\mathbf{p} + \mathbf{q}$.

4 Problem-solving *OPQ* is a triangle.   *B* is the midpoint of *PQ*.
$\overrightarrow{OA} = 3\mathbf{p}$, $\overrightarrow{AP} = \mathbf{p}$, $\overrightarrow{OQ} = 2\mathbf{q}$ and $\overrightarrow{QC} = \mathbf{q}$.

a Find, in terms of **p** and/or **q**, the vectors

i $\overrightarrow{PQ}$ **ii** $\overrightarrow{AC}$ **iii** $\overrightarrow{BC}$

b Hence explain why *ABC* is a straight line.

The length of *AB* is 3 cm.

c Find the length of *AC*.

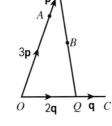

> **Q4b hint**
>
> 'Hence' means you should use your answers to part **a** to help you answer part **b**.

5 *OAYB* is a quadrilateral.
$\overrightarrow{OA} = 3\mathbf{a}$ and $\overrightarrow{OB} = 6\mathbf{b}$.

a Express $\overrightarrow{AB}$ in terms of **a** and/or **b**.

X is the point on *AB* such that
$AX : XB = 1 : 2$
$OX : OY = 1 : 2.5$

b Work out $\overrightarrow{OY}$.

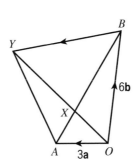

> **Q5b hint**
>
> Start by finding $\overrightarrow{AX}$, then find $\overrightarrow{OX}$. Then
>
> $OX : OY$
>
> $1 : 2.5$
>
> $\square\mathbf{a} + \square\mathbf{b} : \square\mathbf{a} + \square\mathbf{b}$
>
> $\times 2.5$

6 ONM is a straight line.

$\overrightarrow{ON} = 2\mathbf{a} + 3\mathbf{b}$ and $\overrightarrow{OM} = k(\mathbf{a} + \mathbf{b}) - 2\mathbf{a}$, where k is a scalar quantity.

  a Rewrite $\overrightarrow{OM}$ in the form $(\square - \square)\mathbf{a} + \square\mathbf{b}$.

  b Work out k.

  c Find $\overrightarrow{OM}$.

Q6b hint

Since ONM is a straight line, $\overrightarrow{OM}$ is a multiple of $\overrightarrow{ON}$. Coefficients of $\mathbf{a}$ and $\mathbf{b}$ are in the same ratio.
$$\frac{2}{\square - \square} = \frac{3}{\square}$$
Solve to work out k.

Exam-style question

7 OAP, OMB and ANB are straight lines.

$AP = \frac{1}{2}OA$

M is the midpoint of OB.

$\overrightarrow{OA} = 2\mathbf{a}$ $\overrightarrow{OM} = \mathbf{b}$

$\overrightarrow{AN} = k\overrightarrow{AB}$, where k is a scalar quantity.

Given that MNP is a straight line, find the value of k. **(5 marks)**

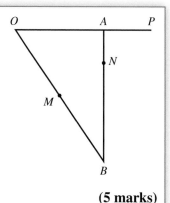

Exam tip

Look carefully at the information you are given. For example, MNP is a straight line, so $\overrightarrow{MP}$ is a multiple of $\overrightarrow{NP}$ and of $\overrightarrow{MN}$.

Exam-style question

8 OAB is a triangle.

OPM and APN are straight lines.

M is the midpoint of AB.

$\overrightarrow{OA} = \mathbf{a}$ $\overrightarrow{OB} = \mathbf{b}$

$OP : PM = 3 : 1$

Work out the ratio $ON : NB$.

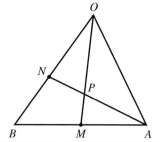

(5 marks)

Working towards A level

9 $\overrightarrow{OA} = \begin{pmatrix} k \\ 3k \end{pmatrix}$ where k is a constant.

Given that $|\overrightarrow{OA}| = 4\sqrt{5}$, find the possible values of k.

> Apply your knowledge of vectors to the information given, then use other mathematical skills to solve the problem.

Q9 hint

The magnitude, $|\overrightarrow{OP}|$, of a vector $\overrightarrow{OP} = \begin{pmatrix} a \\ b \end{pmatrix}$ is given by $\sqrt{a^2 + b^2}$.

10 $\overrightarrow{OB} = \begin{pmatrix} 1 \\ m \end{pmatrix}$ and $\overrightarrow{OC} = \begin{pmatrix} 2m \\ 1 \end{pmatrix}$, where O is the origin.

  a Write $\overrightarrow{BC}$ as a column vector.

  b Given that $|\overrightarrow{BC}| = \sqrt{10}$, find the possible values of m.

> This question requires a combination of different mathematical skills – vector knowledge, Pythagoras' theorem, quadratic equations – and is typical of the demands of A level questions.

Q10a hint
$$\overrightarrow{BC} = \overrightarrow{BO} + \overrightarrow{OC} =$$
$$\begin{pmatrix} -1 \\ -m \end{pmatrix} + \begin{pmatrix} 2m \\ 1 \end{pmatrix} = \begin{pmatrix} \square \\ \square \end{pmatrix}$$

Q10b hint

Use Pythagoras to write and solve a quadratic equation in m.

---

18 Test ready

Summary of key points

To revise for the test:

- Read each key point, find a question on it in the mastery lesson, and check you can work out the answer.

- If you cannot, try some other questions from the mastery lesson or ask for help.

Key points

1 A **vector** is a quantity that has magnitude (size) and direction.
Examples of vectors are force (5 N acting vertically upwards) and velocity (15 km/h due north). → **18.1**

2 **Displacement** is change in position.
A displacement can be written as $\binom{3}{4}$, where 3 is the x-component and 4 is the y-component. → **18.1**

3 The displacement vector from A to B is written $\overrightarrow{AB}$. → **18.1**

4 Vectors are written as **bold** lower case letters: **a**, **b**, **c**
When handwriting, underline the letter: <u>a</u>, <u>b</u>, <u>c</u> → **18.1**

5 **Equal vectors** have the same magnitude and the same direction. → **18.1**

6 The **magnitude** of the vector $\binom{x}{y}$ is its length, i.e. $\sqrt{x^2+y^2}$.

$|a|$ means the magnitude of vector **a**. $|\overrightarrow{OA}|$ means the magnitude of vector $\overrightarrow{OA}$. → **18.1**

7 If $\overrightarrow{AB} = \overrightarrow{CD}$, then the line segments AB and CD are equal in length and are parallel.
$\overrightarrow{AB}$ and $\overrightarrow{CD}$ have the same direction.
$\overrightarrow{BA}$ and $\overrightarrow{DC}$ have the opposite direction. $\overrightarrow{AB} = -\overrightarrow{BA} = \overrightarrow{CD} = -\overrightarrow{DC}$ → **18.2**

8 **2a** is twice as long as **a** and in the same direction.
−a is the same length as **a** but in the opposite direction → **18.2**

9 When a vector **a** is multiplied by a scalar k, the vector k**a** is parallel to **a** and is equal to k times **a**.
A **scalar** is a number, e.g. 3, 2, $\frac{1}{2}$, −1, ... → **18.2**

10 The two-stage journey from A to B and then from B to C has the same starting point and the same finishing point as the single journey from A to C.
So A to B followed by B to C is equivalent to A to C.
$\overrightarrow{AB} + \overrightarrow{BC} = \overrightarrow{AC}$ → **18.2**

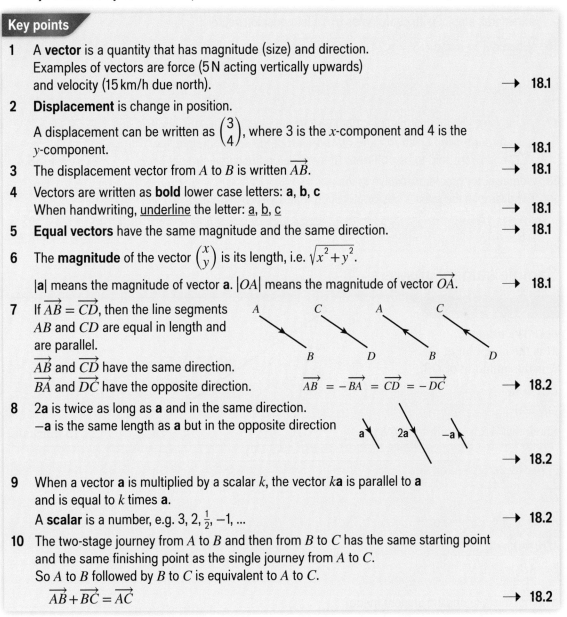

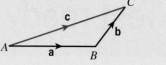

Key points

11 Triangle law for vector addition: Let $\overrightarrow{AB} = \mathbf{a}$, $\overrightarrow{BC} = \mathbf{b}$
and $\overrightarrow{AC} = \mathbf{c}$
Then $\mathbf{a} + \mathbf{b} = \mathbf{c}$ forms a triangle.　　→ **18.2**

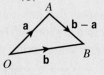

12 $\mathbf{a} + \mathbf{b} = \mathbf{b} + \mathbf{a}$　　→ **18.2**

13 Parallelogram law for vector addition: In parallelogram $PQRS$, where $\overrightarrow{PQ}$ is $\mathbf{a}$
and $\overrightarrow{PS}$ is $\mathbf{b}$, the diagonal $\overrightarrow{PR}$ of the parallelogram is $\mathbf{a} + \mathbf{b}$.　　→ **18.3**

14 When $\mathbf{c} = \mathbf{a} + \mathbf{b}$, the vector $\mathbf{c}$ is called the **resultant vector** of the two vectors $\mathbf{a}$ and $\mathbf{b}$.　→ **18.3**

15 With the origin O, the vectors $\overrightarrow{OA}$ and $\overrightarrow{OB}$ are called the **position vectors** of the points
A and B.
In general, a point with coordinates (p, q) has position vector $\begin{pmatrix} p \\ q \end{pmatrix}$.　　→ **18.4**

16 When $\overrightarrow{OA} = \mathbf{a}$ and $\overrightarrow{OB} = \mathbf{b}$, $\overrightarrow{AB} = \overrightarrow{AO} + \overrightarrow{OB} = \mathbf{b} - \mathbf{a}$.

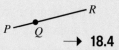

→ **18.4**

17 $\overrightarrow{PQ} = k\overrightarrow{QR}$ shows that the lines PQ and QR are parallel. Also they both
pass through point Q, so PQ and QR are part of the same straight line.
P, Q and R are said to be **collinear** (they all lie on the same straight line).　　→ **18.4**

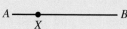

18 Some vector problems involve ratios.
For example, the point X lies on AB such that $AX : XB = 1 : 3$.
$AX = \frac{1}{4}AB$
$AB = 4AX$

$A \; \bullet\!\!\!\!\!-\!\!\!\!\!\underset{X}{\bullet}\!\!\!\!\!-\!\!\!\!\! \bullet \; B$

→ **18.5**

Sample student answers

OAB is a triangle.
M is the midpoint of OA.
N is the midpoint of OB.
$\overrightarrow{OM} = \mathbf{m}$
$\overrightarrow{ON} = \mathbf{n}$
Show that AB is parallel to MN.

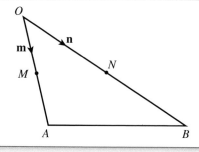

(3 marks)

$\overrightarrow{OA} = 2\underline{m}$
$\overrightarrow{OB} = 2\underline{n}$
$\overrightarrow{AB} = 2\underline{m} + 2\underline{n} = 2(\underline{m} + \underline{n})$
$\overrightarrow{MN} = \underline{m} + \underline{n}$
So $\overrightarrow{AB} = 2\overrightarrow{MN}$

---

a What mistake has the student made?

b How could you improve this answer?

18 Unit test

*Active*Learn
Homework

1 $\mathbf{a} = \begin{pmatrix} 2 \\ -3 \end{pmatrix}$ $\mathbf{b} = \begin{pmatrix} 5 \\ 4 \end{pmatrix}$

Write $2\mathbf{a} - 3\mathbf{b}$ as a column vector. **(2 marks)**

2 The diagram shows two vectors **a** and **b**.

$\overrightarrow{PQ} = \mathbf{a} + 2\mathbf{b}$

Draw the vector $\overrightarrow{PQ}$ on squared paper.

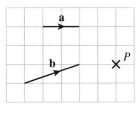

(2 marks)

3 *ABCD* is a parallelogram.
D is the midpoint of *AE*.
$\overrightarrow{AB} = \mathbf{p}$ and $\overrightarrow{AD} = \mathbf{q}$.
Write down in terms of **p** and/or **q**

a $\overrightarrow{AE}$ **(1 mark)**

b $\overrightarrow{AC}$ **(1 mark)**

c $\overrightarrow{CE}$ **(1 mark)**

4 *P* is the point (0, 3). $\overrightarrow{PQ} = \begin{pmatrix} 2 \\ 3 \end{pmatrix}$

a Find the coordinates of *Q*. **(1 mark)**

R is the point (2, 4).

b Express $\overrightarrow{PR}$ as a column vector. **(1 mark)**

5 *A* is the point (3, 4) and *B* is the point (−1, 0).

a Express $\overrightarrow{AB}$ as a column vector. **(1 mark)**

$\overrightarrow{BC} = \begin{pmatrix} 2 \\ 5 \end{pmatrix}$

b Write down the coordinates of point *C*. **(1 mark)**

X is the midpoint of *AB*. *O* is the origin.

c Find $\overrightarrow{OX}$ as a column vector. **(2 marks)**

d What is the magnitude of $\overrightarrow{AB}$?
Give your answer as a surd in its simplest form. **(2 marks)**

6 *ABCDEF* is a regular hexagon.
$\overrightarrow{AB} = \mathbf{a}$, $\overrightarrow{BC} = \mathbf{b}$ and $\overrightarrow{FC} = 2\mathbf{a}$.

a Find in terms of **a** and/or **b**

i $\overrightarrow{FE}$ ii $\overrightarrow{CE}$ **(2 marks)**

$\overrightarrow{CE} = \overrightarrow{EX}$

b Prove that *FX* is parallel to *CD*. **(3 marks)**

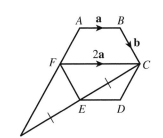

7 *OPQR* is a trapezium.
OR is parallel to *PQ*.
$\overrightarrow{OP} = \mathbf{p}$ and $\overrightarrow{OR} = \mathbf{r}$.
$PQ = 2OR$
X is the point on *PQ* such that
$PX : XQ = 3 : 1$.
Express $\overrightarrow{OX}$ in terms of **p** and/or **q**. **(3 marks)**

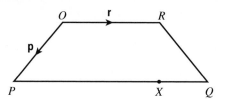

8 *OABC* is a parallelogram.
Y is a point such that $\overrightarrow{AB} = \overrightarrow{BY}$.
The point *X* divides *AC* in the ratio $2 : 1$.
   a Write an expression for $\overrightarrow{OX}$ in terms of **a** and/or **b**. **(3 marks)**
   b Prove that *OXY* is a straight line. **(3 marks)**

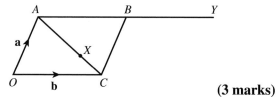

9 *OBC*, *OMA* and *ANB* are straight lines.
$OB : BC = 2 : 3$
M is the midpoint of *OA*.
$\overrightarrow{OA} = \mathbf{a} \qquad \overrightarrow{OB} = \mathbf{b}$
   a Write $\overrightarrow{MC}$ in terms of **a** and/or **b**. **(2 marks)**

$\overrightarrow{NB} = k\overrightarrow{AB}$ where *k* is a scalar quantity.

   b Write $\overrightarrow{NC}$ in terms of *k*, **a** and/or **b**. **(2 marks)**
   c Given that *MNC* is a straight line,
   find the value of *k*.

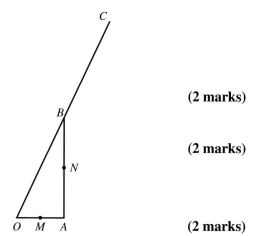

(2 marks)
(TOTAL: 35 marks)

10 Challenge **p** and **q** are two vectors such that
   • $\mathbf{p} \neq \mathbf{q}$
   • magnitude of $\mathbf{p} + \mathbf{q}$ = magnitude of $\mathbf{p} - \mathbf{q}$
   Show that **p** and **q** are perpendicular vectors.

11 Reflect Think back to all units where you have been asked to prove things.
   Choose **A**, **B** or **C** to complete each statement:

I am	**A** good at proof	**B** OK at proof	**C** not very good at proof
I think proof is …	**A** easy	**B** OK	**C** hard
When I think about doing a proof, I feel	**A** confident	**B** OK	**C** unsure

Did you answer mostly **A**s and **B**s?
Are you surprised by how you feel about proof? Why?

Did you answer mostly **C**s?
Find the three questions about proof that you found the hardest.
Ask someone to explain them to you.
Then complete the statements above again.

> **Q11 hint**
>
> You may choose proof questions from this unit or another unit. You could look back at Units 12 and 16 too.

19 Proportion and graphs

Prior knowledge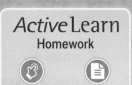

19.1 Direct proportion

- Write and use equations to solve problems involving direct proportion.

*Active*Learn
Homework

Warm up

1 **Fluency** $y = 2z$ and $z = 3x$.
What is y in terms of x?

2 Are x and y in direct proportion? Explain.

x	3	5	8
y	12	20	40

3 **Reasoning** Is cost in direct proportion to the quantity?
If so, write an equation linking cost and quantity.

a 5 apples cost £4.20 and 15 apples cost £12.60.

b 20 GB of data costs £30 and 2 GB of data costs £10.

c 750 screws cost £12 and 1000 screws cost £14.

d 120 units of gas costs £40 and 80 units of gas costs £30.

> **Q3 hint**
>
> Compare ratios.
> cost : quantity

4 The tables show the prices paid for different quantities of euros from two currency exchange websites in July.

travelcash.com

Sterling (£)	50	210	120	400	380	300	250	280
Euros (€)	60	260	150	520	470	390	300	350

currencyexchange.co.uk

Sterling (£)	400	150	30	250	350	300	200	450
Euros (€)	440	160	32	280	390	330	210	500

a Draw a scatter graph for both sets of information on the same axes.

b Draw a line of best fit for each set of data.

c Write a formula for euros, E, in terms of sterling, S, for

　i travelcash.com

　ii currencyexchange.co.uk

> **Q4c hint**
>
> $E = \square S$

d **Reasoning** Which currency exchange website offers better value for money?
Explain your answer.

e **Reflect** Explain how you can tell that two quantities are in direct proportion from

　i their graph　ii an equation

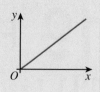

Key point

The symbol $\propto$ means 'is directly proportional to'.
$y \propto x$ means y is directly proportional to x.
In general, if y is directly proportional to x, then $y \propto x$ and $y = kx$,
where k is a number, called the **constant of proportionality**.

Example

y is directly proportional to x. When $y = 20$, $x = 8$.

a Express y in terms of x.

b Find x when $y = 35$.

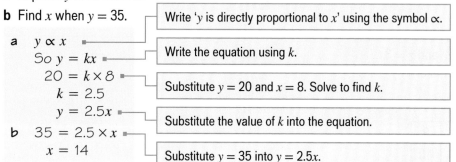

a　$y \propto x$ — Write 'y is directly proportional to x' using the symbol $\propto$.

　　So $y = kx$ — Write the equation using k.

　　$20 = k \times 8$ — Substitute $y = 20$ and $x = 8$. Solve to find k.

　　$k = 2.5$

　　$y = 2.5x$ — Substitute the value of k into the equation.

b　$35 = 2.5 \times x$ — Substitute $y = 35$ into $y = 2.5x$.

　　$x = 14$

5　y is directly proportional to x. $y = 15$ when $x = 3$.

　　a Express y in terms of x.

　　b Find y when $x = 10$.

　　c Find x when $y = 65$.

Q5a hint

Start with the statement $y \propto x$ then write the equation $y = kx$. Use the values of x and y to find the value of k.

6　y is directly proportional to x. $y = 52$ when $x = 8$.

　　a Write a formula for y in terms of x.

　　b Find y when $x = 14$.

　　c Find x when $y = 143$.

7　**Problem-solving** y is directly proportional to x.

　　a $y = 6$ when $x = 4$. Find x when $y = 7.5$.

　　b $y = 31.5$ when $x = 7$. Find x when $y = 45.5$.

　　c $y = 8$ when $x = 5$. Find x when $y = 13$.

8　y is directly proportional to x. When $x = 300$, $y = 10$.

　　a Write a formula for y in terms of x.

　　b Calculate the value of y when x is **i** 5　**ii** 27

　　　Give your answers as fractions in their simplest form.

Exam-style question

9　y is directly proportional to z.
　　When $z = 8$, $y = 60$.
　　z is directly proportional to x.
　　When $x = 4.5$, $z = 90$.
　　Find a formula for y in terms of x.　**(5 marks)**

Exam tip

You need to write two statements of proportionality and two related formulae.

19.2 More direct proportion

*Active*Learn
Homework

- Write and use equations to solve problems involving direct proportion.
- Solve problems involving square and cubic proportionality.

Warm up

1 Fluency Work out **a** 3^3 **b** $\sqrt[3]{8}$ **c** $\sqrt[3]{1000}$

2 $y \propto x$
What are the values of A and B?

x	3	4	B
y	A	18	63

3 $y \propto x$
Use the table of values to find the constant of proportionality, k.

x	6	10	12
y	15	25	30

---

 4 Problem-solving The force, F, on a mass is directly proportional to the acceleration, a, of the mass. When $F = 96$, $a = 12$.

a Express F in terms of a.

b Find F when $a = 20$.

c Find a when $F = 112$.

> **Q4a hint**
> $F \propto a$
> So $F = ka$

 5 Reasoning The table gives information about the perimeter, P, of a shape and the length, l, of one of its sides.

Perimeter, P (cm)	12	24	30	48
Length, l (cm)	5	10	12.5	20

a Show that P is directly proportional to l.

b Given that $P = kl$, work out the value of k.

c Write a formula for P in terms of l.

d Use your formula to work out

  i the value of P when $l = 18$ **ii** the value of l when $P = 42$

> **Q5a hint**
> Use equivalent ratios.
> $12 : 5 = 24 : \square = \square : \square = \square : \square$

---

 6 Reasoning The distance, d (in km), covered by an aeroplane is directly proportional to the time taken, t (in hours).
The aeroplane covers a distance of 1600 km in 3.2 hours.

a Write a formula for d in terms of t.

b Find the value of d when $t = 5$.

c Find the value of t when $d = 2250$.

d Reflect What happens to the distance travelled, d, when the time, t, is

  i doubled   **ii** halved?

---

 7 Reasoning The cost, C (in £), of a newspaper advert is directly proportional to the area, A (in cm^2), of the advert.
An advert with an area of 40 cm^2 costs £2000.

a Sketch a graph of C against A.   **b** Write a formula for C in terms of A.

c Use your formula to work out the cost of an 85 cm^2 advert.

8 y is directly proportional to the square of x.
When $x = 3$, $y = 36$.

    a Write the statement of proportionality.
    b Write an equation using k.
    c Work out the value of k.
    d Find y when $x = 5$.
    e Find x when $y = 25$.

> **Q8a hint**
>
> $\square \propto \square^2$

 9 **Problem-solving** y is directly proportional to the cube of x.
When $x = 2$, $y = 28.8$.

    a Write a formula for y in terms of x.
    b Find y when $x = 4$.
    c Find x when $y = 450$.

> **Q9a hint**
>
> $\square \propto \square^3$

 10 **Problem-solving** y is directly proportional to the square root of x.
When $x = 4$, $y = 50$.

    a Write a formula for y in terms of x.
    b Find y when $x = 9$.
    c Find x when $y = 250$.

> **Exam-style question**
>
> **11** y is directly proportional to the cube root of x.
> When $x = 125$, $y = \frac{1}{20}$.
> Find the value of y when $x = 8000$.   **(3 marks)**

> **Exam tip**
>
> It is important to start your working with a statement of proportionality and then a related formula. Always check you have considered any square, square root, cube or cube root.

12 **Problem-solving** The cost of fuel per hour, C (in £), to propel a boat through the water is directly proportional to the cube of its speed, s (in mph).
A boat travelling at 10 mph uses £50 of fuel per hour.

    a Write a formula for C in terms of s.

    b Calculate C when the boat is travelling at 5 mph.

13 **Reasoning / Future skills** In a factory, chemical reactions are carried out in spherical containers. The time, T (in minutes), the chemical reaction takes is directly proportional to the square of the radius, R (in cm), of the spherical container.
When $R = 120$, $T = 32$.

    a Write a formula for T in terms of R.

    b Find the value of T when $R = 150$.

14 **Reasoning**
In an experiment, measurements of g and h were taken.
Which of these relationships fits the results?

    $g \propto h$ $g \propto h^2$ $g \propto h^3$ $g \propto \sqrt{h}$ $g \propto \sqrt[3]{h}$

h	2	5	7
g	24	375	1029

19.3 Inverse proportion

- Write and use equations to solve problems involving inverse proportion.
- Use and recognise graphs showing inverse proportion

Warm up

1 Fluency What is the cube root of $8\,\text{m}^3$?

2 $a = \dfrac{20}{c^2}$ and $c = 2b$.

Write a in terms of b.
Give your answer in its simplest form.

> **Q2 hint**
>
> $c = 2b$
> So $c^2 = \square$

3 A is directly proportional to B.
$A = 10$ when $B = 40$.

a Write a formula for A in terms of B.

b Use your formula to work out the value of A when $B = 460$.

4 Match each formula to one of the graphs below.

a $y = kx$ **b** $y = kx^2$ **c** $y = \dfrac{k}{x}$

A 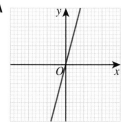   **B** 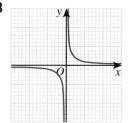   **C**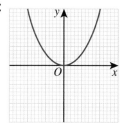

5 Reasoning y is inversely proportional to x.

x	0.25	0.5	1	2	4	8	16	32
y	32	16	8	4	2	1	0.5	0.25

a Draw a graph of y and x.

b $y = \dfrac{k}{x}$ where k is the constant of proportionality. Find k.

c Work out $x \times y$ for each pair of values in the table.

d Reflect How can you tell from a table of values if two variables are inversely proportional?

Key point

When y is **inversely proportional** to x

$y \propto \dfrac{1}{x}$ and $y = \dfrac{k}{x}$

where k is the **constant of proportionality**.

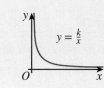

$y = \frac{k}{x}$

Example

y is inversely proportional to x.
When $y = 2$, $x = 3$.

a Write a formula for y in terms of x.

b Calculate the value of y when $x = 8$.

a $\quad y \propto \dfrac{1}{x}$ so $y = \dfrac{k}{x}$

> Write 'y is inversely proportional to x' using the $\propto$ symbol. Then write the equation using k.

$\quad 2 = \dfrac{k}{3}$ so $k = 6$

> Substitute $y = 2$ and $x = 3$. Solve to find k.

$\quad y = \dfrac{6}{x}$

> Substitute k into the equation.

b $\quad y = \dfrac{6}{8} = \dfrac{3}{4}$

> Substitute $x = 8$ into your formula.

6 $\quad y$ is inversely proportional to x.
When $y = 5$, $x = 2$.

a Write a formula for y in terms of x.

b Calculate the value of y when $x = 20$.

c Calculate the value of x when $y = 4$.

 7 $\quad$**Future skills** The pressure, p (in N/m^2), of a gas is inversely proportional to the volume, V (in m^3).
$p = 1500\,\text{N/m}^2$ when $V = 2\,\text{m}^3$.

a Write a formula for p in terms of V.

b Work out the pressure when the volume of the gas is $1.5\,\text{m}^3$.

c Work out the volume of gas when the pressure is $1200\,\text{N/m}^2$.

d What happens to the volume of the gas when the pressure doubles?

 8 $\quad$**Problem-solving / Future skills** The time taken, t (in seconds), to boil water in a kettle is inversely proportional to the power, p (in watts), of the kettle.
A full kettle of power 1500 W boils the water in 400 seconds.

a Write a formula for t in terms of p.

b **Reasoning** A similar kettle has a power of 2500 W.
Can this kettle boil the same amount of water in less than 3 minutes?

9 $\quad$**Reasoning** The time, t (in seconds), it takes an object to travel a fixed distance is inversely proportional to the speed, s (in m/s), at which the object is travelling.
When travelling at 20 m/s it takes an object 40 seconds to travel from A to B.

a Write a formula for s in terms of t.

b Copy and complete the table of values for s and t.

Speed, s (m/s)	4		20	40		160
Time, t (seconds)		80	40		10	

c Sketch a graph to show how s varies with t.

d **Reflect** Explain what happens to the time taken as the speed approaches 0 m/s.

10 Problem-solving The graph shows two variables that are inversely proportional to each other. Find the values of a and b.

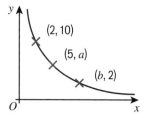

11 Problem-solving y is inversely proportional to the cube of x. When $y = 2$, $x = 3$.

a Write a formula for y in terms of x.

b Calculate y when $x = 5$.

c Calculate x when $y = 6.75$.

Q11a hint

$\square \propto \dfrac{1}{\square^3}$

12 Problem-solving y is inversely proportional to the square root of x. When $y = 2$, $x = 9$.

a Write a formula for y in terms of x.

b Calculate y when $x = 4$.

c Calculate x when $y = 6$.

Q12a hint

$\square \propto \dfrac{1}{\sqrt{\square}}$

Exam-style question

13 The table shows a set of values for r and h.

r	1	2	3	4
h	36	9	4	$2\frac{1}{4}$

h is inversely proportional to the square of r.

a Find an equation for h in terms of r. **(1 mark)**

b Find the positive value of r when $h = 144$. **(2 marks)**

Exam tip

Sometimes there is more than one way to find an answer. For example, you can use any pairs of values for r and h to find an equation. Show your working so that the examiner can clearly understand the method you have used.

---

14 Problem-solving The speed, s (in revolutions per minute), at which each cog in a machine turns is inversely proportional to the square of the radius, r (in cm). When $r = 4$ cm, $s = 212.5$ revolutions per minute.

a Write a formula for s in terms of r.

b Work out the value of s when $r = 4.2$. Round your answer to 2 decimal places.

Q15 hint

Work out k in terms of a. Write a formula for y in terms of x, involving a.

15 Problem-solving y is inversely proportional to x^3. $y = 28$ when $x = a$. Show that $y = 3.5$ when $x = 2a$.

Exam-style question

16 y is inversely proportional to the square of p. When $p = 10$, $y = 6$. p is directly proportional to x. When $x = 4$, $p = 20$. Find a formula for y in terms of x. Give your answer in its simplest form. **(5 marks)**

Exam tip

Read the question carefully. Note whether quantities are in direct or inverse proportion.

19.4 Exponential functions

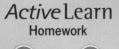

- Recognise graphs of exponential functions.
- Sketch graphs of exponential functions.

Warm up

1 **Fluency** Work out the value of **a** 2^3 **b** 2^4 **c** 3^0 **d** 4^{-1} **e** 4^{-2}

2 A man places a grain of rice on the first square of a chessboard. He then places two grains on the second square, four on the third square, doubling the number of grains each time. How many grains of rice does he place on

a the 10th square **b** the 15th square **c** the 20th square?

3 Find the value of x for each of these equations.

a $2^x = 8$ **b** $3^x = 81$ **c** $10^x = 10\,000$

Key point

Functions of the form $f(x) = a^x$, where a is a positive number, are called **exponential functions**.

4 **a** Copy and complete the table of values for $y = 2^x$.
Give the values correct to 2 decimal places.

x	-4	-3	-2	-1	0	1	2	3	4
y									

b Draw the graph of $y = 2^x$ for $-4 \leqslant x \leqslant 4$.

c Use the graph to find an estimate for

i the value of y when $x = 3.5$ **ii** the value of x when $y = 10$

d **Reflect** Explain what happens to the value of y as the value of x decreases.

Key point

The graph of an exponential function has one of these shapes.

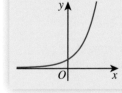

$y = a^x$ where $a > 1$ or
$y = b^{-x}$ where $0 < b < 1$
exponential growth

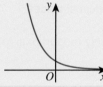

$y = a^{-x}$ where $a > 1$ or
$y = b^x$ where $0 < b < 1$
exponential decay

5 **Reasoning** What is the equation of the line $y = a^x$ when $a = 1$?

6 **Reasoning**

a Draw the graph of each function. Use a grid with x-axis from -2 to 2 and y-axis from 0 to 30.

i $y = 3^x$ **ii** $y = 5^x$

b Predict where the graph of $y = 4^x$ would be. Sketch it on the same axes.

c At which point do all the graphs intersect the y-axis?

d **Reflect** Explain why exponential graphs always cross the y-axis at the same point.

Exam-style question

7 Sketch the graph of $y = \left(\frac{3}{2}\right)^x$.

Give the coordinates of any points of intersection with the axes. **(2 marks)**

Exam tip

For a sketch of a graph, use a coordinate axis like this. Carefully sketch the correct shape for the graph. Write coordinates $(\square, \square)$ on the graph at any point of intersection with the axes.

8 a Copy and complete the table of values for $y = 2^{-x}$.
Give the values correct to 2 decimal places.

x	-4	-3	-2	-1	0	1	2	3	4
y									

b Draw the graph of $y = 2^{-x}$ for $-4 \leqslant x \leqslant 4$.

c Use the graph to find an estimate for
  i the value of y when $x = 3.5$ **ii** the value of x when $y = 10$

9 Reasoning The equation of a curve is $y = a^x$.
P is the point where the curve intersects the y-axis.

a State the coordinates of P.   **b** Does the curve $y = a^{-x}$ also pass through P? Explain.

10 Reasoning / Future skills
The table gives information about the count rate of seaborgium-266.

Time (seconds)	0	30	60	90	120	150	180
Count rate	800	400	200	100	50	25	12.5

The count rate is the number of radioactive emissions per second.

a Draw a graph of the data. Plot time on the horizontal axis and count rate on the vertical axis.

b Is this an example of exponential growth or exponential decay?

c The half-life of a radioactive material is the time it takes for the count rate to halve.
What is the half-life of seaborgium-266?

Example

The sketch shows part of the graph $y = ab^x$, where a and b are constants and $b > 0$.
The points with coordinates $(0, 3)$ and $(2, 12)$ lie on the graph.
Work out the values of a and b.

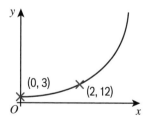

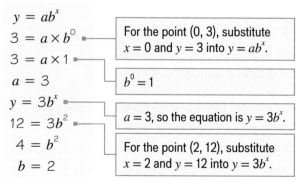

$y = ab^x$

$3 = a \times b^0$ ◂── For the point $(0, 3)$, substitute $x = 0$ and $y = 3$ into $y = ab^x$.

$3 = a \times 1$ ◂──

$a = 3$ ── $b^0 = 1$

$y = 3b^x$ ◂──

$12 = 3b^2$ ◂── $a = 3$, so the equation is $y = 3b^x$.

$4 = b^2$ ── For the point $(2, 12)$, substitute $x = 2$ and $y = 12$ into $y = 3b^x$.

$b = 2$

11 **Problem-solving** The equation of a graph is $y = ka^x$, where k and a are constants and $a > 0$.
The curve passes through (1, 10) and (3, 250).

   a Write two equations in terms of k and a.

   b Solve the equations to work out the value of a.

   c Work out the value of k.

Q11b hint

Divide one equation by the other to eliminate k.

Exam-style question

12 The sketch shows a curve with equation $y = ka^x$, where k and a are constants and $a > 0$.

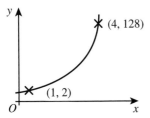

The curve passes through the points (1, 2) and (4, 128).

   a Calculate the value of k and the value of a. **(3 marks)**

   b Find the value of y when $x = -2$. **(1 mark)**

Exam tip

Clearly show every step of your working.

---

13 **Problem-solving** The value, V (in £), of a car depreciates exponentially over time.
The value of the car on 1 January 2018 was £20 000.
The value of the car on 1 January 2020 was £16 200.
The sketch graph shows how the value of the car changes over time.
The equation of the graph is $V = ab^t$,
where t is the number of years after 1 January 2018, and a and b are positive constants.

   a Use the information to find the values of a and b.

   b Use your values of a and b in the formula $V = ab^t$ to estimate the value of the car on 1 January 2021.

14 **Problem-solving** The population of a country is currently 4 million, and is growing at a rate of 5% a year.
The expected population, p (in millions), in t years' time, is given by the formula $p = 4 \times 1.05^t$.

   a Use a table of values to draw the graph of p against t for the next 6 years.

   b Use your graph to estimate

      i the size of the population after 2.5 years

      ii the time taken for the population to reach 5 million

   c **Reasoning** Explain how you know that the growth in population is exponential.

15 **Future skills** £10 000 is invested in a savings account paying 2% compound interest a year.

   a Write a formula for the value of the savings account (V) and the number of years (t).

   b Draw a graph of V against t for the first 10 years.

   c Use the graph to estimate when the investment will reach a value of £11 000.

19.5 Non-linear graphs

- Match equations to graphs.
- Calculate the gradient of a tangent at a point.
- Estimate the area under a non-linear graph.

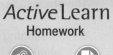
Active Learn
Homework

Warm up

1 **Fluency** Calculate the area of this shape.

7 cm

5 cm

3 cm

2 This flask fills up with water.
h is the height of the water after time t.
Which graph shows the correct relationship between h and t?

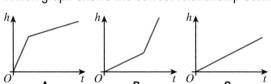

A   B   C

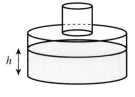

h

3 The velocity–time graph shows the
motion of a particle.
Given that the particle travelled in
a straight line, calculate the distance
travelled between $t = 0$ and $t = 12$.

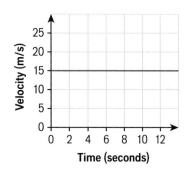

4 Points A and B are connected by a straight line.
Write the gradient of the line AB for

  a $A(0, 0)$ and $B(2, 6)$ **b** $A(3, 5)$ and $B(7, 11)$ **c** $A(-2, 4)$ and $B(2, 0)$

Exam-style question

5 Here are some graphs.

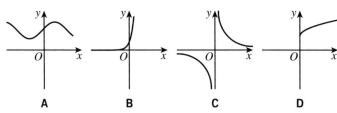

A   B   C   D

Match each equation to a graph.

  i $y = \sqrt{x} + 4$ **ii** $y = \sin x + 4$ **iii** $y = 4^x$ **iv** $y = \dfrac{4}{x}$ **(3 marks)**

6 **Reasoning** The diagrams show two students' tangents to the graph $y = x^2 - x + 1$ at point P, where $x = 1$.

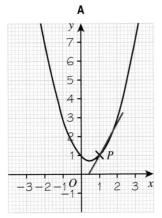

A

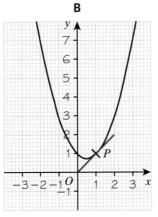

B

a Which student has drawn the more accurate tangent? Explain.

b Calculate an estimate for the gradient of the graph at point P.

Example

Water is poured into the container at a constant rate.

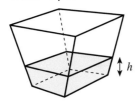

The graph shows the height, h (in cm), of the water after time t (in seconds).
Estimate the rate at which h is increasing after 10 seconds.

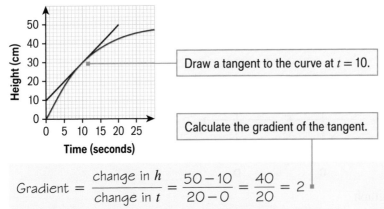

Draw a tangent to the curve at $t = 10$.

Calculate the gradient of the tangent.

$$\text{Gradient} = \frac{\text{change in } h}{\text{change in } t} = \frac{50 - 10}{20 - 0} = \frac{40}{20} = 2$$

At $t = 10$ the height of the water is increasing at 2 cm per second.

7 **Problem-solving / Reasoning** The graph shows the relationship between the temperature, T (in °C), of a cup of coffee and time, t (in seconds).

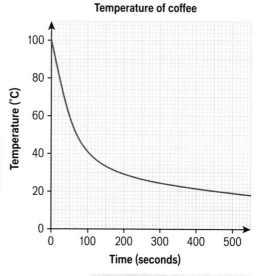

Temperature of coffee

a Describe how the temperature changes over time.

b Work out the average rate of temperature reduction over the first 300 seconds.

> **Q7b hint**
>
> First work out the temperature decrease from 0 to 300 seconds. Give your answer in °C/second.

c Copy the graph. Draw a tangent to the curve at $t = 300$. Use the tangent to estimate the rate of temperature reduction at exactly 300 seconds.

d **Reflect** Write a sentence comparing the average rate of temperature reduction over the first 300 seconds with the rate of temperature reduction at exactly 300 seconds.

> **Q7d hint**
>
> Which is faster?

> **Key point**
>
> On a distance–time graph, the gradient of the tangent at any point gives the speed at that time.

8 **Reasoning** The distance–time graph shows information about a runner in a 100 m race.

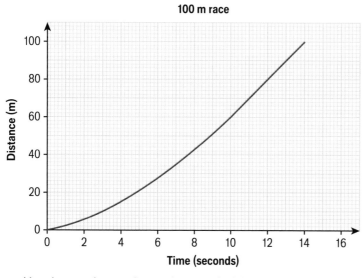

100 m race

> **Q8 hint**
>
> Carefully read any question that asks for a speed from a distance–time graph. Are you being asked for an average speed over time or an estimate of speed at an exact time?

a Use the graph to estimate the speed of the runner 6 seconds into the race.

b After how many seconds is the runner running at full speed? Explain how you know.

c If she could maintain this speed, how long would it take to run a 200 m race?

> **Key point**
>
> The straight line that connects two points on a curve is called a **chord**. The gradient of the chord gives the average rate of change and can be used to find the average rate of change between two points. For example, on a distance–time graph, the gradient of a chord gives the average speed between two times.
>
>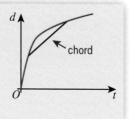

9 **Problem-solving / Reasoning** A car drives away from a set of traffic lights. The velocity–time graph gives some information about the motion of the car.

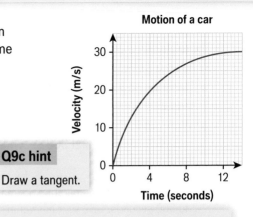
Motion of a car

a Copy the graph. Draw a chord from $t = 0$ to $t = 10$.

b Use the chord from part **a** to calculate the average acceleration of the car over the first 10 seconds.

c Estimate the acceleration at time $t = 2$ seconds.

Q9c hint
Draw a tangent.

d Describe how the acceleration changes over the 12 seconds.

Key point

The area under a velocity–time graph shows the displacement, or distance from the starting point. To estimate the area under a part of a curved graph, draw a chord between the two points you are interested in, and straight lines down to the horizontal axis to create a trapezium. The area of the trapezium is an estimate for the area under this part of the graph.

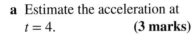

---

10 **Problem-solving** For the car in **Q9**, estimate the displacement from

a $t = 0$ to $t = 4$ **b** $t = 4$ to $t = 8$ **c** $t = 8$ to $t = 12$

11 **Problem-solving** The graph shows the speed of a van, in metres per second, during the first 15 seconds of a journey.

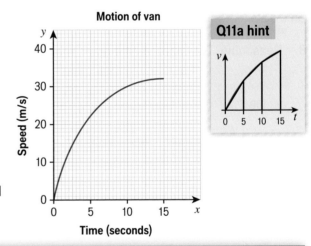
Motion of van

a Work out an estimate for the distance the van travelled in the first 15 seconds.
Use three strips of equal width.

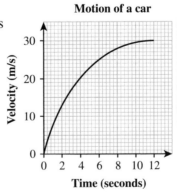
Q11a hint

b **Reasoning** Is your answer to part **a** an underestimate or an overestimate of the actual distance the van travelled in the first 15 seconds?
Give a reason for your answer.

Exam-style question

12 The velocity–time graph describes the motion of a car.
Velocity, v, is measured in metres per second (m/s) and time, t, is measured in seconds. The car travels in a straight line.

Motion of a car

Exam tip

Clearly show any marks on a graph that you have used to answer exam questions (for example, any tangents, chords or trapezia).

a Estimate the acceleration at $t = 4$.   **(3 marks)**

b Estimate the distance travelled between $t = 6$ and $t = 8$.   **(2 marks)**

The acceleration at exact time T is equal to the average acceleration for the first 8 seconds.

c Find an estimate for the value of T.   **(2 marks)**

19.6 Translating graphs of functions

*Active*Learn
Homework

- Understand the relationship between translating a graph and the change in its function notation.

Warm up

1 **Fluency** Find the new coordinates when

  a the point $(1, 2)$ is translated by $\begin{pmatrix} 3 \\ 0 \end{pmatrix}$
  b the point $(-3, 4)$ is translated by $\begin{pmatrix} 0 \\ -2 \end{pmatrix}$

2 $f(x) = 3x + 1$ and $g(x) = 2x^2$.
  Find the value of
  a $f(2)$
  b $f(-3)$
  c $g(1)$
  d $f(0)$
  e $g(0)$

3 $g(x) = 5x + 2$
  a Find the value of
    i $g(3) + 2$
    ii $g(3 + 2)$
  b Write out in full
    i $y = g(x) + 2$
    ii $y = g(x + 2)$

4 Match each equation to one of the graphs below.
  a $y = \sin x + 0.5$
  b $y = \sin(x + 30)$

A
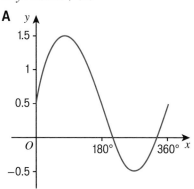

B
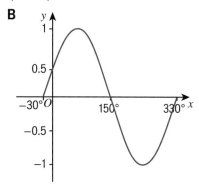

5 $f(x) = x^2$

  a Draw the graph of $y = f(x)$ on a coordinate grid with x-axis from -5 to $+5$ and y-axis from -10 to $+30$.

  b On the same set of axes draw the graphs of
    i $y = f(x) - 5$
    ii $y = f(x + 1)$

  c The minimum point of $y = f(x)$ is $(0, 0)$.
    Write the coordinates of the minimum point of
    i $y = f(x) - 5$
    ii $y = f(x + 1)$

  d Describe the transformation that maps the graph of $y = f(x)$ to the graph of
    i $y = f(x) - 5$
    ii $y = f(x + 1)$

Q5a and b hint

Create a table of values for each graph.

Q5b i hint

$y = x^2 - 5$

Q5b ii hint

$y = (x + 1)^2$

6 **Problem-solving** Here is the graph of $y = f(x) = x^2$.

Copy the graph.
On the same axes sketch the graphs of

a $y = f(x) + 1$

b $y = f(x) - 2$

c $y = f(x + 2)$

d $y = f(x - 4)$

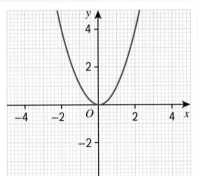

7 **Reasoning** Write the vector that translates $y = f(x)$ to

a $y = f(x) + 2$ b $y = f(x) - 3$ c $y = f(x + 1)$ d $y = f(x - 4)$ e $y = f(x + 5) - 2$

Example

Graph A is a translation of the graph of $y = f(x)$.
Write the equation of graph A.

Find corresponding points on the two graphs that are easy to read, for example where each graph is horizontal.

Write your final answer in function notation.

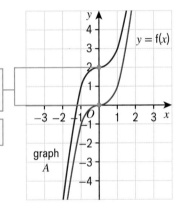

The graph of $y = f(x)$ has been translated by $\begin{pmatrix} 0 \\ 2 \end{pmatrix}$.

The equation of graph A is $y = f(x) + 2$.

Exam-style question

8 The graph of $y = f(x)$ is shown on the grid.

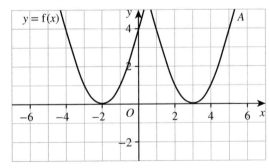

Graph A is a translation of the graph of $y = f(x)$.
Write down the equation of graph A.

(1 mark)

9   The graph of $y = f(x)$ is shown on the grid.

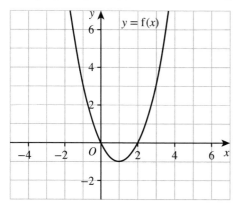

Copy the diagram and sketch the graph of $y = f(x-1)$.   **(1 mark)**

10 **Problem-solving** Here is a sketch of the graph of $y = f(x) = x^3$.

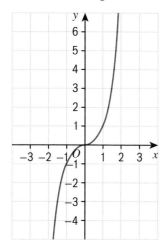

a Draw sketches of the graphs of
   i $y = f(x) + 2$
   ii $y = f(x - 3)$
b Write the coordinates of the point which $(0, 0)$ is mapped to for both graphs.

11 **Reasoning** $f(x) = 3x + 2$
   a Draw the graph of $y = f(x)$.
   b Draw the graph of $y = f(x + 1)$.
   c Write the algebraic equation of $y = f(x + 1)$.

12 **Problem-solving** $f(x) = \dfrac{1}{x}$
   a Sketch the graph of $y = f(x + 2) - 3$.
   b Write the equation of each asymptote.

Q12 hint

An **asymptote** is a line that a curve approaches but never reaches.

19.7 Reflecting graphs of functions

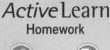
- Understand the effect reflecting a curve in one of the axes has on its function form.

Warm up

1 **Fluency** Find the image of the point $(-1, -2)$ when it is

   a reflected in the x-axis   **b** reflected in the y-axis   **c** translated by $\begin{pmatrix} 0 \\ 5 \end{pmatrix}$

2 $f(x) = 6x - 4$
Write out in full

   a $f(-x)$ **b** $-f(x)$

3 $g(x) = 2x + 5$
Find

   a $g(2)$ **b** $-g(-3)$

4 Here is the graph of $y = \cos x$.

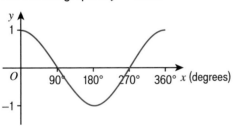

Sketch the graph of

   a $y = -\cos x$ **b** $y = -\cos x + 1$

---

5 $f(x) = 4x - 2$

   a Copy and complete the table.

x	−2	−1	0	1	2
$f(x)$					
$-f(x)$					
$f(-x)$					

   b On the same set of axes, draw the graphs of

     i $y = f(x)$

     ii $y = -f(x)$

     iii $y = f(-x)$

   c Describe the transformation that maps $f(x)$ to $-f(x)$.

   d Describe the transformation that maps $f(x)$ to $f(-x)$.

6   **Problem-solving** The diagram shows the graph of $y = f(x)$.

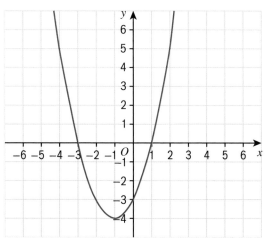

a   Copy the sketch.
    On the same axes sketch the graphs of
    i $y = -f(x)$
    ii $y = f(-x)$

b   **Reasoning** Finley says, 'The graphs of $y = f(x)$ and $y = -f(x)$ always intersect the y-axis in the same place.'
    Is Finley right? Explain your answer.

7   **Reasoning** The diagram shows the graph of $y = f(x)$.
    The turning point of the curve is $A(2, 4)$.

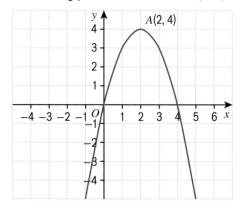

Write the coordinates of the turning point of the curve with equation
a $y = -f(x)$
b $y = f(-x)$
c $y = -f(-x)$

8 **Problem-solving** Here is the graph of $y = f(x) = x^3 + 2$.

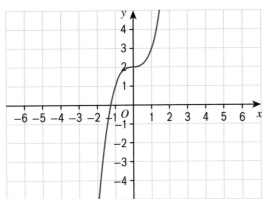

a Copy the graph and on the same axes sketch the graphs of

   i $-f(x)$ **ii** $f(-x)$ **iii** $-f(-x)$

b Describe the transformation that maps $f(x)$ to $-f(-x)$.

9 On the grid, graph G has been reflected to give graph A.
Graph A has been translated to give graph B.
The equation of graph G is $y = g(x)$.

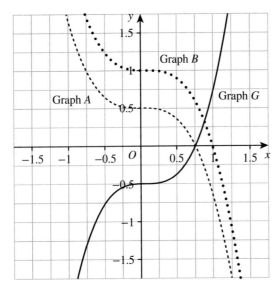

From the box, choose the equation of

a graph A **(1 mark)**

b graph B **(1 mark)**

$y = -g(x)$	$y = -g(-x)$	$y = g(-x)$
$y = -g(x+\frac{1}{2})$	$y = -g(x)+\frac{1}{2}$	$y = g(-x+\frac{1}{2})$
$y = g(-x)+\frac{1}{2}$	$y = -g(-x+\frac{1}{2})$	$y = g(-x+\frac{1}{2})$

10 The graph of the curve C with equation $y = f(x)$ is transformed to give the graph of the curve S with equation $y = f(-x) + 5$.
Point P on curve C with coordinates $(2, -1)$ is mapped to point Q on curve S.
Find the coordinates of Q.   **(2 marks)**

19 Check up

Active Learn
Homework

Proportion

1 In a circuit with a fixed resistance, the current, I, is directly proportional to the voltage, V.
When the current is 10 amps, the voltage is 4 volts.

   a Write a formula for I in terms of V.

   b Calculate the current when the voltage is 10 volts.

   When the voltage is constant and the resistance is allowed to vary, the current is inversely proportional to the resistance, R. When the resistance is 20 ohms, the current is 2 amps.

   c Write a formula for I in terms of R.

   d Calculate the current when the resistance is 4 ohms.

2 y is directly proportional to $\sqrt{x}$.
When $y = 48$, $x = 4$.

   a Write a formula for y in terms of x.

   b Find y when $x = 9$.

   c Find x when $y = 168$.

3 c is inversely proportional to d^3.
When $c = 5.5$, $d = 4$.

   a Write a formula for c in terms of d.

   b Find c when $d = 5$.

Exponential and other non-linear graphs

4 On the same axes, sketch the graphs of

   a $y = 2^x$ **b** $y = 2^{-x}$ **c** $y = 3^x$

5 The diagram shows a sketch of the curve $y = ab^x$.
The curve passes through the points $A(0, 3)$ and $B(1, 6)$.

   a Find the values of

   i a **ii** b

   b Find the value of y when $x = 4$.

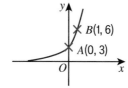

6 The velocity–time graph shows a car travelling in a straight line away from a junction.
The time after the junction, t, is measured in seconds, s.
The velocity, v, is measured in metres per second, m/s.

   a Calculate the average acceleration of the car between $t = 20$ and $t = 30$.

   b Estimate the acceleration at $t = 40$.

   c Estimate the distance travelled between $t = 40$ and $t = 60$.

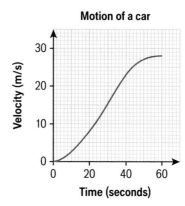

Motion of a car

Unit 19 Proportion and graphs   273

Transformations of graphs of functions

7 The function $y = f(x)$ is shown in the diagram.
Sketch the graph of

a $y = f(x) + 3$

b $y = f(x - 2)$

c $y = f(-x)$

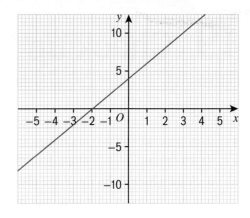

8 $y = f(x)$
The graph of $y = f(x)$ is shown on the grid.

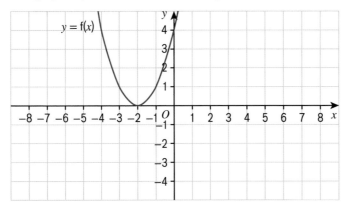

Copy the diagram and sketch the graph of

a $y = f(x - 3)$ **b** $y = -f(x)$

9 **Reflect** How sure are you of your answers? Were you mostly

Just guessing 😞   Feeling doubtful 😐   Confident 🙂

What next? Use your results to decide whether to strengthen or extend your learning.

Challenge

10 Antony transforms the graph $y = 4x + 3$ using all four transformations shown on the cards.

| Reflect in the x-axis | Translate $\begin{pmatrix} 0 \\ 4 \end{pmatrix}$ | Reflect in the y-axis | Translate $\begin{pmatrix} 1 \\ 0 \end{pmatrix}$ |

a Does the order of the transformations change the final result?
There is more than one order that maps the graph to itself.

b Find the different orders of transformations that map the graph to itself.

c Explain why applying the transformations in different orders can lead to the same outcome.

> **Q10a hint**
> Draw a sketch and try different orders.

> **Q10c hint**
> Use a combination of algebraic and function notation.

19 Strengthen

Active Learn
Homework

Proportion

1 Reasoning

a i Write the lengths, l, and widths, w, of different rectangles with an area, A, of $24\,\text{cm}^2$.

ii Is l directly or inversely proportional to w?

iii Draw a set of axes with l on the horizontal axis and w on the vertical axis. Draw a graph of possible lengths and widths of a rectangle with area $24\,\text{cm}^2$.

b i Write the areas, A, and lengths, l, of different rectangles with a width of $5\,\text{cm}$.

ii Is A directly or inversely proportional to l?

iii Draw a set of axes with A on the vertical axis and l on the horizontal axis. Draw a graph of possible areas and lengths of a rectangle with width $5\,\text{cm}$.

> **Q1a ii hint**
> $l = \square w$ or $l = \dfrac{\square}{w}$?

> **Q1b ii hint**
> $A = \square l$ or $A = \dfrac{\square}{l}$?

2 Write **i** the statement of proportionality **ii** the formula for each of these. The first two are started for you.

a A is directly proportional to B. **i** $A \propto \square$ **ii** $A = k\square$

b C is inversely proportional to D. **i** $C \propto \dfrac{1}{\square}$ **ii** $C = \dfrac{k}{\square}$

c M is directly proportional to the square of N.

d F is inversely proportional to the cube of G.

e H is inversely proportional to the square root of T.

f R is directly proportional to the cube of S.

> **Q2c hint**
> $M \propto \square^2$

3 F is directly proportional to a. $F = 20$ when $a = 2$.
$F \propto a$ so $F = ka$, where k is the constant of proportionality.

a Substitute $F = 20$ and $a = 2$ into $F = ka$.

b Find the value of k.

c Use your value of k from part **b** to write a formula for F in terms of a.

d Work out the value of F when $a = 4$.

e Work out the value of a when $F = 60$.

4 a is inversely proportional to b. When $a = 10$, $b = 2$.
$a \propto \dfrac{1}{b}$ so $a = \dfrac{k}{b}$, where k is the constant of proportionality.

a Substitute $a = 10$ and $b = 2$ into the formula $a = \dfrac{k}{b}$.

b Find the value of k.

c Use your value of k from part **b** to write a formula for a in terms of b.

d Calculate **i** a when $b = 5$ **b** b when $a = 5$

5 d is directly proportional to the square of t.

$d = 80$ when $t = 4$.

$d \propto t^2$ so $d = kt^2$, where k is the constant of proportionality.

a Find the value of k.

b Write a formula for d in terms of t.

c Calculate the value of d when $t = 7$.

d Calculate the positive value of t when $d = 45$.

Q5 hint

The 'square of t' means t^2.

Exponential and other non-linear graphs

1 **Future skills** The number of bacteria, n, in a Petri dish doubles every minute.
At time $t = 0$ minutes there is 1 bacterium in the Petri dish.

a Copy and complete the table of values.

b Use a set of axes with n on the vertical axis and t on the horizontal axis to draw a graph of t and n.

t	0	1	2	3	4
n	1				

c Which of these graphs have you plotted?

A $t = 2^n$ **B** $n = 2^t$ **C** $n = t^2$ **D** $n = 2^{-t}$ **E** $n = 8000 \times 2^{-t}$ **F** $n = 8000^{-t}$

d As time, t, increases, does the number of bacteria, n, grow (increase) or decay (decrease)?

e Use your answer to part **d** to state if this is an example of exponential growth or decay.

f The number of bacteria, n, in another Petri dish halves every minute.
At time $t = 0$ minutes there are 8000 bacteria in the Petri dish.
Answer parts **a** to **e** for these bacteria.

2 **Reasoning** The diagram shows a sketch of the curve $y = ab^x$.
The curve passes through the points $A(0, 4)$ and $B(1, 8)$.

a Substitute the values of x and y from point A into $y = ab^x$.

b Find the value of a.

c Substitute the values of x and y from point B and your value of a from part **b** into $y = ab^x$.

d Find the value of b.

e Substitute your values of a and b into $y = ab^x$.

f Find the value of y when $x = 3$.

Q2b hint

$b^0 = 1$

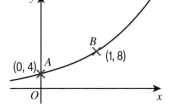

3 **Reasoning** The velocity–time graph shows a train pulling away from a station.
The train travels in a straight line.

a What is the velocity of the train at

   i point A

   ii point B

   iii point C?

b Use the formula

$$\text{acceleration} = \frac{\text{change in velocity}}{\text{time taken}}$$

to estimate the acceleration between points A and B.

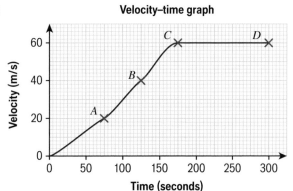

Velocity–time graph

4 Look at the velocity–time graph in **Q3**.

a Estimate the distance travelled between C and D.

b Estimate the distance travelled between B and C.

Q4a hint

Distance travelled = area under the graph

Q4b hint

Read off the values and draw a trapezium. Find the area of the trapezium underneath the line BC.

Transformations of graphs of functions

1 $f(x) = x^2$

a Copy and complete the table of values for $f(x)$.

x	−4	−3	−2	−1	0	1	2	3	4
$f(x)$	16								

b Sketch the graph of $y = f(x)$.

c Copy and complete the table of values for $-f(x)$.

x	−4	−3	−2	−1	0	1	2	3	4
$-f(x)$	−16								

d Sketch the graph of $y = -f(x)$.

e Describe how $y = f(x)$ is transformed into $y = -f(x)$.

2 $f(x) = x + 1$

a Copy and complete the table of values for $f(x)$.

b Sketch the graph of $y = f(x)$.

x	−3	−2	−1	0	1	2	3
$f(x)$							

c Copy and complete the table of values for $f(-x)$.

x	−3	−2	−1	0	1	2	3
$f(-x)$							

Q2c hint

$f(-x) = (-x) + 1$

d Sketch the graph of $y = f(-x)$.

e Describe how $y = f(x)$ is transformed into $y = f(-x)$.

3 $f(x) = 2x - 4$

a Copy and complete the table of values for $f(x)$.

b Sketch the graph of $y = f(x)$.

x	−3	−2	−1	0	1	2	3
$f(x)$							

c Draw a table of values for $f(x) + 3$.

d Copy and complete the sentence.
For the same values of x, the values of $f(x) + 3$ are always _____ than the values of $f(x)$.

e Sketch the graph of $y = f(x) + 3$.

f Describe how $y = f(x)$ is transformed into $y = f(x) + 3$.

g Draw a table of values for $f(x + 2)$.

h Sketch the graph of $y = f(x + 2)$.

i Describe how $y = f(x)$ is transformed into $y = f(x + 2)$.

Q3g hint

$f(x + 2) = 2(x + 2) - 4$

19 Extend

1 **Reasoning** The distance–time graph describes the distance, d, of a tennis ball from a fixed point as it is thrown vertically upwards.

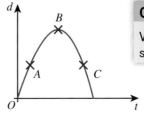

Q1 hint

Velocity is a vector. It is speed in a given direction.

  a Describe the motion of the ball at points A, B and C.

  b Compare the speed of the ball at A and C.

  c Compare the velocity of the ball at A and C.

Exam-style question

2 These graphs represent four different types of function, f.

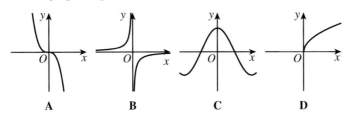

| A | B | C | D |

Exam tip

Make sure you are familiar with the shapes of graphs, including linear, quadratic, exponential as well as those shown here.

Match each description of a function to one of the graphs.

  i f(x) is inversely proportional to x.

  ii f(x) is directly proportional to $\sqrt{x}$.

  iii f(x) is a cubic function.

  iv f(x) is a trigonometrical function.

(**3 marks**)

Exam-style question

3 Lee invests a sum of money for 30 years at 4% per annum compound interest.

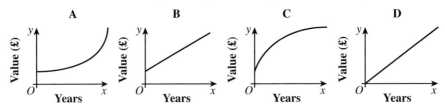

  a Which graph best shows how the value of Lee's investment changes over the 30 years?

(**1 mark**)

Hamish invested an amount of money in an account paying 5% per annum compound interest.

After 1 year the value of his investment was £2775.

  b Work out the amount of money that Hamish invested.

(**2 marks**)

4 **Reasoning** When 20 litres of water are poured into any cylinder, the depth, D (in cm), of the water is inversely proportional to the square of the radius, r (in cm), of the cylinder.
When $r = 15\,\text{cm}$, $D = 28.4\,\text{cm}$.

   a Write a formula for D in terms of r.

   b Find the depth of the water when the radius of the cylinder is $25\,\text{cm}$.

   c Find the radius of the cylinder when the depth is $64\,\text{cm}$.

   d Cylinder A has radius $x\,\text{cm}$ and is filled with water to a depth of $d\,\text{cm}$.
   This water is poured into cylinder B with radius $2x\,\text{cm}$.
   What is the depth of water in cylinder B?

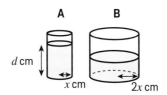

5 **Problem-solving / Future skills** As part of a science experiment, Michael places different-sized spheres into a measuring jug of water.
He estimates the radius, r, of the spheres and measures the amount of water displaced, W.
The table shows the results from his experiment.

Radius, r (cm)	2	8	6	10	5	1	3	4.5
Water displaced, W (litres)	0.03	2.1	0.9	4.1	0.5	0.004	0.11	0.38

   a Draw a scatter graph of Michael's results.

   b Which rule best describes the relationship between r and W?

   c Write a formula for estimating the relationship between r and W.

   d Estimate the amount of water displaced by a sphere with radius $16\,\text{cm}$.

   $\boxed{W \propto r}$ $\boxed{W \propto r^2}$ $\boxed{W \propto r^3}$

   $\boxed{W \propto \dfrac{1}{r}}$ $\boxed{W \propto \dfrac{1}{r^2}}$ $\boxed{W \propto \dfrac{1}{r^3}}$

6 The graph shows a company's profits over a 6-month period.

   a Describe how the level of profit changed over this period.

   b What does the area under the graph represent?

Profits January to June

(graph: y-axis Profit (£ millions) from 0 to 4, x-axis Month from 0 to 6)

7 **Reasoning** $f(x) = x^2 + 2x - 8$

   a Write down the coordinates of the points where $y = f(x)$ intersects the x-axis.

   b Write down the minimum value of y.

   c Sketch the graph of

   i $y = f(x)$

   ii $y = x^2 + 2x - 2$

   iii $y = (x+1)^2 + 2(x+1) - 8$

> **Q7a hint**
> Factorise $x^2 + 2x - 8$.

> **Q7b hint**
> Complete the square.

8 Here is the graph of $y = f(x) = \dfrac{1}{x}$.

a Sketch the graph of
$y = f(x+2) - 3$. **(2 marks)**

b Write the equations of the two
asymptotes. **(2 marks)**

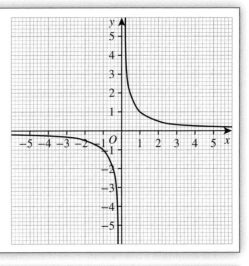

9 This is the graph of $y = 2^x$.

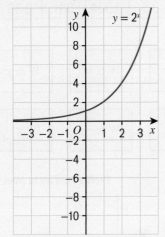

Q9 hint

Use your knowledge of the transformations
$y = f(-x)$, $y = -f(-x)$, $y = f(x) - a$ and
$y = f(x-a)$.

This question combines knowledge of the
exponential function and transformations
of graphs – both essential elements of
work at A level.

Copy the diagram and on the same axes sketch the graphs of

a $y = 2^{-x}$ **b** $y = -(2^{-x})$ **c** $y = 2^x - 3$ **d** $y = 2^{(x-3)}$

For each graph

  i state the coordinates of the point where the graph crosses the y-axis

  ii state the equation of the asymptote

10 A particle is travelling in a straight line.
Its speed, v m/s, at time t seconds is given by
$v = t^2 + t + 4$, where $t \geqslant 0$.
Work out an estimate of the distance travelled by
the particle in the first 5 seconds of its motion.
Use 5 strips of equal width.

Using the trapezium rule and the motion
of particles are both included in A level
Mathematics. This question connects
these two ideas.

19 Test ready

Summary of key points

To revise for the test:

- Read each key point, find a question on it in the mastery lesson, and check you can work out the answer.

- If you cannot, try some other questions from the mastery lesson or ask for help.

Key points

1   When a graph of two quantities is a straight line through the origin, one quantity is **directly proportional** to the other. 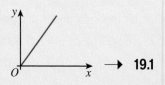 → 19.1

2   The symbol $\propto$ means 'is directly proportional to'.
$y \propto x$ means y is directly proportional to x. → 19.1

3   If y is directly proportional to x, then $y \propto x$ and $y = kx$, where k is a number, called the **constant of proportionality**. → 19.1

4   A quantity can be directly proportional to the *square*, the *square root*, the *cube* or the *cube root* of another quantity:

- if y is proportional to the square of x, then $y \propto x^2$ and $y = kx^2$
- if y is proportional to the cube of x, then $y \propto x^3$ and $y = kx^3$
- if y is proportional to the square root of x, then $y \propto \sqrt{x}$ and $y = k\sqrt{x}$
- if y is proportional to the cube root of x, then $y \propto \sqrt[3]{x}$ and $y = k\sqrt[3]{x}$. → 19.2

5   When y is **inversely proportional** to x, then $y \propto \dfrac{1}{x}$ and $y = \dfrac{k}{x}$.

→ 19.3

6   Functions of the form $f(x) = a^x$, where a is a positive number, are called **exponential functions**. → 19.4

7   The graph of an exponential function has one of these shapes.

$y = a^x$ where $a > 1$ or
$y = b^{-x}$ where $0 < b < 1$
exponential growth

$y = a^{-x}$ where $a > 1$ or
$y = b^x$ where $0 < b < 1$
exponential decay

→ 19.4

8   The **tangent** to a curved graph is a straight line that touches the graph at a point. The gradient at a point on a curve is the gradient of the tangent at that point. → 19.5

9   On a distance–time graph, the gradient of the tangent at any point gives the speed at that time. → 19.5

10 The straight line that connects two points on a curve is called a **chord**. The gradient of the chord gives the average rate of change and can be used to find the average rate of change between two points. For example, on a distance–time graph, the gradient of a chord gives the average speed between two times.

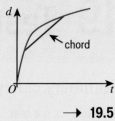

→ **19.5**

11 The area under a velocity–time graph shows the displacement, or distance from the starting point. To estimate the area under a part of a curved graph, draw a chord between the two points you are interested in, and straight lines down to the horizontal axis to create a trapezium. The area of the trapezium is an estimate for the area under this part of the graph.

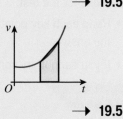

→ **19.5**

12 The graph of $y = f(x)$ is transformed into the graph of:

• $y = f(x) + a$ by a translation of a units parallel to the y-axis or a translation by $\begin{pmatrix} 0 \\ a \end{pmatrix}$ → **19.6**

• $y = f(x + a)$ by a translation of $-a$ units parallel to the x-axis or a translation by $\begin{pmatrix} -a \\ 0 \end{pmatrix}$ → **19.6**

• $y = f(-x)$ by a reflection in the y-axis → **19.7**

• $y = -f(x)$ by a reflection in the x-axis. → **19.7**

Sample student answer

Exam-style question

This is a graph of the function $y = f(x + 2)$.

Sketch the curve of the function $y = f(x)$. **(2 marks)**

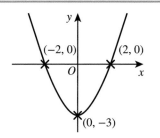

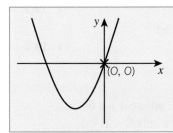

a The student has only marked the coordinates of one point on the sketch. Which other points should the student mark?

b Has the student translated the graph in the correct direction?

c Sketch the correct graph with all the relevant points marked.

19 Unit test

*Active*Learn
Homework

1 The time, T (in seconds), it takes a water heater to boil some water is directly proportional to the mass of water, m (in kg), in the water heater.
When $m = 250$, $T = 600$.

a Find T when $m = 400$. **(3 marks)**

The time, T (in seconds), it takes a water heater to boil a constant mass of water is inversely proportional to the power, P (in watts), of the water heater.
When $P = 1400$, $T = 360$.

b Find the value of T when $P = 900$. **(3 marks)**

2 Here are four graphs.

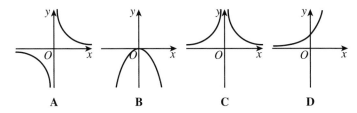

A   B   C   D

Write the letter of the graph that could have equation

a $y = \dfrac{4}{x}$ **(1 mark)**

b $y = 3^x$ **(1 mark)**

3 y is inversely proportional to x^3.
Given that $y = 2.5$ when $x = 2.4$

a find an equation for y in terms of x **(2 marks)**

b find the value of x when $y = 0.54$ **(2 marks)**

4 y is directly proportional to the square root of x.
$y = \frac{1}{3}$ when $x = 2\frac{1}{4}$.
Find the value of y when $x = 24$. **(3 marks)**

5 The sketch shows the graph of $y = f(x)$.

The maximum turning point at A has coordinates $(-3, 9)$.
The graph intersects the x-axis at $B(-6, 0)$ and $C(0, 0)$.
Write the coordinates of A, B and C for the graph of

a $y = -f(x)$ **(1 mark)**

b $y = f(-x)$ **(1 mark)**

c $y = f(x - 2)$ **(1 mark)**

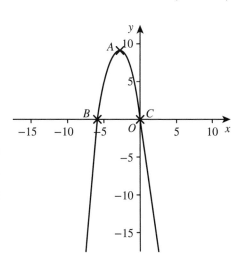

6 The velocity–time graph shows the first 60 seconds of a space flight.
Time, t, is measured in seconds.
Velocity, v, is measured in metres per second.

a Estimate the acceleration at
$t = 30$. **(2 marks)**

b Given that the spacecraft travelled in a straight line, estimate the distance travelled between $t = 20$ and $t = 30$. **(2 marks)**

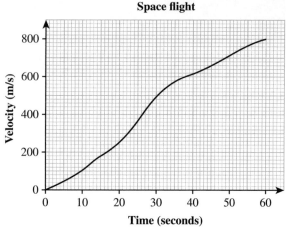

Space flight

7 Sketch the graph of $y = 3^{-x}$.
Give the coordinates of any points of intersection with the axes. **(2 marks)**

8 The diagram shows a sketch of $y = f(x)$.
Sketch the graph of

a $y = f(x) + 5$ **(1 mark)**

b $y = -f(x) - 5$ **(1 mark)**

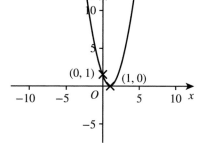

9 The diagram shows a sketch of the curve $y = ab^x$.
The curve passes through the points $A(1, 12)$ and $B(3, 108)$.

a Find the value of

  i a **ii** b **(3 marks)**

b Find the value of y when $x = 4$. **(1 mark)**

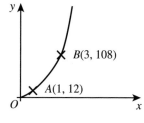

(TOTAL: 30 marks)

10 Challenge The graphs in the diagram make a pattern with reflective symmetry in the x- and y-axes.

a Write possible equations for graphs A, B, C and D to make this pattern.

b Write other equations of graphs that make a pattern with reflective symmetry in the x- and y-axes.

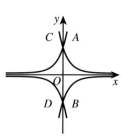

11 Reflect

a Write down a word that describes how you feel:

- before a maths test

- during a maths test (when you know how to answer a question)

- during a maths test (when you don't immediately know how to answer a question)

- after a maths test

b Discuss with a classmate what you could do to change negative feelings to positive feelings.

Mixed exercise 6

1 **Problem-solving** y is directly proportional to x.
Copy and complete the table.

x	10	15	24	
y	8			9

2 **Reasoning** Here are some graphs.

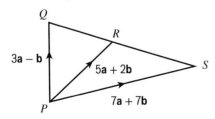

A	B	C	D
E	F	G	H

Match each equation with the letter of its graph.

  i $y = \cos x$ **ii** $y = 3^x$ **iii** $y = \dfrac{3}{x}$ **iv** $y = x^3$

3 **Reasoning** $\mathbf{a} = \begin{pmatrix} -3 \\ 3 \end{pmatrix}$, $\mathbf{b} = \begin{pmatrix} -4 \\ 10 \end{pmatrix}$, $\mathbf{c} = \begin{pmatrix} -10 \\ -2 \end{pmatrix}$

  a Work out the magnitude of vector $\mathbf{c}$. Give your answer correct to 3 significant figures.

  b Show that $3\mathbf{a} - \mathbf{b}$ is parallel to $\mathbf{c}$.

  c Work out the magnitude of vector $\mathbf{a} + \mathbf{b}$. Give your answer correct to 3 significant figures.

Exam-style question

4 Yesterday it took 20 cleaners $6\frac{1}{2}$ hours to clean all the wards in a hospital.
There are only 15 cleaners to clean all the wards in the hospital today.
Each cleaner is paid £8.60 for each hour or part of an hour they work.
How much will each cleaner be paid today?   **(3 marks)**

---

5 Write $\dfrac{x+4}{3} - \dfrac{2x}{5}$ as a single fraction in its simplest form.

6 **Reasoning** The diagram shows vectors $\overrightarrow{PQ}$, $\overrightarrow{PR}$ and $\overrightarrow{PS}$.

Is QRS a straight line? Show working to explain your answer.

7 Write $\dfrac{4}{x+3}+\dfrac{5}{2x}$ as a single fraction in its simplest form. **(2 marks)**

8 The ratio $x+y : x-y$ is equivalent to $a : 1$.

Show that $x = \dfrac{y(a+1)}{a-1}$. **(3 marks)**

9 Simplify fully $\dfrac{2x^2-7x-4}{4x^2-1}$.

10 Reasoning Felix is asked to sketch the curve with equation $y = 2^x$.
He draws this sketch.

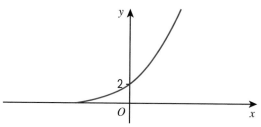

Felix does not receive any marks for his sketch.
Explain why.

11 Show that $(2x+3)(3x+1)(x-4) = 6x^3 - 13x^2 - 41x - 12$. **(3 marks)**

12 Problem-solving $f(x) = x+5$ and $g(x) = 2x-1$.
Write $f(3x) + g(x+7)$ as a simplified expression.

13 Prove algebraically that the sum of the squares of any two consecutive odd numbers is always an even number. **(3 marks)**

14 Problem-solving y is directly proportional to the cube of x.
When $x = 6$, $y = 972$.
Find x when $y = 7.776$.

15 Problem-solving $f(x) = \cos x$ and $g(x) = x + 45°$.
Sketch the graph of the composite function $y = fg(x)$ for $-180° \leqslant x \leqslant 180°$.

16 q is inversely proportional to the square of p.
$q = 10$ when $p = 4.5$.
Find the negative value of p when $q = \frac{10}{11}$.
Give your answer as a simplified surd. **(3 marks)**

17 Reasoning $ABCDEF$ is a hexagon.
Show that $BCEF$ is a parallelogram.

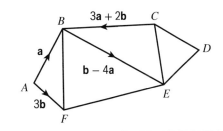

Exam-style question

18 The function f is given by $f(x) = 3x^2 - 4$.

a Show that $f^{-1}(71) = 5$. **(2 marks)**

The functions g and h are given by $g(x) = x + 3$ and $h(x) = x^2 + 1$.

b Find the values of x for which $hg(x) = 4x^2 + 7x + 8$. **(4 marks)**

19 Reasoning The diagram shows triangle OPQ.

$$\overrightarrow{OP} = 4\mathbf{p} \qquad \overrightarrow{OQ} = 6\mathbf{q}$$

M is the midpoint of PQ.

a Write $\overrightarrow{OM}$ in terms of $\mathbf{p}$ and $\mathbf{q}$.
Give your answer in its simplest form.

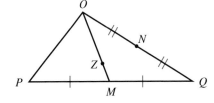

N is the midpoint of OQ.

Z is the point on OM such that $OZ : ZN = 2 : 1$.

b Show that PZN is a straight line.

Exam-style question

20 A triangle has vertices A, B and C.
The coordinates of A are $(-4, 5)$.
The coordinates of B are $(6, 1)$.
The coordinates of C are $(2, -5)$.
M is the midpoint of AB.
N is the midpoint of BC.
Prove MN is parallel to AC.
You must show each stage of your working. **(4 marks)**

Exam-style question

21 A ball is thrown upwards from a height of 1.5 metres
above the ground.
The graph shows the height of the ball above
the ground.
Work out an estimate of the speed of the ball,
in m/s, 1 second after it was thrown. **(3 marks)**

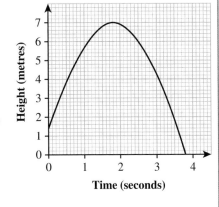

22 Reasoning Prove that $3n^3\left(\dfrac{2}{n} + 4\right) + 6n(5n - 2n^2)$ is a square number when n is a positive integer.

Exam-style question

23 The graph of $y = f(x)$ is shown on the grid.

  a Copy the grid and draw the graph with
    equation $y = f(x+2) - 3$.   **(2 marks)**

Point $P(-1, 3)$ lies on the graph
of $y = f(x)$.

When the graph of $y = f(x)$ is transformed
to the graph with equation $y = f(-x)$, point
P is mapped to point Q.

  b Write down the coordinates of point Q.
                          (1 mark)

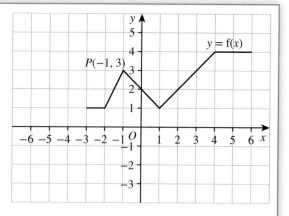

24 **Reasoning** Here is a sketch of the graph of $y = x^2 - 4$ and
curve C. Curve C is the graph of $y = x^2 - 4$ after a translation
of 3 units to the right.

  a Write the equation of curve C in the form $y = x^2 + bx + c$.

  b The graph $y = x^2 - 4$ is reflected in the x-axis to give
    graph D. Write the equation of graph D.

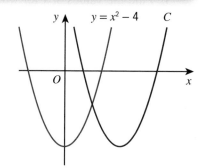

Exam-style question

25 Show that $\dfrac{5x-35}{x^2-3x-28} \div \dfrac{x-4}{x^3-16x}$ simplifies to ax, where a is an integer.   **(4 marks)**

26 Solve $\dfrac{x}{5} - \dfrac{3x}{x+4} = 2$. Give your answer correct to 2 decimal places.

27 **Reasoning** Show that $\dfrac{1}{1+\sqrt{2}} - \dfrac{3}{4+\sqrt{2}}$ can be written in the form $\dfrac{a+b\sqrt{2}}{c}$,

  where a, b and c are integers.

Exam-style question

28 $OABC$ is a parallelogram.
$\overrightarrow{OA} = \mathbf{a}$ $\overrightarrow{OB} = \mathbf{b}$

P is the point on OA such that $OP : PA = 2 : 1$.
Q is the point on OB such that $OQ : QB = 2 : 3$.
Work out, in its simplest form, the ratio $CP : CQ$.
You must show all your working.
                        (5 marks)

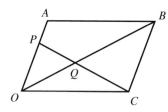

Answers

UNIT 11 Multiplicative reasoning

11.1 Growth and decay

1 a 1.5   **b** 0.15   **c** 0.015   **d** 0.0015

2 a 1.3^4 **b** 101.3   **c** 1.013

3 125

4 a 1.2   **b** 0.86   **c** 1.072   **d** 0.975

5 £28

6 a 0.65   **b** £4225   **c** 0.85   **d** £3591.25

e

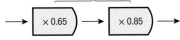

× 0.5525

→ ×0.65 → ×0.85 →

f Students' own checks.

7 a 0.68   **b** 1.0815   **c** 1.0246

8 £36 949.50

9 £146 664

10 No; $0.85 \times 1.22 = 1.037$, which is equivalent to a 3.7% increase.

11 a 1.405 (3 d.p.)   **b** 0.522 (3 d.p.)
    c 1.12^3, 0.85^4

12 a 18%   **b** 4 years

13 a $A_1 = 1000 \times 1.02$
     $A_2 = 1000 \times 1.02 \times 1.02$
     $A_3 = 1000 \times 1.02 \times 1.02 \times 1.02$
     $A_4 = 1000 \times 1.02 \times 1.02 \times 1.02 \times 1.02$
     $A_5 = 1000 \times 1.02 \times 1.02 \times 1.02 \times 1.02 \times 1.02$
    b 1000×1.02^5 **c** £1104.08 (nearest penny)

14 £3692.42

15 Students' own answers.

16 £96.77 (nearest penny)

17 Cash savings account (£203.31 total interest compared with £137.90 from Bonus savings account)

18 £3753.67 (nearest penny)

19 $R = 1.4$

20 $x = 1.3$

21 a 1263.5   **b** 4 hours

22 a 302
    b Nearest whole number. The answer must be a whole number.

23 450

11.2 Compound measures

1 a £10 per hour   **b** 15 km per litre

2 a 6   **b** −30   **c** −0.125

3 a 16 km/h   **b** 30 km   **c** 3 hours
    d i About $50 \div 5 = 10$ km/h
      ii Underestimate

4 a 7 hours 30 minutes   **b** 6 hours 12 minutes

5 a i 4.5 km   **ii** 5.5 km
    b i 59.5 minutes   **ii** 1 hour 0.5 minutes

6 3 hours

7 a i 1.5 litres   **ii** 3.75 litres
    b 40 hours

8 a 16 km/litre
    b The exact rate will vary depending on the speed of the car, whether it is going uphill or downhill, etc.

9 a More   **b** 5400 m/h
    c Fewer   **d** 90 m/min   **e** 1.5 m/s
    f Estimate first; if the answer needs to be smaller, divide; if the answer needs to be larger, multiply.

10 a 6.5 km/h   **b** 256 000 m/h   **c** 43 200 m/h
    d 1 m/s   **e** 22.5 m/s   **f** 72 km/h

11 a 3600 s   **b** 4 s

12 a About $3600 \div 240 = 15$ s
    b Underestimate; the speed was rounded up

13 a $\frac{1000x}{3600}$ m/s   **b** $\frac{3600y}{1000}$ km/h

14 1.8 km/h

15 1.3 km/min

16 53.8 km/h (1 d.p.)

17 72.5 km/h

18 a About 9 days   **b** 10 hours exactly

19 a $u = 0$, $a = 5$, $s = 200$ **b** 44.7 m/s

20 2 m/s^2

21 2.5 m/s

11.3 More compound measures

1 a $A = \pi r^2$
    b $V = $ area of cross-section × length

2 a 7500 g   **b** 1000 mg   **c** $10\,000 \text{ cm}^2$
    d 6.25 m^2 **e** $1\,000\,000 \text{ cm}^3$ **f** 0.095 m^3

3 a $m = 30$ **b** $v = 16$

4 a i 4.05 cm   **ii** 4.15 cm
    b i 407.5 g   **ii** 412.5 g
    c i 41.05 kg   **ii** 41.15 kg

5

Metres per second	Kilometres per hour
15	54
20	72
30	**108**
45	**162**

6 Falcon is faster: car speed = 350 km/h = 97.2 m/s (1 d.p.); falcon speed = 388.8 km/h = 108 m/s

7 26.4 km

8 8.3 g/cm^3

9 2.4 g/cm^3

10 a 480 cm^3 **b** 288 g

11 2047.5 g

12 675 cm^3

13 2700 kg/m^3

14 $1000x \text{ kg/m}^3$

15 Sodium is denser: lithium density = 0.53 g/cm^3; sodium density = 0.97 g/cm^3

16 Maximum 8.1 g/cm^3, minimum 7.8 g/cm^3 (1 d.p.)

17 1.01 g/cm^3 (2 d.p.)

18 a 23.1 N/m^2 **b** 73.0 N
    c 8.33 m^2 **d** $0.002\,31 \text{ N/cm}^2$

19 0.153 N/cm^2

20 12 000 N

21 a 784 N **b** 49 N/cm^2

   c 18 375 N/m^2 **d** Sitting

11.4 Ratio and proportion

1 a B **b** A **c** C

2 a i

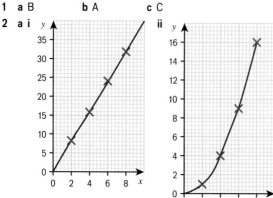

b Graph **i**; straight line through origin

   c $y = 4x$

3 1 : 1.33 (2 d.p.)

4 a $A = \frac{3}{5}B$ **b** $\frac{P}{Q} = \frac{7}{4}$, $P = \frac{7}{4}Q$

   c $\frac{X}{Y} = \frac{9}{5}$, 9 : 5

5 a 1 : 1.6

   b

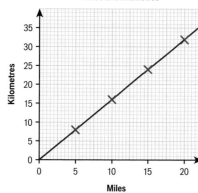

Miles and kilometres

   c Yes; straight line through origin

   d 1.6 **e** Kilometres = 1.6 × miles

   e The constant in the formula is the gradient of the graph, which is the same as the ratio in part **a**.

6 a Yes; values are in the same ratio, $\frac{8}{10} = \frac{16}{20} = \frac{24}{30} = \frac{32}{40} = \frac{40}{50}$

   b $t = 1.25s$ or $s = 0.8t$ **c** 20 miles

7 a

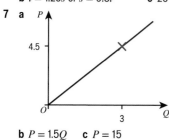

   b $P = 1.5Q$ **c** $P = 15$

8 a

x	0	5	10	20
C	0	6.5	13	26

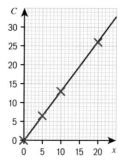

Graph plotted from the table of values is a straight line through origin so in direct proportion.

   b $C = 1.3x$ **c** £71.50

9 660 s or 11 minutes

10 When $d = 8$, $P = 0.8$, so $P = 0.1d$.
When $d = 75$, the pressure will be $75 \times 0.1 = 7.5$ bars.
This is less than 8.5 bars, so the watch will still work.

11 Students' own answers, for example, it is likely that D and E are not in direct proportion – the more electricians there are, the quicker they should do the job.

12 a $H = \frac{40}{B}$

   b i 6 hours 40 minutes **ii** 13 hours 20 minutes
   iii 4 hours 27 minutes (nearest minute)

   c The answers are the same; $H \times B = 40$.

13 a No; if the number of typists doubles, the time taken does not double.

   b $H = \frac{15}{T}$ **c** 2 hours 9 minutes

   d The same report; all type at the same speed.

14 $W = 7$, $X = 10$, $Y = 2$, $Z = 5$

15 a Direct **b** Inverse **c** Neither
   d Inverse **e** Direct

16 96 hours

17 15 amps

18 a $r = \frac{4.5}{t}$ **b** $r = 1.125$

19 a

x	1	2	5	10
y	10	5	2	1

   b

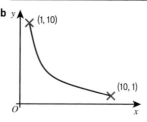

   c $y = 0.5$ **d** Yes **e** $y = 20$; yes

20 Yes; $T = 296.47$ s (2 d.p.), which is less than 300 s.

11 Check up

1 £7200

2 £3869.28

3 6060

4 5 years

5 320 seconds

6 0.8 g/cm^3

7 **a** $8050 \, \text{kg/m}^3$

b $1006.25 \, \text{g}$ or $1.00625 \, \text{kg}$

8 $93.75 \, \text{N/m}^2$

9 Usain Bolt is faster: Usain Bolt: $12.3 \, \text{m/s} = 44.3 \, \text{km/h}$; Great white shark: $11.1 \, \text{m/s} = 40 \, \text{km/h}$

10 **a** Yes; values are in same ratio.

b $E = 1.3P$　　**c** €32.50

11 **a** $c = 8d$　　**b** $c = 96$

12 10.5 amps

14 **a**

$$\times 12 \begin{cases} 5280 \text{ feet} & 1 \text{ hour} \\ 63\,360 \text{ inches} & 1 \text{ hour} \end{cases}$$

$$\div 60 \begin{cases} 63\,360 \text{ inches} & 60 \text{ minutes} \\ 1056 \text{ inches} & 1 \text{ minute} \end{cases} \div 60$$

b i

$$\qquad\qquad \text{distance} \qquad \text{time}$$

$$\times 1000 \begin{cases} 1 \text{ km} & 1 \text{ hour} \\ 1000 \text{ m} & 1 \text{ hour} \end{cases}$$

$$\times 1000 \begin{cases} 1000 \text{ m} & 1 \text{ hour} \\ 1000\,000 \text{ mm} & 1 \text{ hour} \end{cases}$$

$$\div 60 \begin{cases} 1000\,000 \text{ mm} & 60 \text{ minutes} \\ 16\,666.7 \text{ mm} & 1 \text{ minute} \end{cases} \div 60$$

ii

$$\qquad\qquad \text{volume} \qquad \text{time}$$

$$\times 1000 \begin{cases} 1 \text{ litre} & 1 \text{ minute} \\ 1000 \text{ ml} & 1 \text{ minute} \end{cases}$$

$$\times 60 \begin{cases} 1000 \text{ ml} & \frac{1}{60} \text{ hour} \\ 60\,000 \text{ ml} & 1 \text{ hour} \end{cases} \times 60$$

iii

$$\qquad\qquad\qquad \text{mass} \qquad \text{volume}$$

$$\div 1000 \begin{cases} 1 \text{ mg} & 1 \text{ m}^3 \\ 0.001 \text{ g} & 1 \text{ m}^3 \end{cases}$$

$$\div 1000 \begin{cases} 0.001 \text{ g} & 1 \text{ m}^3 \\ 0.000\,001 \text{ kg} & 1 \text{ m}^3 \end{cases}$$

$$\div 1\,000\,000 \begin{cases} 0.000\,001 \text{ kg} & 1 \text{ m}^3 \\ 0.000\,000\,000\,001 \text{ kg} & 1 \text{ cm}^3 \end{cases} \div 1\,000\,000$$

c Students' own conversions.

15 5 km race leader board: Allia, Chaya, Hafsa, Billie
10 km race leader board: Fion, Daisy, Gracie, Ellie

11 Strengthen

Percentages

1 **a** 1.2　　**b** 1.09　　**c** 1.12　　**d** 1.037

2 **a** 0.77　　**b** 0.94　　**c** 0.92　　**d** 0.925

3 1.308

4 0.7238

5 1.0304

6

Year	Amount at start of year	Amount plus 3% interest	Total amount at end of year
4	£437.09	437.09×1.03 $= 400 \times 1.03^4$	£450.20
5	£450.20	450.20×1.03 $= 400 \times 1.03^5$	£463.71
6	£463.71	463.71×1.03 $= 400 \times 1.03^6$	£477.62

7 £6144

8 5 years

9 558

Compound measures

1 **a**

$$\times 60 \searrow \begin{array}{l} 50 \text{ ml in 1 second} \\ 3000 \text{ ml in 60 seconds} \\ \text{(1 minute)} \\ 3 \text{ litres in 1 minute} \end{array} \nwarrow \times 60$$

$$\div 1000$$

b 4 minutes

2 **a** $\text{Speed} = \dfrac{\text{Distance}}{\text{Time}}$　　**b** $\text{Time} = \dfrac{\text{Distance}}{\text{Speed}}$

c $\text{Distance} = \text{Speed} \times \text{Time}$

d $\text{Force} = \text{Pressure} \times \text{Area}$

e $\text{Area} = \dfrac{\text{Force}}{\text{Pressure}}$　　**f** $\text{Pressure} = \dfrac{\text{Force}}{\text{Area}}$

3 **a**

Metal	Mass (g)	Volume (cm^3)	Density (g/cm^3)
copper	**1090**	122	8.96
lead	450	**39.8**	11.3
mercury	110	8.15	**13.5**

b

Force (N)	Area (cm^2)	Pressure (N/cm^2)
104	13	8
48	12	**4**
65	**5**	13

4 **a** $1000 \, \text{g}$　　**b** $10\,000 \, \text{cm}^2$　　**c** $1\,000\,000 \, \text{cm}^3$

5 **a**

$$\begin{array}{l} 270 \text{ kg per m}^2 \\ 270\,000 \text{ g per m}^2 \\ 27 \text{ g per cm}^2 \end{array} \begin{array}{l} \times 1000 \\ \div 10\,000 \end{array}$$

b

$$\begin{array}{l} 50 \text{ g per cm}^3 \\ 0.050 \text{ kg per cm}^3 \\ 50\,000 \text{ kg per m}^3 \end{array} \begin{array}{l} \div 1000 \\ \times 1\,000\,000 \end{array}$$

6

km/h	m/h	m/min	m/s
18	**18 000**	300	5
36	**36 000**	600	10
24	**24 000**	400	**6.67**
57.6	57 600	960	16

Ratio and proportion

1 $W = 24$, $X = 22.5$, $Y = 30$, $Z = 18$

2 200 seconds

3 $W = 6$, $X = 3$, $Y = 4$, $Z = 4$

4 64 N

11 Extend

1 **a** 375 N　　**b** 24 cm

2 1100 N

3 **a** $16 \, \text{m/s}^2$　　**b** $\dfrac{v}{20} \, \text{m/s}^2$

4 **a** 43 minutes 13 seconds (nearest second)

b Time would increase.

5 $n = 4$

6 $x = 1.65$ (2 d.p.)

7 363 g

8 C

9 **a** Yes; exterior angle × number of sides = constant, so if number of sides is doubled, exterior angle is halved.

b 18°

10 Both parts of the journey took the same time.

11 a Yes; upper bound for speed is
$285 \div 122.5 = 2.32... \text{ km/min} = 139.59... \text{ km/h}$.

   b Yes; upper bound is now
$276 \div 122.5 = 2.25... \text{ km/min} = 135.18... \text{ km/h}$,
which is less than 136 km/h.

12 Yes, he could be wrong; upper bound for petrol consumption
$= 100 \times 11.25 \div 151.5 = 7.43$ litres per 100 km.

13 a $k = 0.975$ **b** 1084 litres (nearest litre)

11 Test ready

Sample student answers

1 a i 1.25×4 is a whole number.

   ii Student B multiplies numerator and denominator by 100 to get rid of the decimal in the denominator.

   iii Students' own answers, e.g. Student B's working shows more explanation of what is being calculated at each step.

  b Student B has written the formula being used, and included the units at more stages.

11 Unit test

1 a 13.2 m/s   **b** 47.4 km/h

2 1390 N/m² (3 s.f.)

3 6.6 kg

4 a $D = 0.08t$ **b** 9.6 km

  c 138 minutes or 2 hours 18 minutes (3 s.f.)

5 5.57 m/s

6 £743.62 (nearest penny)

7 a $P = \dfrac{48}{A}$

  b No; $P = 48 \div 2 = 24 \text{ N/m}^2$, which is less than the limit.

8 £8365.43 (nearest penny)

9 £4086.87 (nearest penny)

10 A

11 a Compound interest is better because Antonia will get interest on the interest gained each year.

  b ROI = 6.93%

12 Students' own answers, e.g.
Using a decimal multiplier to work out a percentage increase or decrease; working out the amount of interest earned; working out the pressure when force and area are known.

13 Students' own answers. e.g.
Both pay interest (usually each year), calculated as a percentage of the amount in the account.
For compound interest, the amount includes interest previously earned; for simple interest, it does not.

UNIT 12 Similarity and congruence

12.1 Congruence

1 180°

2 $a = 134°$ (vertically opposite angles are equal)
$b = 134°$ (alternate angles are equal)
$c = 180 - 134 = 46°$ (angles on a straight line add up to 180°)

3 a GHI and PQR **b** A and C

4 a 17.0 cm   **b** 15.0 cm

5 a SAS   **b** RHS   **c** SSS   **d** AAS

6 a Congruent (SAS); A corresponds to R, B corresponds to Q, C corresponds to P.

  b Not congruent

7 DEF (SSS) and GHI (SAS)

8 a Yes, congruent (SAS or RHS, using Pythagoras' theorem to find the missing sides)

  b Yes, congruent (SSS or RHS, using the sine ratio to find the missing sides)

9 No. All the triangles with a 12 cm hypotenuse will be congruent, and all the triangles where 12 cm is not the hypotenuse will be congruent.

10 a $\angle EBA = 50°$, $\angle EAB = 20°$, $\angle EDC = 20°$, $\angle CED = 110°$

  b Two angles and a corresponding side are equal (AAS).

11 JL is common so all three sides are equal, so JKL and JML are congruent (SSS).

12.2 Geometric proof and congruence

1 SSS, SAS, AAS or RHS

2 a

3 a i EM **ii** FM

  b $\angle EMG$

4 a

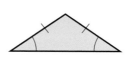

  b Students' own proofs, e.g. $\angle WXY = \angle WZY = 90°$;
WY is common (hypotenuse);
$WZ = XY$ (opposite sides of a rectangle are equal).
Therefore the triangles are congruent (RHS).

  c $\angle WYZ$

5 Students' own proofs, e.g. $ABCD$ is a rhombus, so $\angle ABC = \angle ADC$, $BA = AD$ and $BC = CD$.
Therefore the triangles are congruent (SAS).

6 a Students' own proofs, e.g. SQ is common; $PQ = SR$; $PS = QR$. Therefore the triangles are congruent (SSS).

  b i 123°   **ii** 28.5°

  c $\angle SQP = \angle QSR$ (alternate angles are equal);
$\angle SQR = \angle QSP$ (alternate angles are equal);
QS is common. Therefore the triangles are congruent (SAS).

7 Students' own proofs, e.g. $LX = XM$; $XK = XJ$;
$\angle JXL = \angle KXM$ (vertically opposite angles are equal).
Therefore the triangles are congruent (SAS).

8 $FG = GH$; EG is common; $\angle FEG = \angle GEH = 90°$.
Therefore the triangles are congruent (RHS).

9 Students' own proofs, e.g.

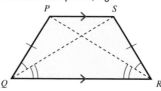

$PQ = SR$; $\angle PQR = \angle SRQ$; QR is common.
Therefore triangles PQR and SQR are congruent (SAS).
Since the triangles are congruent, then $PR = QS$.

10 *SM* is common, $RS = ST$; $\angle SMR = \angle SMT = 90°$.
Therefore triangles *RSM* and *MST* are congruent (RHS).
Since the triangles are congruent, then $RM = MT$ so the line *SM* bisects the base.

11 $PQ = ST$; $\angle QPR = \angle RTS$ (alternate angles are equal); $\angle PQR = \angle RST$ (alternate angles are equal).
Therefore triangles *PQR* and *RST* are congruent (AAS).
Since the triangles are congruent, then $PR = RT$ so *R* is the midpoint of *PT*.

12 a Students' own proofs, e.g. $\angle GDE = \angle GFC$ (alternate angles are equal); $DE = CF$; $\angle DEG = \angle FCG$ (alternate angles are equal). Therefore triangles *DEG* and *CFG* are congruent (AAS).

b Students' own proofs, e.g. $\angle DCG = \angle GEF$ (alternate angles are equal); $\angle GDC = \angle GFE$ (alternate angles are equal); $DC = EF$. Therefore triangles *CDG* and *EFG* are congruent (AAS).

c Triangles *DEG* and *CFG* are congruent, so $DG = FG$.
Triangles *CDG* and *EFG* are congruent, so $CG = GE$.
Therefore *G* is the midpoint of *CE* and of *DF*.
$DC = DE$; $CG = GE$; *DG* is common.
Therefore triangles *CDG* and *DEG* are congruent (SSS).
Since the triangles are congruent, then $\angle DGC = \angle DGE = 180 \div 2 = 90°$ (angles on a straight line add up to 180°) so the diagonals of a rhombus intersect at right angles.

13 a, b

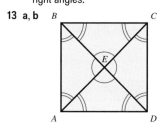

c Triangles *BEC*, *CED*, *DEA* and *AEB* are congruent, and triangles *BCD*, *ACD*, *ABD* and *ABC* are congruent.

d They are all equal; therefore they are all $360 \div 4 = 90°$.

e Since all four triangles are congruent, $AE = EC$ and $BE = ED$; therefore *E* is the midpoint of both *AC* and *BD*, which intersect at right angles (from part **d**).

14 $XY = XZ$ (triangle *XYZ* is isosceles); $XA = XB$; $\angle BXA$ is common. Therefore triangles *XAZ* and *XBY* are congruent (SAS).

12.3 Similarity

1 Same: all sides in each pentagon are equal length, all angles are equal.
Different: the sides of the two pentagons are different lengths.

2 a 2   **b** $\frac{1}{2}$

3 a *PQ* and *XY*, *PR* and *XZ*, *QR* and *YZ*
b $PQ = 1.5\,\text{cm}$; $XY = 3\,\text{cm}$; $PR = 1.9\,\text{cm}$; $XZ = 3.8\,\text{cm}$; $QR = 1.6\,\text{cm}$; $YZ = 3.2\,\text{cm}$
$\angle QPR = \angle YXZ = 54°$; $\angle PQR = \angle XYZ = 76°$; $\angle QRP = \angle YZX = 50°$
c Corresponding angles are the same.
d $\frac{PQ}{XY} = \frac{1}{2}$; $\frac{PR}{XZ} = \frac{1}{2}$; $\frac{QR}{YZ} = \frac{1}{2}$
e They are all $\frac{1}{2}$.

4 $\frac{100}{225} = \frac{4}{9}$

5 a i *TU*   **ii** *UV*
b i $\frac{4}{9}$ **ii** $\frac{5}{12}$
c The ratios of corresponding sides are not the same, therefore the parallelograms are not similar.

6 a Not similar **b** Similar   **c** Not similar **d** Similar

7 $\angle DCB = \angle YXW = 155°$; all the angles in the pentagons are the same; the angles in triangles *ABC* and *VWX* are the same.
$CD = AE$ and $XY = VZ$
$\frac{CD}{XY} = \frac{12}{8} = 1.5$
$\frac{DE}{YZ} = \frac{9}{6} = 1.5$
Corresponding sides are in the same ratio.
Therefore the shapes are similar.

8 55 cm

9 18 cm

10 14 cm

11 a 10 m
b It does not matter, provided that the ratios are consistent.

12 12.65 m

13 a Corresponding angles are all equal, 30°, 90°, 60°.
b 15 cm
c All right-angled triangles with one other angle the same are similar.
d 0.5   **e** Sine

14 a Yes; corresponding sides are all in the same ratio.
b No; corresponding sides are not in the same ratio.
c No; corresponding sides are not in the same ratio.

15 a Each shape has 6 equal sides and 6 equal angles, so they are similar.
b Yes; all regular hexagons have 6 equal sides and 6 equal angles so they are all similar.

12.4 More similarity

1 1 : 1.5

2 a 1 : 3   **b** 1 : 3   **c** 1 : 9

3 a 1.5   **b** $\frac{2}{3}$

4 $a = 62°$ (corresponding angles are equal)
$b = a = 62°$ (alternate angles are equal)
$c = b = 62°$ (vertically opposite angles are equal)
$d = 62°$ (base angles of an isosceles triangle are equal)

5 a $\angle AEB = 94°$ (vertically opposite angles are equal)
$\angle EAB = 39°$ (alternate angles are equal)
$\angle ABE = 47°$ (alternate angles are equal)
b Yes; $\angle EDC = \angle EBA$, $\angle DCE = \angle EAB$ and $\angle CED = \angle AEB$. Therefore all angles are equal so the triangles are similar.

6 a $\angle RPQ = \angle RTS$ (alternate angles are equal); $\angle PQR = \angle RST$ (alternate angles are equal); $\angle PRQ = \angle SRT$ (vertically opposite angles are equal). Therefore all angles are equal so the triangles are similar.
b 10 cm

7 a $\angle F$ is common; $\angle FGH = \angle FJK$ (corresponding angles are equal); $\angle FHG = \angle FKJ$ (corresponding angles are equal). Therefore all angles are equal so the triangles are similar.
b 60 mm   **c** 64 mm

8 a $\angle PQN = \angle MNL = 52°$ (corresponding angles are equal); $\angle LMN = \angle LPQ = 102°$ (corresponding angles are equal); $\angle L$ is common. Therefore all angles are equal so the triangles are similar.
b 44 cm   **c** 18 cm

9 a, b 4 cm (assuming that *BD* and *AE* are parallel, so *AC* is corresponding side to *BC*)
24 cm (assuming that *BD* and *AE* are not parallel, and *AC* is corresponding side to *DC*)

10 a 54 m   **b** 135 m^2

11 $28800\,\text{cm}^2$

12 a $1:9$ **b** 9 **c** $1:3$
  d 3 **e** $7.2\,\text{cm}$

13 a $7.5\,\text{cm}$ **b** $18\,\text{cm}$

14 $12\,\text{cm}, 15\,\text{cm}, 19.2\,\text{cm}$

15 a 4 **b** 2 **c** $21\,\text{cm}$

16 $5\,\text{cm}$

17 a $24\,\text{cm}^2$ **b** $54\,\text{cm}^2$

12.5 Similarity in 3D solids

1 Volume $= 27\,\text{cm}^3$, surface area $= 54\,\text{cm}^2$

2 1.125

3 a 5 **b** $\frac{4}{3}$

4 a

	Linear scale factor	Volume A	Volume B	Volume scale factor
i	2	2	16	8
ii	k	24	$24k^3$	k^3

  b Each volume scale factor is the cube of the linear scale factor.

5 $96\,\text{cm}^3$

6 $405\,\text{cm}^3$

7 $60\,\text{cm}^3$

8 a 1.3 **b** 11 litres

9 a $1:8$ **b** 8 **c** $1:2$
  d 2 **e** $12\,\text{cm}$

10 $7.5\,\text{cm}$

11 a $21\,\text{cm}$ **b** $6\,\text{cm}$

12 a $1:1.331$ **b** $50\,\text{cm}$

13 a 125 **b** 5 **c** 25 **d** $1500\,\text{cm}^2$

14 $563\,\text{cm}^2$ (3 s.f.)

15 $810\,\text{ml}$

16 Area scale factor $= \frac{92}{207} = \frac{4}{9}$
  So linear scale factor $= \frac{2}{3}$ and volume scale factor $= \frac{8}{27}$
  Volume of cone $B = 837 \times \frac{8}{27} = 248\,\text{cm}^3$

17 a $4:9$ **b** $3:10$ **c** $4:9:30$

12 Check up

1 A and C are congruent (SAS).

2 a, b Students' own proofs, e.g.
  $\angle JHK = \angle HKL = 90°$; $\angle JKH = \angle LHK$
  (alternate angles are equal); HK is common.
  Therefore the triangles are congruent (AAS).
  $\angle JHK = \angle HKL = 90°$; $JK = HL$ (opposite sides of parallelogram are equal); HK is common.
  Therefore the triangles are congruent (RHS).

3 a Missing angles are $58°$ and $35°$, therefore all angles are equal and the triangles are similar.
  b All corresponding sides are in the same ratio (scale factor 3.5).

4 a $20\,\text{cm}$ **b** $18\,\text{cm}$

5 a The angle at A is common; $\angle EBA = \angle DCA$
  (corresponding angles are equal); $\angle EDC = \angle AEB$
  (corresponding angles are equal). All angles are equal so the triangles are similar.
  b $CD = 32\,\text{cm}$

6 a $\angle PQR = \angle RST$ (alternate angles are equal);
  $\angle QPR = \angle RTS$ (alternate angles are equal);
  $\angle PRQ = \angle SRT$ (vertically opposite angles are equal).
  All angles are equal so the triangles are similar.
  b $x = 40\,\text{cm}$; $y = 19.5\,\text{cm}$

7 a $55.2\,\text{cm}$ **b** $200\,\text{cm}^2$

8 $2580.5\,\text{cm}^3$

9 $519\,\text{cm}^3$

11 a Accurate constructions of the triangles

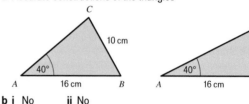

  b i No **ii** No

12 Strengthen

Congruence

1 D

2 A and C

3 Rectangle, parallelogram (2 ways), kite, isosceles triangle (2 ways)

4 a SSS **b** RHS **c** AAS

5 a KM is a common side; $LM = NM$; $KL = KN$
  b SSS

Similarity in 2D shapes

1 a i [diagram of triangles ABC and EDF]

  ii AB and ED, BC and DF, AC and EF
  iii $\angle ABC = \angle EDF$, $\angle ACB = \angle EFD$, $\angle CAB = \angle FED$
  b i [diagram of triangles IJH and IKG]

  ii IJ and IK, JH and KG, HI and GI
  iii $\angle IJH = \angle IKG$, $\angle JHI = \angle KGI$, $\angle HIJ = \angle GIK$

2 a

C	D	$\dfrac{C}{D}$
3	6	$\frac{1}{2}$
5	x	$\frac{5}{x}$
y	8	$\frac{y}{8}$

  b $x = 10$, $y = 4$

3 $1.5\,\text{cm}$

4 **a** 1 **b** 1 **c** $\frac{PQ}{1} = PQ$

5 $\sin 30° = \frac{PQ}{1} = \frac{0.5}{1} = 0.5$

6 **a** 150° **b** 210°, 330°

7 **a** 0.6 **b** 0.4 **c** −0.2 **d** −0.8

8 **a** 1 **b** −1
c 0°, 180°(any multiple of 180°)

9 **a** Decreases from 1 to 0 **b** Decreases from 0 to −1
c Increases from −1 to 0

10 **a i** 1 **ii** 0.97
b $x = 90°$
c i 0.87 **ii** 120°
d i 180° **ii** 135° **iii** 165°

11 **a** Rotational symmetry of order 2 about (180°, 0)
b Students' own checks.

12
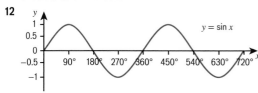

13 **a** 0 **b** 1

14 **a i** $\frac{\sqrt{3}}{2}$ **ii** $\frac{\sqrt{3}}{2}$
b The graph repeats every 360°, so sin 420° is the same as
sin 60°.
The graph is symmetrical about $x = 450°$,
so sin 480° = sin 420°.

15 **a i** 210°, 330°, 570°, 690°
ii 240°, 300°, 600°, 660°
b Students' own checks.

16 $A(90°, 1)$, $B(180°, 0)$, $C(270°, −1)$, $D(540°, 0)$

17 **a** $\sin x = \frac{3}{5}$ **b** $x = 36.9°$ (1 d.p.)
c $x = 36.9°$, 143.1°, 396.9°, 503.1° (all 1 d.p.)

18 **a** $x = 18.2°$, 161.8°, 378.2°, 521.8° (all 1 d.p.)
b $x = 56.4°$, 123.6°, 416.4°, 483.6° (all 1 d.p.)

13.3 Graph of the cosine function

1 **a** $\frac{\sqrt{3}}{2}$ **b** $\frac{1}{\sqrt{2}}$ **c** $\frac{1}{2}$ **d** 0

2 **a** 0.96 **b** 16.3° (1 d.p.)

3 **a** 1 **b** $\frac{OQ}{1} = OQ$

4 **a** $\cos 60° = \frac{OQ}{1} = \frac{0.5}{1}$ **b** 0.5
c i 300° **ii** 120°, 240°

5 **a** 330° **b** 150°

6 **a** 0.4 **b** −0.6 **c** −0.6 **d** 0.4

7 **a** Decreases from 0 to −1 **b** Increases from −1 to 0
c Increases from 0 to 1

8 **a i** −1 **ii** −0.87
b Reflection symmetry with mirror line $x = 180°$
c i 0.5 **ii** 300°
d i 270° **ii** 240° **iii** 360°
e Answers between 105° and 110° and between 250°
and 255°

9
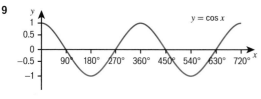

10 **a i** 0.5 **ii** −0.5
b 60°, 300°, 420°, 660°

11 **a i** $\frac{\sqrt{3}}{2}$ **ii** $-\frac{\sqrt{3}}{2}$
b 150°, 210°, 510°, 570°

12 Students' own checks.

13 **a** $\theta = 117.0°$ (1 d.p.)
b $\theta = 117.0°$, 243.0°, 477.0°, 603.0° (all 1 d.p.)

14 **a** $x = 35.9°$ (1 d.p.)
b $x = 35.9°$, 324.1°, 395.9°, 684.1° (all 1 d.p.)

13.4 Graph of the tangent function

1 **a** $\frac{1}{\sqrt{3}}$ **b** 1 **c** $\sqrt{3}$

2 **a** 0.225 **b** 12.7° (1 d.p.)

3 $\frac{\sqrt{3}}{3}$

4 **a** 240° **b** 120°, 300°

5 **a** 315° **b** 135°
c Students' own answers, e.g.
Drawing the diagram makes it easier to see when to add
180° and when to subtract from 180°.

6 **a** 0.6 **b** 0.6 **c** −1.2 **d** 1.5

7 **a i** Decreases from 0 to minus infinity
ii Increases from 0 to infinity
iii Decreases from 0 to minus infinity
b Students' own answers, e.g.
The value of $\tan \theta$ is where the line OP crosses the vertical
line but when θ is exactly 90°, the two lines are parallel
and never meet.
c i 270°, 450°, ... **ii** Students' own checks.

8 **a** Every 180°
b i 1.7 **ii** −1.7
c Rotational symmetry of order 2 about (180°, 0)
d i 240° **ii** 280° **iii** 300°

9
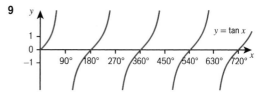

10 **a i** 0 **ii** 1
b i $\sqrt{3}$ **ii** $-\sqrt{3}$
c The graph repeats every 180°, so tan 240° is the same as
tan 60°.
The graph is symmetrical about $x = 90°$,
so tan 120° = −sin 60°.

11 **a i** 45°, 225°, 405°, 585° **ii** 135°, 315°, 495°, 675°
b Students' own checks.

12 **a** $x = 74.7°$ (1 d.p.)
b $x = 74.7°$, 254.7°, 434.7°, 614.7° (all 1 d.p.)

13 **a**
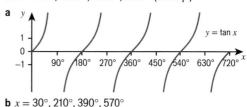

b $x = 30°$, 210°, 390°, 570°

13.5 Calculating areas and the sine rule

1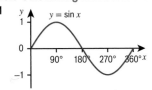

2 **a** $9\,\text{cm}^2$ **b** $20\,\text{cm}^2$ **c** $78.5\,\text{cm}^2$
d $77.0\,\text{cm}^2$ **e** $67.0\,\text{cm}^2$

3 **a** $x^2 + 3x - 28 = 0$
b $x = 4, x = -7$

4 $3.38\,\text{cm}$

5 **a** $h = p\sin\theta$ **b** Area $= \frac{1}{2}pq\sin\theta$

6 **a i, ii**

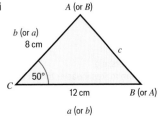

a (or *b*)

iii $36.8\,\text{cm}^2$
b i, ii

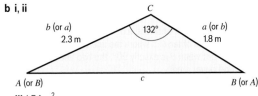

iii $1.54\,\text{m}^2$

7 **a** $7.99\,\text{cm}$ **b** $68.4°$
8 $11.5\,\text{cm}$ (1 d.p.)
9 **a** $12.45\,\text{cm}^2$ **b** $20.73\,\text{cm}^2$ **c** $8.27\,\text{cm}^2$
10 $67.1\,\text{m}^2$
11 **a** $118.0°$ **b** $28.8\,\text{cm}^2$
12 **a** $\frac{1}{2}x(x+1)\sin30° = 5$ so $\frac{1}{4}x^2 + \frac{1}{4}x - 5 = 0$
    or $x^2 + x - 20 = 0$
b $x = 4$ (or $x = -5$)
c x cm and $(x+1)$ cm are both lengths and cannot be negative.
13 **a** $22.9\,\text{cm}$ **b** $25.5\,\text{mm}$ **c** $14.7\,\text{m}$
14 $\frac{\sin\theta}{15} = \frac{\sin110°}{18}$

$\sin\theta = \frac{15\sin110°}{18}$

$\theta = \sin^{-1}\left(\frac{15\sin110°}{18}\right) = 51.5°$ (1 d.p.)

15 **a** $48.2°$ **b** $19.8°$ **c** $68.9°$ **d** $55.2°$
16 **a** $11.3\,\text{cm}$ **b** $38.7°$
17 **a** 0.857 **b** $59.0°, 121.0°$
18 $75.4°, 104.6°$
19 $15.8°, 164.2°$

13.6 The cosine rule and 2D trigonometric problems

1 **a** 8π **b** $2\pi + 8$ **c** $6\pi + 8$
2 **a** $067°$ **b** $247°$
3 5.05
4 **a** $23.4\,\text{cm}^2$ **b** $32.1°$
5 **a** $8.43\,\text{cm}$ **b** $8.15\,\text{cm}$ **c** $21.1\,\text{cm}$ **d** $12.5\,\text{m}$

6 **a i, ii**

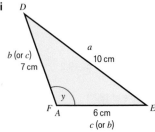

b $100.3°$
7 **a** $59.6°$ **b** $151.3°$ **c** $99.1°$ **d** $82.4°$
8 $106.3°$
9 **a** $15.4\,\text{cm}$ **b** $26.6°$ **c** $93.7\,\text{cm}^2$
10 **a** $7.90\,\text{cm}$
b Yes; $\frac{OD}{\sin39°} = \frac{12}{\sin107°}$ so $OD = \frac{12\sin39°}{\sin107°} = 7.90$
c $43.8\,\text{cm}$
11 $113°$
12 **a** $16.6\,\text{km}$ **b** $111°$
13 **a** Sine rule; $56.3°$ **b** Cosine rule; $110.8°$
14 $14.2\,\text{cm}$

13.7 Solving problems in 3D

1 **a** Sine rule **b** Cosine rule
2 **a** $65.0°$ **b** $36.7°$ **c** $60.9°$ **d** $7.5\,\text{cm}$

3 **a**

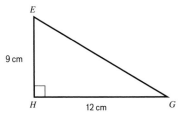

b $15\,\text{cm}$ **c** $17\,\text{cm}$ **d** $28.1°$
4 **a i** $15\,\text{cm}$ **ii** $20.5\,\text{cm}$ **iii** $16.6\,\text{cm}$ **iv** $20.5\,\text{cm}$
b $43.0°$ **c** $43.0°$ **d** $35.8°$
5 $19.1°$
6 **a i** $22.6\,\text{cm}$ **ii** $11.3\,\text{cm}$ **iii** $21.2\,\text{cm}$
b $62°$
7 **a** $10.3\,\text{cm}$ **b** $6.65\,\text{cm}$ **c** $109.6°$
8 **a** $67.6°$ **b** $238\,\text{cm}^2$
9 **a** $11°$ **b** $32°$
10 $55.2°$

13.8 Transforming trigonometric graphs 1

1 **a** $(3, -5)$ **b** $(-3, 5)$ **c** $(-3, -5)$
2 **a** $\frac{\sqrt{3}}{2}$ **b** $\frac{1}{\sqrt{3}}$ **c** $\frac{1}{\sqrt{2}}$
d 0 **e** 0 **f** $\sqrt{3}$

3 a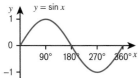

b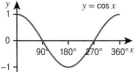

c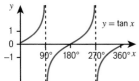

4 a i, ii

	x	$\sin x$	$-\sin x$
A	$-180°$	0	0
B	$-150°$	-0.5	0.5
C	$-90°$	-1	1
D	$-30°$	-0.5	0.5
E	$30°$	0.5	-0.5
F	$90°$	1	-1
G	$150°$	0.5	-0.5
H	$180°$	0	0

b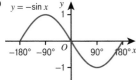

c Reflection in x-axis

5 a

	x	$\sin x$	$-\sin x$	$\sin(-x)$
A	$-180°$	0	0	0
B	$-150°$	-0.5	0.5	0.5
C	$-90°$	-1	1	1
D	$-30°$	-0.5	0.5	0.5
E	$30°$	0.5	-0.5	-0.5
F	$90°$	1	-1	-1
G	$150°$	0.5	-0.5	-0.5
H	$180°$	0	0	0

b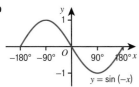

c Reflection in y-axis

6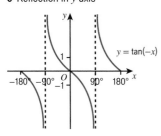

7 a Reflection in y-axis

b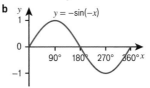

Same as graph of $y = \sin x$

8 Students' own answers, e.g.
A reflection in the x-axis followed by a reflection in the y-axis is equivalent to a rotation through $180°$ about the origin, and the sine curve has rotational symmetry of order 2 about $(0°, 0)$.

9 a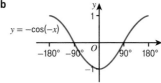

b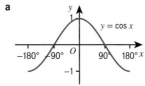

10 a Reflection in x-axis

b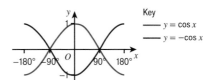

11 a Rotation through $180°$ about the origin

b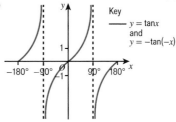

12 a $P(90°, -1)$, $Q(360°, 0)$, $R(450°, -1)$

b Q

13 A and F, B and D, C and E

13.9 Transforming trigonometric graphs 2

1 a B   **b** C   **c** A

2 a $(30, 2.5)$ **b** $(15, 1.5)$

3 a, b, c, e, f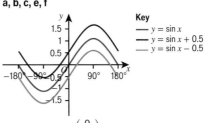

d Translation by $\begin{pmatrix} 0 \\ 0.5 \end{pmatrix}$

f Translation by $\begin{pmatrix} 0 \\ -0.5 \end{pmatrix}$

4 a, b

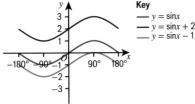

Key
— $y = \sin x$
— $y = \sin x + 2$
— $y = \sin x - 1$

5 **a** $y = \cos x - 1$
  b $y = \sin x + 1$
  c $y = \tan x + 2$

6 a

x	0°	30°	60°	90°
$\cos(x + 30°)$	$\frac{\sqrt{3}}{2}$	$\frac{1}{2}$	0	$-\frac{1}{2}$

b

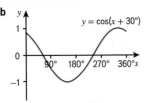

$y = \cos(x + 30°)$

c Translation by $\begin{pmatrix} -30 \\ 0 \end{pmatrix}$

7 a Translation by $\begin{pmatrix} -60 \\ 0 \end{pmatrix}$ **b** Translation by $\begin{pmatrix} -20 \\ 0 \end{pmatrix}$

  c Translation by $\begin{pmatrix} 30 \\ 0 \end{pmatrix}$

8 a Translation by $\begin{pmatrix} -40 \\ 0 \end{pmatrix}$ **b** Translation by $\begin{pmatrix} -30 \\ 0 \end{pmatrix}$

  c Translation by $\begin{pmatrix} 60 \\ 0 \end{pmatrix}$

9 a C   **b** B   **c** A

10 Yes, with reasoning, e.g.
From the unit circle definitions, $\sin\theta = \cos(90° - \theta)$ so
$\sin(x + 30°) = \cos(90° - (x + 30°)) = \cos(60° - x)$
and $\cos(60° - x) = \cos(x - 60°)$ by symmetry.

11

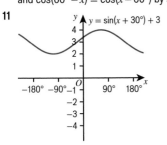

$y = \sin(x + 30°) + 3$

12 $a = 90, b = -2$

13 Check up
Key: UB = upper bound; LB = lower bound
1 $7.96\,\text{m}^2$

2 a 16.0 cm   **b** 4.61 cm

3 a 71.7°   **b** 20.0°

4 UB: $x = 4.5939$ m (4 d.p.); LB: $x = 4.4151$ m (4 d.p.);
  $x = 4.5$ m (to the nearest 0.5 metre)

5

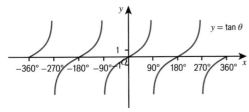

$y = \tan\theta$

---

6 a

y

b −1

7 a $x = 66°, 294°$ **b** B

8 a C   **b** B   **c** D   **d** A

9 $x = 19.5°, 160.5°, 379.5°, 520.5°$

10 23.3 cm

11 a 15.2 cm   **b** 11.3 cm

13 Triangle A 12 cm, 12 cm, 17.8 cm
  Triangle B 17.8 cm, 17.8 cm, 26.5 cm
  Triangle C 26.5 cm, 26.5 cm, 39.4 cm
  Triangle D 39.4 cm, 39.4 cm, 58.6 cm

13 Strengthen

Accuracy and 2D problem-solving
Key: UB = upper bound; LB = lower bound
1 a i, ii

C
b (or a) 12 cm   82° a (or b) 18 cm
A (or B) c B (or A)

b 107 cm²

2 a 39.6 cm²   **b** 166 m²

3 a Triangle correctly labelled
  b $\dfrac{x}{\sin 35°} = \dfrac{16}{\sin 56°}$ **c** $x = 11.1$

4 a $x = 25.4$ **b** $x = 13.3$

5 a Triangle correctly labelled
  b $\dfrac{\sin\theta}{31} = \dfrac{\sin 146°}{74}$ **c** $\theta = 13.5°$

6 a 50.9°   **b** 25.0°

7 a Triangle correctly labelled
  b $x^2 = 23^2 + 37^2 - 2 \times 23 \times 37 \times \cos 48°$
  c $x = 27.6$

8 a $x = 68.0$ **b** $x = 61.6$

9 a Triangle correctly labelled
  b $\cos\theta = \dfrac{25^2 + 32^2 - 41^2}{2 \times 25 \times 32}$ **b** $\theta = 91.1°$

10 a 120.9°   **b** 27.7°

11 a

	Upper bound value	Lower bound value
5.7	5.75	5.65
23	23.5	22.5

b UB = 0.3987; LB = 0.3827
c 5.65, 23.5   **d** 5.75, 22.5

12 UB = 10.7294; LB = 9.6295

Trigonometric graphs

1 a

x	0°	10°	20°	30°	40°	50°	60°	70°	80°	90°
$\sin x$	0	0.17	0.34	0.5	0.64	0.77	0.87	0.94	0.98	1

b, c, d

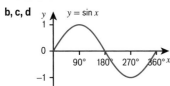

e $x = 270°$

2 a

x	0°	10°	20°	30°	40°	50°	60°	70°	80°
$\tan x$	0	0.18	0.36	0.58	0.84	1.19	1.73	2.75	5.67

b

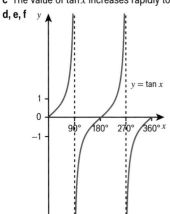

c The value of $\tan x$ increases rapidly towards infinity.

d, e, f

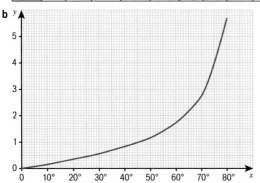

g 135°, 315°

3 $x = 126°$, $x = 234°$

4

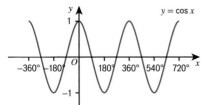

5 a No   **b** Yes   **c** No

6 a i y direction

    ii 4 up   **iii** $\begin{pmatrix} 0 \\ 4 \end{pmatrix}$

  b i x direction

    ii 4 left   **iii** $\begin{pmatrix} -4 \\ 0 \end{pmatrix}$

7 a $\cos x = \dfrac{3}{4}$

b

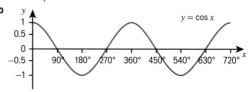

c $x = 41.4°$ **d** $x = 41.4°, 318.6°, 401.4°, 678.6°$

3D problem-solving

1 a

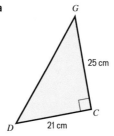

b 32.6 cm

c

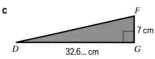

d 12.1°

2 a

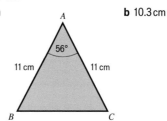

b 10.3 cm

c

d 72.3°

13 Extend

1 a $\dfrac{1}{2}ab \sin C$ **b** $\dfrac{1}{2}ac \sin B$, $\dfrac{1}{2}bc \sin A$

  c The three expressions must all have the same value, so

$$\dfrac{1}{2}ab \sin C = \dfrac{1}{2}ac \sin B$$
$$ab \sin C = ac \sin B$$
$$b \sin C = c \sin B$$
$$\dfrac{b}{\sin B} = \dfrac{c}{\sin C}$$

and similarly for $\dfrac{a}{\sin A}$

2 $d^2 = a^2 + b^2 + c^2$

3 a $a^2 = h^2 + b^2 - 2bx + x^2$ **b** $c^2 = h^2 + x^2$

  c Substituting for $h^2 + x^2$ in part **a** gives $a^2 = b^2 + c^2 - 2bx$

  d $x = c \cos A$, so $a^2 = b^2 + c^2 - 2bc \cos A$

4 $A = \frac{1}{2} \times 1 \times \sqrt{3} \times \sin\theta$

$\frac{3}{4} = \frac{1}{2} \times \sqrt{3} \times \sin\theta$

$3 = 2\sqrt{3} \times \sin\theta$

$\sin\theta = \frac{\sqrt{3}}{2}$

$\theta = 60°$

5 $\frac{\pi r^2}{3} - \frac{\sqrt{3}\,r^2}{4}$

6 68.7%

7 $x = 6$

8 a, c

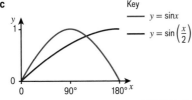

Key
—— $y = \sin x$
—— $y = \sin\left(\frac{x}{2}\right)$

b

x	0°	60°	90°	120°
$\sin\left(\frac{x}{2}\right)$	0	0.5	0.7	0.9

9

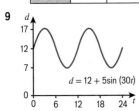

$d = 12 + 5\sin(30t)$

10 1130 cm³

11 a $x = 5$ **b** $k = 10$

12 Using the cosine rule in triangle BDE

$DE^2 = x^2 + x^2 - 2 \times x \times x \times \cos 30°$

$= 2x^2 - 2x^2 \times \frac{\sqrt{3}}{2}$

$= x^2(2 - \sqrt{3})$

Angle DAC = angle ACE = 90°, so $AC = DE$ and $AC^2 = DE^2$

Using the cosine rule in triangle ABC

$\cos ABC = \frac{AB^2 + BC^2 - AC^2}{2 \times AB \times BC}$

$= \frac{8^2 + 8^2 - x^2(2 - \sqrt{3})}{2 \times 8 \times 8}$

$= \frac{128 - x^2(2 - \sqrt{3})}{128}$

$= 1 - \frac{(2 - \sqrt{3})}{128}x^2$

13 a Area $= \frac{1}{2}(x+2)(x+4)\sin 120° = 11\sqrt{3}$

$\frac{\sqrt{3}}{4}(x+2)(x+4) = 11\sqrt{3}$

$(x+2)(x+4) = 44$

$x^2 + 6x + 8 - 44 = 0$

$x^2 + 6x - 36 = 0$

b $(x+3)^2 - 9 - 36 = 0$

$(x+3)^2 = 45$

$x + 3 = \pm\sqrt{45} = \pm 3\sqrt{5}$

$x = -3 \pm 3\sqrt{5}$

But $x + 2$ is a length and must be positive, so $x = -3 + \sqrt{5}$

14 a $k = 60$ **b** $(60°, 2)$ **c** $p = -120°, r = 240°$

d $q = 1.5$

Sample student answers

a Yes. Student A is using the formula Area $= \frac{1}{2}ab\sin C$ in the equilateral triangle.
Student B has divided the equilateral triangle into two congruent right-angled triangles and is using the formula Area $= \frac{1}{2} \times$ base $\times$ height to find the area of one of these. They will double this to obtain the area of the equilateral triangle.

b The required expression for the area is in terms of π.

c Student A:
Area $= \frac{1}{2} \times \frac{16}{\pi^2} \times \frac{\sqrt{3}}{2} = \frac{4\sqrt{3}}{\pi^2}$ cm²
Student B:
Area of triangle $XYZ = \frac{1}{2} \times XZ \times YZ$

$= \frac{1}{2} \times \frac{4}{\pi} \times \sin 30° \times \frac{4}{\pi} \times \sin 60°$

$= \frac{1}{2} \times \frac{4}{\pi} \times \frac{4}{\pi} \times \frac{1}{2} \times \frac{\sqrt{3}}{2}$

$= \frac{16 \times \sqrt{3}}{8\pi^2} = \frac{2\sqrt{3}}{\pi^2}$ cm²

So area of equilateral triangle $= 2 \times \frac{2\sqrt{3}}{\pi^2} = \frac{4\sqrt{3}}{\pi^2}$ cm²

13 Unit test
Key: UB = upper bound; LB = lower bound

1 UB: $11.25 \times \sin 48.5° = 8.4258$ m (4 d.p.)
LB: $11.15 \times \sin 47.5° = 8.2206$ m (4 d.p.)
$BC = 8$ m (nearest whole metre)

2 $A = \frac{1}{2} \times 4x \times 3x \times \sin 30° = \frac{1}{2} \times 12x^2 \times \frac{1}{2} = 3x^2$

$x^2 = \frac{A}{3}$ so $x = \sqrt{\frac{A}{3}}$

3 41.1 cm²

4 22.0 cm²

5 49.6 m

6 49.9°, 130.1°

7 a 31.0 km **b** 078°

8

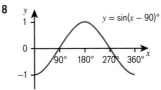

$y = \sin(x - 90)°$

9 a $y = -\cos x$ **b** $y = \sin(-x)$

10 $A(0, 1)$, $B(90, 2)$, $C(270, 0)$

11 a $\tan^{-1}\left(\frac{7}{5}\right) = 54.462... = 54.5°$ (1 d.p.)

b $\theta = 54.5°, 234.5°, 414.5°, 594.5°$ (all 1 d.p.)

12 Students' own answers, e.g.
$\tan x = \frac{1}{\sqrt{3}}$

Mixed exercise 4

1 **i** B, **ii** C, **iii** F, **iv** A, **v** E, **vi** D → 13.2, 13.3, 13.4

2 a False → 12.3
b True → 12.3
c True → 12.3
d False → 12.3

3 33.8 cm → 12.4

4 No. Original pressure $= 90 \div 40 = 2.25$ N/cm²
New pressure $= 110 \div 60 = 1\frac{5}{6}$ N/cm²
Percentage decrease $= 18.5\%$ (to 3 s.f.) → 11.3

5 Length PQ = length SR because opposite sides in a parallelogram are equal
Length QR = length PS because opposite sides in a parallelogram are equal
Length PR is common to both triangles
So triangle PQR is congruent to triangle RSP (SSS). → 12.2

6 70.4 km/h → 11.2

7 No, 15 m/s = 54 000 m/h = 54 km/h, or
No, 53 km/h = 53000 ÷ 3600 ≈ 14.7 m/s → 11.2

8 a Direct proportion → 11.4
b A → 11.4

9 2.2% → 11.1

10 0.001 82 N/cm^2 and 0.001 71 N/cm^2 → 11.3

11 1090 g → 11.3

12 5 hours 20 minutes → 11.4

13 5 times → 12.5

14 a A v, B vi, C ii → 13.8, 13.9
b i $y = \sin x + 2$

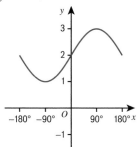

→ 13.8, 13.9

iii $y = -\sin x - 1$

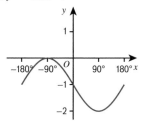

→ 13.8, 13.9

iv $y = \sin(x + 45°)$

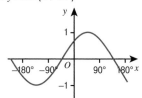

→ 13.8, 13.9

15 6.36 cm → 13.5, 13.6

16 15.5 cm^2 → 13.5

17 a Students' own answers, e.g. $y = -\cos x$
or $y = \sin(x - 90°)$ → 13.8, 13.9
b Students' own answers, e.g. $y = -\sin x + 1$
or $y = \cos(x + 90°) + 1$ → 13.8, 13.9

18 4.03 g/cm^3 → 11.3

19 4.25 cm → 13.5, 13.6

20 a 121 cm^3 → 13.7
b Angle $PTR = 90°$ → 13.7

UNIT 14 Further statistics

14.1 Sampling

1 a $\frac{9}{20} = \frac{63}{140}$ **b** $N = 55$

2 a No; it is biased to people who shop in the supermarket.
b Yes; each person is equally likely to be selected.
c Students' own answers, e.g.
It would take too long (or be impractical in other ways) to survey the whole population.
The results would be very similar to those from a representative sample.

3 a 13, 48, 09, 32, 02, 31, 50 **b** 86, 13, 60, 78, 48, 80

4 46, 12, 48, 06, 24, 14, 37, 39

5 a Mean = 2.04, median = 2, mode = 1, range = 8
b Mean = 3.2, median = 2, no mode, range = 7
c Ruby's sample gives a better estimate for the mean and mode, because they are closer to the population mean and mode. Both samples give the same estimate for the median and range.

6 a It would take too long to survey 240 students.
b $\frac{13}{40}$ **c** 78

7 Mark: If the sample does not represent the population, the proportion of students choosing salad in the population may not be the same as the proportion in the sample.
Dan: If the 40 students are not a random sample, the proportion choosing salad in the biased sample may not represent the proportion of the population who would choose salad.

8 a 1200
b Students' own answers, e.g.
The proportions choosing the different holidays this year will be the same as last year.

9 a $\frac{1}{8}$ **b** 320

10 a 342 or 343
b Students' own answers, e.g.
The population has not changed between the release and recapture times (no ants have died or have been born).
The probability of being captured is the same for each ant.
Marks have not rubbed off.

14.2 Cumulative frequency

1 a i 4 **ii** 17 **iii** 13
b 10 m; we do not know the exact lengths of the longest and shortest lengths of wood.

2 a 21
b i 15 **ii** 26

3

Mass, m (kg)	Cumulative frequency
$3 < m \leqslant 4$	4
$3 < m \leqslant 5$	16
$3 < m \leqslant 6$	33
$3 < m \leqslant 7$	43
$3 < m \leqslant 8$	50

4

Height, h (m)	Cumulative frequency
$4.0 < h \leqslant 4.2$	2
$4.0 < h \leqslant 4.4$	5
$4.0 < h \leqslant 4.6$	10
$4.0 < h \leqslant 4.8$	18
$4.0 < h \leqslant 5.0$	30
$4.0 < h \leqslant 5.2$	48
$4.0 < h \leqslant 5.4$	63
$4.0 < h \leqslant 5.6$	70

5 a

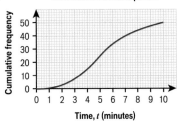

Height of giraffes

b 5.05 m

6 a

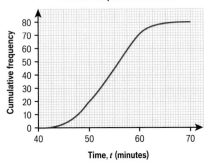

Time to solve a maths puzzle

b 5 minutes

c We do not know the exact values of the longest and shortest times.

7 a

Time to complete a 10k fun run

b 54 minutes **c** 50 minutes **d** 58 minutes

e i $\frac{1}{4}$ **ii** $\frac{3}{4}$

f 8 minutes

8 a

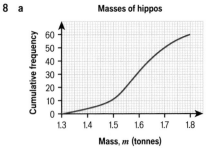

Masses of hippos

b Median = 1.59 tonnes, LQ = 1.52 tonnes, UQ = 1.67 tonnes, IQR = 0.15 tonnes

c About 20 **d** 1.64

9 a

Time, t (minutes)	Cumulative frequency
$0 < t \leqslant 10$	16
$10 < t \leqslant 20$	50
$20 < t \leqslant 30$	82
$30 < t \leqslant 40$	96
$40 < t \leqslant 50$	100

b

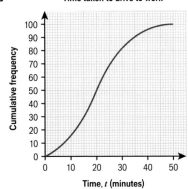

Time taken to drive to work

c About 60%

14.3 Box plots

1 a 25.5th **b** 16

2 a 25.5 **b** 38 **c** 50%

3

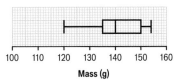

Masses of tomatoes

4 a 25 minutes

b LQ = 19 minutes, UQ = 28 minutes

c

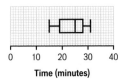

Time to complete an essay

d The difference between the $\frac{n+1}{4}$th and the $\frac{n}{4}$th values is small, and a graph is not accurate enough to be able to distinguish between them.

5 a

Height of trees

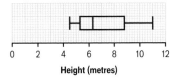

Height (metres)

b 25%

6 a

Age of people on bus

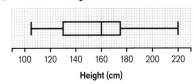

Age (years)

b 23

7 a

Height of sunflowers

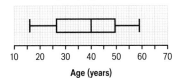

Height (cm)

b 15

c No; a quarter of the sunflowers (15 sunflowers) are between 175 cm (the UQ) and 220 cm but they could all be between 200 cm and 220 cm, for example.

8 a, b

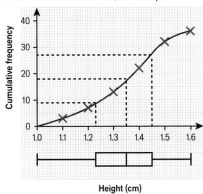

Height (cm)

c $1.23 \leqslant h \leqslant 1.45$

9 a Club A   **b** 13   **c** 75%

d Club A: 6; Club B: 3   **e** Club A: 12; Club B: 8

10 a Less variation in temperatures

b Higher average temperature

11 a Masses of different species of birds

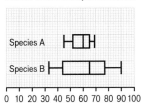

Mass (grams)

b Species A   **c** Species B

12 a Medians: 6.6 female, 7.1 male
LQs: 6 female, 6.2 male
UQs: 7.3 female, 7.9 male

b

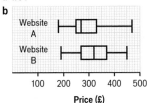

Mass (kg)

c Females had a lower median mass and a smaller interquartile range, so the spread in masses of the middle 50% of the females was lower.

13 a £80

b

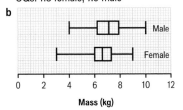

Price (£)

c Yes; median price on website B is higher than median price on website A.

14.4 Drawing histograms

1 a 3.2   **b** 7.9   **c** 150   **d** 160

2 a

Masses of birds

b $14 < m \leqslant 16$ **c** Frequency

3 a 5, 10, 10   **b** 2.4, 3.5, 1.5

4 5.5, 11.5, 15.5, 10, 8

5 a

Time, t (minutes)	Frequency	Class width	Frequency density
$40 < t \leqslant 45$	4	5	0.8
$45 < t \leqslant 50$	17	5	3.4
$50 < t \leqslant 60$	22	10	2.2
$60 < t \leqslant 80$	12	20	0.6

b

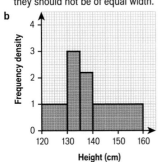

Time taken to complete a fun run

6 a Vertical axis and values plotted are frequency, and should be frequency density.
Horizontal axis scale should be labelled 120, 130, 140, ...
The bars should not be labelled with a class interval and they should not be of equal width.

b

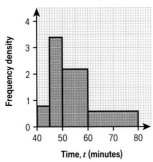

c On a histogram, the bar with largest area represents the modal class, and this may not be the tallest bar, e.g. in part **b**, modal class is $140 < h \leqslant 160$ but tallest bar is for $130 < h \leqslant 135$.

7

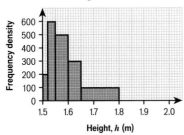

Heights of students

14.5 Interpreting histograms

1 a $\frac{1}{2}$ **b** $\frac{3}{4}$ **c** $\frac{1}{4}$

2 a 120 g (3 s.f.) **b** $120 < m \leqslant 130$

3 a Frequency **b** 15

c

Mass, m (g)	Frequency
$150 < m \leqslant 200$	$50 \times 0.3 = \mathbf{15}$
$200 < m \leqslant 300$	$\mathbf{100 \times 0.45 = 45}$
$300 < m \leqslant 400$	$\mathbf{100 \times 0.55 = 55}$
$400 < m \leqslant 600$	$\mathbf{200 \times 0.08 = 16}$

d 100   **e** 131

4 a 40
 b i Height 24, width 5   **ii** 120
 c i 80   **ii** 14   **iii** 94

5 a

Distance to work, x (km)	Frequency
$0 < x \leqslant 10$	140
$10 < x \leqslant 25$	**60**
$25 < x \leqslant 30$	**30**
$30 < x \leqslant 40$	100

b

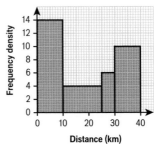

Distance to work

6 a 30

b

Time, t (mins)	Frequency
$15 < t \leqslant 16$	6
$16 < t \leqslant 18$	14
$18 < t \leqslant 20$	20
$20 < t \leqslant 25$	15
$25 < t \leqslant 30$	5

c 19.8 minutes

7 165th value = 25th value in the $10 < x \leqslant 25$ class.
$\frac{25}{60} \times 15 = 6.25$ so median is approx. 16 km

8 a 46   **b** 4.92 tonnes   **c** 6

9 a 100

b

Height, h (m)	Frequency
$1.40 < h \leqslant 1.45$	5
$1.45 < h \leqslant 1.48$	15
$1.48 < h \leqslant 1.50$	20
$1.50 < h \leqslant 1.55$	20
$1.55 < h \leqslant 1.60$	15
$1.60 < h \leqslant 1.70$	12

c 87   **d** 1.51 m   **e** 1.527 m (3 d.p.)
f 36   **g** 1.482 m

14.6 Comparing and describing distributions

1 a Range, IQR   **b** Mean, median

2 a 294.1, 267.3 (1 d.p.)   **b** 30, 14

3 On average, the males weigh more and have a larger spread of masses.

4 a, b Female African elephants are on average taller and have a greater spread of heights.

5 a 23.55 mins **b** 18 mins   **c** 85, 8
d Median and IQR; they are not affected by the outlier 95, and are more representative of the data.

6 a Median and IQR (unaffected by extreme values: 120 in males, for example)
b The males completed the race quicker on average (medians are 78 and 84.5) and had a smaller spread of times (IQRs are 8 and 14).

7 a, b

Train delays at Stratfield station

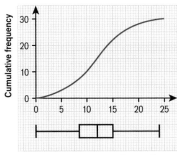

Length of delay, x (minutes)

c The median length of delay was longer at Westford.
The spread (IQR) of the delays was greater at Westford.

8 **a** Checkpoint B has higher average speeds (median 38 compared to 32).

b The speeds are more spread out at checkpoint A (IQR 21 compared to 9).

9 The females have a higher average age and a larger spread of ages.

10 **a** By drawing cumulative frequency graph:
'Before' times (seconds): LQ $\approx$ 19.5, median $\approx$ 23, UQ $\approx$ 27, IQR $\approx$ 7.5
'After' times (seconds): LQ $\approx$ 18.5, median $\approx$ 21.5, UQ $\approx$ 24.5, IQR $\approx$ 6

b Yes; the median 'after' time is lower than the median 'before' time, showing that the time to solve the puzzle has improved; the 'before' times are more spread out than the 'after' times, so the 'after' times are less varied.

14 Check up

1 048, 157, 164, 023, 125

2 **a** 90

b The population has not changed between the release and recapture times (no squirrels have been born or died).
The probability of being captured is the same for each squirrel.
Marks have not rubbed off.

3 **a**

Mass, m (kg)	Cumulative frequency
$20 < m \leqslant 23$	1
$20 < m \leqslant 26$	5
$20 < m \leqslant 29$	13
$20 < m \leqslant 32$	34
$20 < m \leqslant 35$	66
$20 < m \leqslant 38$	84
$20 < m \leqslant 41$	90

b

Masses of emperor penguins

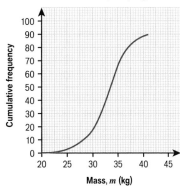

Mass, m (kg)

c 33 kg

d LQ = 30.5 kg, UQ = 35 kg, IQR = 4.5 kg

e **i** 74   **ii** 71

f

Masses of emperor penguins

Mass (kg)

4 **a**

Heights of pine trees

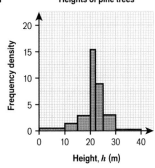

Height, h (m)

b 36

5 The girls have a higher average time and a bigger spread of times.

6 **a** 28   **b** 19

c First party had a higher average age and a greater spread of ages.

8 About 57

14 Strengthen

Sampling

1 **a** 02, 79, 21, 51, 21, 08, 01, 57, 01, 87, 33, 73, 17, 70, 18, 40, 21, 24, 20, 66, 62

b 02, 21, 21, 08, 01, 01, 17, 18, 21, 24, 20

c 02, 21, 08, 01, 17, 18, 24, 20

d 01, 02, 08, 17, 21

2 02, 21, 51, 08, 01, 57, 33, 17

3 a $\frac{1}{4}$

b In the sample: 4 tagged ducks $= \frac{1}{4}$ of the sample

In the population: 20 tagged ducks $= \frac{1}{4}$ of the population
Total population = 80 ducks

4 a 45

b The sample is representative of the population (and/or no geese have died, no geese have lost their tags).

Graphs and charts

1 a

Mass, m (kg)	Cumulative frequency
$70 < m \leqslant 75$	1
$70 < m \leqslant 80$	9
$70 < m \leqslant 85$	28
$70 < m \leqslant 90$	43
$70 < m \leqslant 95$	53
$70 < m \leqslant 100$	60

b

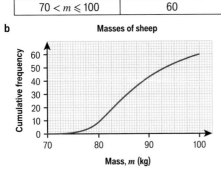

Masses of sheep

2 a Median = £4.40, LQ = £3.50, UQ = £5.10

b £1.60   **c** 36   **d** 4

3 a–d

Pocket money

Pocket money (£)

4 a i 5   **ii** 1.6

Length, l (mm)	Frequency	Class width	Frequency density
$10 < l \leqslant 15$	2	$15 - 10 = 5$	$\frac{2}{5} = 0.4$
$15 < l \leqslant 20$	8	$20 - 15 = 5$	$\frac{8}{5} = 1.6$
$20 < l \leqslant 30$	15	$30 - 20 = 10$	$\frac{15}{10} = 1.5$
$30 < l \leqslant 40$	12	$40 - 30 = 10$	$\frac{12}{10} = 1.2$
$40 < l \leqslant 60$	5	$60 - 40 = 20$	$\frac{5}{20} = 0.25$

b

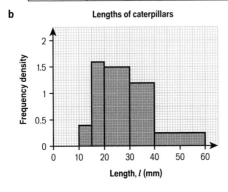

Lengths of caterpillars

5 a 3, 6, 15, 4, 2   **b** 9

c 15   **d** $\frac{3}{5}$ **e** 9   **f** 18

Comparing data

1 a

	Lower quartile	Median	Upper quartile	Interquartile range
Boys	4	6	7	3
Girls	3	5	8	5

b i higher   **ii** boys, girls

2 a 23   **b** 12, 6, 18

c 67 kg, 59 kg, 75 kg   **d** 67 kg, 59 kg, 75 kg

e 16 kg

f Female wild boars have a lower median mass.
The masses of female wild boars are more varied/spread out (have a larger IQR) than the masses of the males.

14 Extend

1 a

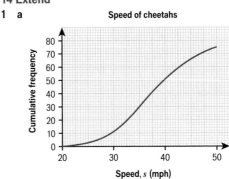

Speed of cheetahs

b i 37 mph   **ii** 9 mph

2 a The 342 customers on the list

b 40 customers

c £26.50 × 200 = £5300
That the mean of the sample is a good estimate for the mean of the population. That all the customers would spend close to the mean amount.

3 a 4   **b** 4 minutes, 9.6 minutes

4 2

5 a 14.9 s   **b** LQ = 14.3 s, UQ = 15.5 s, IQR = 1.2 s

c 1.8 s   **d** 28.6 s

6 a

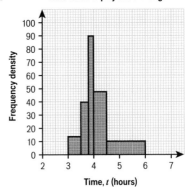

Time taken to play a round of golf

b 23

7 a

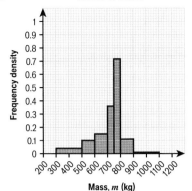

Masses of camels

b 707.5 kg   **c** 747 kg   **d** 783 kg

8 a

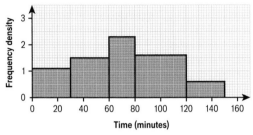

Time in car park

The missing number in the frequency table is 46.

b $T \approx 48$

9 a 5.5, 7.5   **b** Width 2 cm, height 3.5 cm

14 Test ready

Sample student answers

a Student B

b Student A has compared a measure of spread rather than a measure of average.
Student C has not given a reason to support their answer.

14 Unit test

1 a Cheaper, less time-consuming

  b No; it is biased to people at the school so it is not representative.

2 a 624

  b The sample is representative of the population/a random sample.

  c Answer in the range 120–160; it is 10% of the population (or larger) but still a small enough sample to be manageable.

3 a

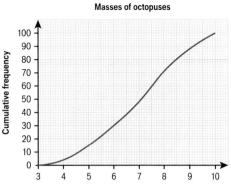

Masses of octopuses

b 7.1 kg   **c** 2.5 kg

d From graph, about 39% weigh between 3 kg and 6.5 kg, so about $100 - 39 = 61$ weigh more than 6.5 kg; $\frac{61}{100} = 61\%$, which is over 60%.

4 a

Median	22
Lower quartile	17
Upper quartile	24
Shortest time	12
Longest time	32

b Yes; the women's median time was 22 seconds, which is faster than the men's median time of 24 seconds.

c No; the IQR of the men's times is 4 seconds, which is smaller than the IQR of the women's times (IQR = 7 seconds).

5 a

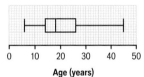

Ages of cast in show

b 45

6 a 42   **b** Median $\approx$ 304 g, range $\approx$ 43 g

c Assumptions for part **a**: probability of being captured is the same for all squirrels; population has not changed between the release and recapture times, and no tags have fallen off.
Assumption for part **b**: the tagged squirrels are representative of the population.

7 a

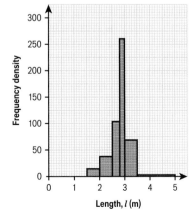

Lengths of dolphins

b 2.87 m (2 d.p.)   **c** 24

8 a $\frac{13}{50}$

b The calculation assumes that the lengths of the seahorses are evenly distributed in the 15 cm to 17 cm class interval.

9

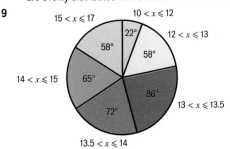

Angles do not add to 360° due to rounding.

10 Students' own answers.

UNIT 15 Equations and graphs

15.1 Solving simultaneous equations graphically

1 a $x^2 + y^2 = 9$

b

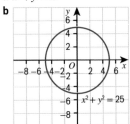

2 a i, ii

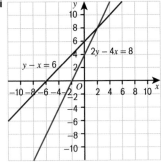

b $x = 2, y = 8$ **c** Students' own checks.

3 a i B ** ii** A ** iii** C
b i $x = 1, y = 4$ ** ii** $x = 4, y = 1$
 iii $x = 0, y = 2$

4 Answers in ranges $x = 1.3$ to 1.4, $y = -0.7$ to -0.6

5 a $x = 1, y = 6$ **b** $x = 3, y = 9$
 c $x = -1, y = 7$ **d** $x = -2, y = -3$
 e Algebraic solutions should be the same as graphical solutions.

6 $x = 1.3, y = 0.3$ and $x = -2.3, y = -3.3$

7 a, b $x = 3, y = -2$ and $x = -1.5, y = 4.75$
 c Algebraically, because values read from a graph are estimates.

8 a, b

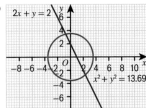

b $x = -0.8, y = 3.6$ and $x = 2.4, y = -2.8$

9 $x = -1.6, y = 2.6$ and $x = 2.6, y = -1.6$

15.2 Representing inequalities graphically

1 a $\{x : x \leqslant 6\}$ **b** $\{x : x > -2\}$ **c** $\{x : x \leqslant 4\}$

2 a i $x < 2$ ** ii** $y \leqslant 2$ ** iii** $-1 \leqslant x \leqslant 3$

b i

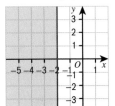

ii

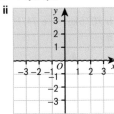

iii

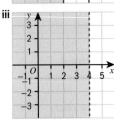

iv

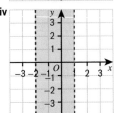

v

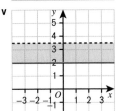

vi

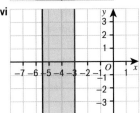

vii

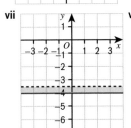

viii

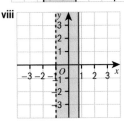

3 a, b, d

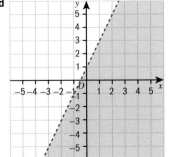

c $y = 3$ and $2x + 1 = 5$; yes

4 **a**

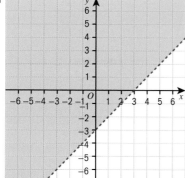

b

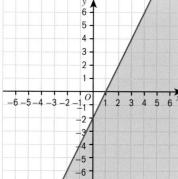

c

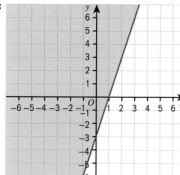

d

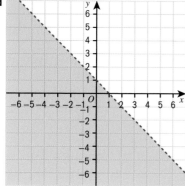

5

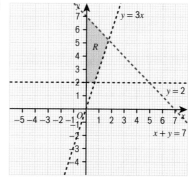

6 **a i** $y = 3, y = 2x - 1, x = -2$
   ii $y < 3, y > 2x - 1, x \geqslant -2$
   b i $x = 2, y = x + 2, y = -x - 2$
   ii $x < 2, y \leqslant x + 2, y \geqslant -x - 2$

7 $y > 2x - 3, y > -3x, y \leqslant 5$

8 15 points

9 **a, b**

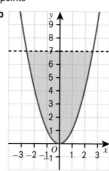

10 a $-1 \leqslant x \leqslant 2$ **b** $x < -1$ and $x > 2$

11 a

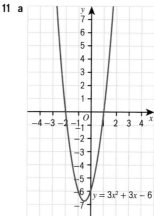

b i $-2 < x < 1$ **ii** $\{x : -2 < x < 1\}$

12 a

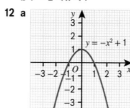

b $\{x : x \leqslant -1\} \cup \{x : x \geqslant 1\}$

13 a

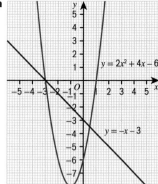

$y = 2x^2 + 4x - 6$

$y = -x - 3$

b $\{x : x \leqslant -3\} \cup \left\{x : x \geqslant \frac{1}{2}\right\}$

15.3 Quadratic equations

1 a $x = -2, x = -1$ **b** $x = 1.5, x = -4$

2 a $(x+1)^2 - 6$ **b** $2(x+2)^2 - 4$

3 a $(1, 7)$ **b** $x = 1$

4

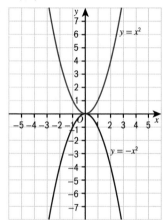

$y = x^2$

$y = -x^2$

5 a $x = -1$ and $x = -3$ **b** $(-2, -1)$

6 a

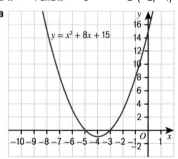

$y = x^2 + 8x + 15$

b $x = -3$ and $x = -5$

c Line of symmetry is halfway between the roots.

d $(-4, -1)$ **e** $(0, 15)$ **f** Minimum

g A positive coefficient of x^2 gives a curve with a minimum; a negative coefficient of x^2 gives curve with a maximum.

7 a, b, c, g

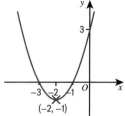

$(-2, -1)$

b $x = -3$ and $x = -1$ **c** $y = 3$

d Minimum **e** $x = -2$

f $y = -1$

8 a

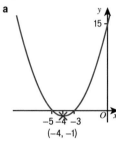

$(-4, -1)$

b

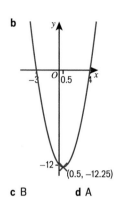

$(0.5, -12.25)$

9 a C **b** D **c** B **d** A

10 $(-1.5, -6.25)$

11 a Minimum, $(1, 3)$ **b** Maximum, $(-3, -2)$

c Minimum, $(5, -2)$ **d** Minimum, $(-3, -5)$

e Minimum, $(2, 1)$ **f** Maximum, $(-1, 4)$

g When a quadratic is written in completed square form, $y = a(x+b)^2 + c$, the coordinates of the turning point are $(-b, c)$.

12 a $(4, 0)$ and $(-2, 0)$ **b** $(0, -8)$

c $(x-1)^2 - 9$ **d** $(1, -9)$

e Minimum; coefficient of x^2 is positive.

f

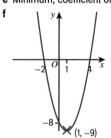

$(1, -9)$

13 a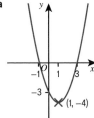

b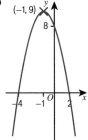

c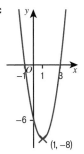

d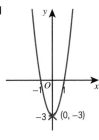

14 a $a = 2, b = 3$ **b** $(2, 3)$

15 a $(-3, -5)$ **b** $(-3 \pm \sqrt{5}, 0)$

16 a $x = -1 \pm \sqrt{3}$ **b** $x = -2$ and $x = 4$

  c $x = -2 \pm \dfrac{2}{\sqrt{3}}$

17 a $x = -2 \pm \sqrt{7}$ **b** $x = 2 \pm \dfrac{3}{\sqrt{2}}$

  c $x = -3 \pm \sqrt{13}$

18 The graph should have a maximum since the coefficient of x^2 is negative.
The graph should cross the y-axis at $(0, 6)$.
The graph should have roots at $x = 3$ and $x = -1$.
The graph should have a turning point at $(1, 8)$.

19 $y = x^2 + 4x + 1$

15.4 Using quadratic graphs

1 a $x = 5.3$ and $x = -1.3$ **b** $x = 2.4$ and $x = -0.9$
  c $x = -1.0$ and $x = 1.7$

2 a $x = -5$ and $x = -1$ **b** $x = -1 \pm \dfrac{\sqrt{6}}{2}$

  c $x = -1 \pm \dfrac{2}{\sqrt{3}}$

3 a i A   **ii** B
  b i $x = -1.3$ and $x = 5.3$
    ii $x = -1.1$ and $x = 1.5$
  c i $x = -1.32$ and $x = 5.32$
    ii $x = -1.14$ and $x = 1.47$

4 a

x	-4	-3	-2	-1	0	1	2
y	14	4	-2	-4	-2	4	14

b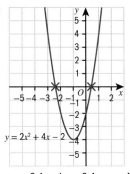

  c $x = -2.4$ and $x = 0.4$ **d** $2(x+1)^2 - 4$

5 a $y = x^2 - x - 2$ **b** $y = x^2 - 10x + 21$
  c $y = x^2 + 4x + 4$

6 a i $(-1, 1)$ **ii** $(0, 2)$
  iii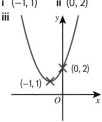

  b i $(-2, -3)$ **ii** $(0, -7)$
    iii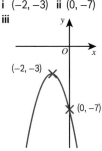

  c i $(1, 1)$ **ii** $(0, 3)$ **iii**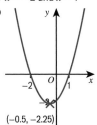

  d The graphs of the equations do not cross the x-axis.

7 a $(3, 5)$
  b The turning point is at $(3, 5)$. Since this is a minimum, the whole graph is above the x-axis.

8 a 0 roots   **b** 2 roots   **c** 2 roots   **d** 1 root
  e 0 roots   **f** 2 roots   **g** 1 root   **h** 2 roots

9 a $x = 2 \pm \sqrt{7}$
  b $x^2 - 7x + 13 = 0$
  $\left(x - \dfrac{7}{2}\right)^2 - \dfrac{49}{4} + 13 = 0$
  $\left(x - \dfrac{7}{2}\right)^2 = -\dfrac{3}{4}$
  There are no real roots of a negative number.
  OR
  The graph has a turning point at $\left(\dfrac{7}{2}, \dfrac{3}{4}\right)$; since this is a minimum, the whole graph is above the x-axis.

10 a $x = -2$ and $x = 1$
  b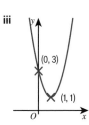

  c $\{x : -2 < x < 1\}$ **d** $\{x : x \leqslant -2\} \cup \{x : x \geqslant 1\}$

11 a $\{x : -1 < x < 3\}$ **b** $\{x : -5 < x < 2\}$
  c $\{x : x < -4\} \cup \{x : x > -1\}$

12 a $\{x : -5 < x < -2\}$ **s** $\{x : x < 2\} \cup \{x : x > 4\}$
  c $\{x : 1 < x < 5\}$

13 a

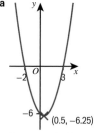

b $\{x : x < -2\} \cup \{x : x > 3\}$

14 a $\{x : -4 < x < 3\}$　　**b** $\{x : x \leqslant -1\} \cup \{x : x \geqslant 2\}$
　c $\{x : x \leqslant -3\} \cup \{x : x \geqslant 3\}$

15 $\{x : x < -6\} \cup \{x : x > 6\}$

16 $\{n : 2 < n < 4\}$

17 a $x < -2$ and $x > 2$; $-8 \leqslant x \leqslant 8$

　b

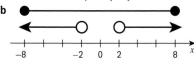

　c $\{x : -8 \leqslant x < 2\} \cup \{x : 2 < x \leqslant 8\}$

18 $\{m : -5 < m < -3\} \cup \{m : 3 < m < 5\}$

15.5 Cubic equations

1 a $x^2 + 5x + 6$　　**b** $x^2 + x - 12$
　c $2x^2 - 9x - 5$　　**d** $3x^2 - 10x + 8$

2 a $x = 5$ or $x = -2$　　**b** $x = 1$ or $x = -1$
　c $x = 3$ or $x = -1$　　**d** $x = -2$ or $x = \frac{1}{3}$

3 $x^3 + 6x^2 + 9x + 2$

4 $x^3 + 9x^2 + 26x + 24$

5 a $x^3 + 8x^2 + 17x + 10$　　**b** $x^3 - x^2 - 14x + 24$
　c $x^3 + 4x^2 + x - 6$　　**d** $x^3 + x^2 - 20x$
　e $x^3 + x^2 - x - 1$　　**f** $x^3 + 9x^2 + 27x + 27$
　g $2x^3 - 4x^2 - 30x$　　**h** $2x^3 - 3x^2 - 50x + 75$
　i $2x^3 - 3x^2 - 50x + 75$
　j The answers are the same. The order in which you expand
　brackets doesn't matter. Students' own preference as to
　which they found easier.

6 $x^3 + 2x^2 - x - 2$

7 a $x = -4, x = -1, x = 2$　　**b** $(0, -8)$

8 a $x = -1, x = -2, x = -5$　　**b** $(0, 10)$
　c

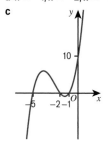

9 a D　　　**b** C　　　**c** A
　d E　　　**e** F　　　**f** B

10 a 3 roots　　**b** 1 repeated root
　c 3 roots　　**d** 2 roots (one repeated)
　e 1 root　　**f** 3 roots

11 a

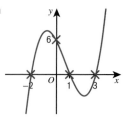

b

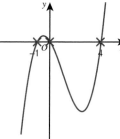

c

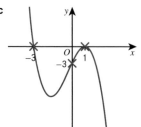

d

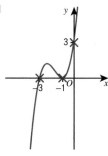

e

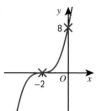

f

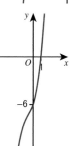

12

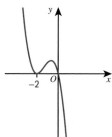

13 $a = 2, b = -9, c = -18$

14 $a = 3, b = 6, c = -8$

15.6 Using iteration to solve equations

1 a $x = \pm 8$　　**b** $x = 5$　　**c** $n = \pm 4.2$　　**d** $m = 3.1$

2 a $x^2 = y - 4$　　　　**b** $x^2 = x + 3$
　c $x^2 = 2x + 5$　　　　**d** $x^2 = 4 - 3x$

3 a When $x = 2$, $x^2 - x = 2 < 3$;
　　when $x = 3$, $x^2 - x = 6 > 3$
　　so $2 < x < 3$
　b When $x = 1$, $2x^2 - x = 1 < 2$;
　　when $x = 2$, $2x^2 - x = 6 > 2$
　　so $1 < x < 2$
　c When $x = 0$, $x^2 + 2x = 0 < 1$;
　　when $x = 1$, $x^2 + 2x = 3 > 1$
　　so $0 < x < 1$

4 a $x = \sqrt{x - 3}$　**b** $x = \sqrt{7 + 2x}$　**c** $x = \sqrt[3]{4 - x}$
　d $x = \sqrt[3]{11 + x}$　**e** $x = \sqrt{5x + 1}$　**f** $x = \sqrt[3]{2x^2 - 3}$

5 a 3.233 879 677　　　　**b** 5.694 010 531
　c $-1.322 033 898$

6 a When $x = 1$, $x^3 - x = 0 < 3$;
when $x = 2$, $x^3 - x = 6 > 3$
so solution where $y = 3$ is for x between 1 and 2

b $x^3 - x = 3$; $x^3 = x + 3$; $x = \sqrt[3]{x + 3}$

c $x_3 = 1.670\,492\,373$

7 a $x(x - 3) = -7$; $x = \dfrac{-7}{x - 3}$

b $x(x^2 + 5) - 8 = 0$; $x(x^2 + 5) = 8$; $x = \dfrac{8}{x^2 + 5}$

8 a $x^2 + 4x - 2 = 0$; $x(x + 4) = 2$; $x = \dfrac{2}{x + 4}$

b $x^3 - 2x + 5 = 0$; $x(x^2 - 2) = -5$; $x = \dfrac{-5}{x^2 - 2}$

9 a $x_3 = 0.608\,672\,037\,7$

b $0.008\,15\ldots$, which is very close to zero.

10 a $x_3 = 1.409\,45$

b $0.028\,25\ldots$ (using unrounded value from part **a**), which is close to zero, so the solution is fairly accurate.

11 a When $x = 1$, $x^3 - 5x + 3 = 1^3 - 5 + 3 = -1$;
when $x = 2$, $x^3 - 5x + 3 = 2^3 - 10 + 3 = 1$
For $x = 1$, $y < 0$ and for $x = 2$, $y > 0$
So solution to $y = 0$ lies between $x = 1$ and $x = 2$

b $x^3 - 5x + 3 = 0$; $x(x^2 - 5) = -3$; $x = \dfrac{-3}{x^2 - 5}$

c $x_3 = 0.660\,364\ldots$

d $-0.013\,848\ldots$, which is very close to zero.

12 $a = 3$, $b = -6$

13 a $x_1 = -\dfrac{4}{7}$, $x_2 = -\dfrac{49}{180}$, $x_3 = -0.254\,718\,983\,6$

b $x = \dfrac{1}{x^2 - 4}$ rearranges to $x^3 - 4x - 1 = 0$ so x_1, x_2 and x_3 are estimates for a solution to $x^3 - 4x - 1 = 0$

14 $x = 1.4562$

15 $x = -1.912\,23$

15 Check up

1

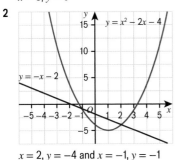

$x = 8$, $y = 0$

2

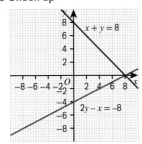

$x = 2$, $y = -4$ and $x = -1$, $y = -1$

3 $y < 2x - 1$, $x \leqslant 2$, $y \geqslant -2$

4 $x = 2.7$ and $x = -0.7$

5 a, b

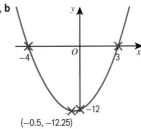

$(-0.5, -12.25)$

c $\{x : -4 < x < 3\}$

6 a $(0, 1)$ **b** $(-3, -8)$

7 a $y = x^2 - 4x + 3$ **b** $(2, -1)$

8 $\{x : x \leqslant -4\} \cup [x : x \geqslant 4]$

9 $x^3 + 5x^2 - 17x - 21$

10

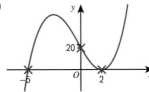

11 a $x^2 + x = 1$; $x^2 = 1 - x$; $x = \sqrt{1 - x}$

b $x_3 = 0.627\,491\ldots$

13 Any equations of these forms:
$y = a(x + 1)(x - 3)$
$y = a(x + 1)^2(x - 3)$
$y = a(x - 3)^2(x + 1)$
$y = a(x + b)(x + 1)(x - 3)$

15 Strengthen

Simultaneous equations and inequalities

1 a $(-2, 2)$ **b** $x = -2$, $y = 2$

2 a

x	-2	-1	0	1	2	3	4
y	11	5	1	-1	-1	1	5

b, c

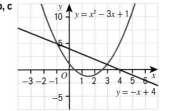

d $(3, 1)$ and $(-1, 5)$

e $x = 3$, $y = 1$, and $x = -1$, $y = 5$

3 a $x = 1$, $y = 6$

b $x = 3$, $y = 4$ and $x = -1$, $y = 0$

4 a $x = -2$, $y = -3$, $y = x + 2$

b $x = -2$: $\leqslant$ or $\geqslant$; $y = -3$: $<$ or $>$; $y = x + 2$: $\leqslant$ or $\geqslant$

c $x \leqslant -2$, $y > -3$, $y \leqslant x + 2$

5 $y < 1$, $y \geqslant x - 3$, $y \geqslant -x$

Graphs of quadratic equations

1 a i $x = -1$ and $x = 3$ **ii** $y = -6$

  b i $x = -3$ and $x = 1$ **ii** $y = -9$

  c i $x = -2$ and $x = 2$ **ii** $y = 4$

  d i $x = 1$ and $x = 4$ **ii** $y = -4$

2 a When $y = 0$
$x^2 + 2x - 8 = 0$
So $(x - 2)(x + 4) = 0$
There are two possible solutions:
$x - 2 = 0$ hence $x = 2$
$x + 4 = 0$ hence $x = -4$
So the roots are $x = 2$ and $x = -4$

b When $x = 0$
$y = x^2 + 2x - 8$
$y = 0^2 - 0 - 8$
$y = -8$
So the intercept with the y-axis is at $y = -8$

3 a $x = 5$ and $x = -3$; $y = -15$
b $x = -8$ and $x = 2$; $y = 16$
c $x = -1$ and $x = 3$; $y = -6$
d $x = 1$ and $x = -5$; $y = -15$

4 a $y = (x + 1)^2 - 1 - 8$; $y = (x + 1)^2 - 9$
b $x = -1$, $y = -9$ **c** Minimum

5 a $(1, -16)$, minimum   **b** $(-3, 25)$, maximum
c $(1, -8)$, minimum   **d** $(-2, -27)$, minimum

6 a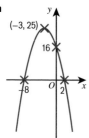
b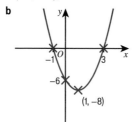
c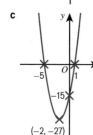

7 a $0 = 1 - b + c$; $-2 = c$ **b** $b = -1$, $c = -2$
c $y = x^2 - x - 2$

8 a $x = -3$ and $x = 5$
b i $-3 < x < 5$ **ii** $x < -3$ and $x > 5$
c i $\{x : -3 < x < 5\}$ **ii** $\{x : x < -3\} \cup \{x : x > 5\}$

9 a $x^2 - 36 \leqslant 0$
b

c $-6 \leqslant x \leqslant 6$

Cubic equations

1 a $x^2 - 2x - 3$
b

×	x^2	$-2x$	-3
x	x^3	$-2x^2$	$-3x$
$+5$	$5x^2$	$-10x$	-15

c $x^3 + 3x^2 - 13x - 15$

2 a i $x = -1$, $x = 1$, $x = -2$ **ii** $y = -2$
b i $x = -1$, $x = -2$, $x = -5$ **ii** $y = 10$
c i $x = -1$, $x = -2$, $x = 3$ **ii** $y = 6$

3 a $x - 3 = 0$ hence $x = 3$
$x + 4 = 0$ hence $x = -4$
$x - 1 = 0$ hence $x = 1$
So the roots are $x = 3$, $x = -4$ and $x = 1$

b When $x = 0$
$y = (x - 3)(x + 4)(x - 1)$
$y = (0 - 3)(0 + 4)(0 - 1)$
$y = 12$

4 a $x = -3$, $x = 7$, $x = -2$; $y = -42$
b $x = 10$, $x = -2$, $x = -1$; $y = 20$
c $x = 2.5$, $x = -2$, $x = 1$; $y = 10$
d $x = 3$, $x = 2$; $y = -54$

Iteration

1 a $y = x^2 - 4x + 1$
$0 = x^2 - 4x + 1$
$x^2 = 4x - 1$
$x = \sqrt{4x - 1}$
$x_{n+1} = \sqrt{4x_n - 1}$

b $x_1 = 3.605\,551\,275$
$x_2 = 3.663\,632\,774$
$x_3 = 3.695\,203\,796$
$x_4 = 3.712\,252\,037$
$x_5 = 3.721\,425\,553$

2 a $x_{n+1} = \sqrt{8x_n - 1}$ **b** 7.873

15 Extend

1 a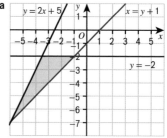

b 6.25 units^2

2 a 1 root   **b** 0 roots   **c** 2 roots
d 0 roots   **e** 2 roots   **f** 0 roots

3 a i $y = -a(x + 3)^2 + 4$ **ii** $y = a(x - 4)^2 - 3$
b $y = ax(x + 3)$

4 $6\sqrt{5}$ units

5 RHS $= x(x - 2)(x + 2)$
$= x(x^2 - 2x + 2x - 4)$
$= x(x^2 - 4)$
$= x^3 - 4x$
$=$ LHS

6 $-n^2 + 4n + 20 = -(n - 2)^2 + 24$
The largest value this can take is when $n - 2 = 0$, when $n = 2$

7 4

8 a $x = 0.6$, $y = 4.2$ and $x = -3.2$, $y = 2.7$
b $x = -0.4$, $y = -2.6$ and $x = 0.6$, $y = 1.6$

9 5

10 a $\{x : -1 < x < 1.5\}$ **b** $\{x : -2 < x < 5\}$

11 $x > 2$

12 $(x + 2)(x - 1.5)(x - 4) = x^3 - 3.5x^2 - 5x + 12$
The curve passes through $(0, 24)$, so the value of y when $x = 0$ has to be 24, hence all the coefficients obtained from the first expansion must be doubled.
So the equation of the cubic curve is
$y = 2x^3 - 7x^2 - 10x + 24$, i.e. $a = 2$, $b = -7$, $c = -10$, $d = 24$

13 a $f(1) = 1 + 1 - 3 - 2 = -3 < 0$
and $f(2) = 8 + 4 - 6 - 2 = 4 > 0$
so the graph of $f(x)$ will cross the x-axis between $x = 1$
and $x = 2$
b $f(x) = x^2(x + 1) - 3x - 2$
So $f(x) = 0$ can be written as $x^2(x + 1) = 3x + 2$
hence $x = \sqrt{\dfrac{3x + 2}{x + 1}}$
c $x_1 = 1.612\,451\,550$
$x_2 = 1.617\,781\,747$
$x_3 = 1.618\,022\,615$
$x_4 = 1.618\,033\,476$
Since x_3 and x_4 both round to 1.6180 to 4 decimal places,
$\alpha = 1.6180$ (4 d.p.)

15 Test ready

Sample student answers

Student B gives the best answer as they have used the information on the graph and the equation given.
Student A has calculated c correctly, but has incorrectly identified b.
Student C has incorrect values for b and c.

15 Unit test

1

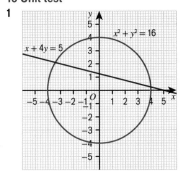

$x \approx 4$, $y \approx 0.25$ and $x \approx -3.4$, $y \approx 2.1$

2 a C   **b** A   **c** B   **d** D

3

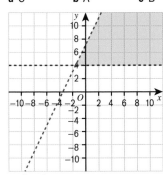

4 a $x^3 - x^2 - 14x + 24$ **b** Cubic
  c $x = 3$, $x = -4$, $x = 2$ **d** $(0, 24)$

5

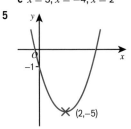

6 Any equation of the form $y = -a(x - 3)^2 - 4$

7 $25 - 6.25\pi$

8 $4\sqrt{2}$

9 $x < -3$ and $x > 2$

10 a No   **b** Yes   **c** Yes

11 6, 7, 8

12 a $x_1 = -2.168\,702\,885$
$x_2 = -2.166\,482\,503$
$x_3 = -2.166\,324\,805$
b $x_{n+1} = \sqrt[3]{x_n - 8}$ is a rearrangement of $x^3 - x + 8 = 0$ so x_1,
x_2 and x_3 are iterations that are estimates of a solution
to $x^3 - x + 8 = 0$

13 a Stream Speed: $y = 20 + 1.5x$, ONline: $y = 2x$
  b 40 GB   **c** Above 40 GB

14 a $37°$, $143°$ **b** $11°$, $191°$

15 Students' own answers.

UNIT 16 Circle theorems

16.1 Radii and chords

1 $x = 110°$ (base angles of an isosceles triangle are equal)
$y = 145°$ (angles on a straight line add up to $180°$)

2 AD is common; $\angle ABD = \angle ACD$ and $AB = AC$
(triangle ABC is isosceles); $\angle BAD = \angle CAD = 180° - 90°$
$-\angle ABD$ (angles in a triangle add up to $180°$).
Therefore the triangles are congruent (SAS).

3 a, b, c Students' own diagrams, e.g.

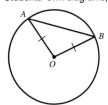

  d Isosceles
  e AO and OB are radii, so are equal in length. So triangle
AOB has two equal sides and is isosceles.

4 a $i = 30°$ **b** $j = 21°$ **c** $k = 64°$; $l = 116°$; $m = 32°$

5 a OM is common; $OA = OB$ (radii of same circle);
$\angle OMA = 90°$ (angles on a straight line add up to $180°$).
Therefore the triangles are congruent (RHS).
  b Triangles OAM and OBM are congruent. $OA = OB$,
$OM = OM$, so $AM = MB$.
  c $AM = MB$ so M is in the middle between A and B.
Therefore M is the midpoint of AB.

6 a, b, c Students' own diagrams, e.g.

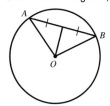

  d $90°$

7 a

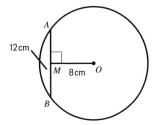

  b $AM = 6$ cm (the perpendicular from the centre of a circle
to a chord bisects the chord)
  c $AO = 10$ cm

8 $\angle OMA = 90°$ (line from centre of circle to midpoint of chord is perpendicular to the chord) so triangle OAM is right-angled, and $OM^2 = OA^2 - AM^2 = 17^2 - 8^2$ (Pythagoras' theorem). Hence $OM = 15\,\text{cm}$.

9 $AB = 20\,\text{cm}$

10 a 90° (line from centre of circle to midpoint of chord is perpendicular to the chord)

    b 65° (angles in a triangle add up to 180°)

    c 130° (triangles OAM and OBM are congruent)

16.2 Tangents

1 a 58°   **b** 11.3 cm (1 d.p.)

2 6.5 cm (1 d.p.)

3 a i 90°   **ii** 90°

    (angle between tangent and radius is 90°)

    b i 20°   **ii** 20°

    (base angles of an isosceles triangle are equal)

    c 70°

4 a OC is common; $AO = OB$ (radii of same circle); $\angle OAC = \angle OBC = 90°$.

    Therefore the triangles are congruent (RHS).

    b $AC = BC$ because the two triangles are congruent.

5 a $\angle OAP = \angle OBP = 90°$ (angle between tangent and radius is 90°)

    $a = 360 - 90 - 90 - 20 = 160°$ (angles in a quadrilateral add up to 360°)

    b $OA = OB$ (radii of same circle)

    $\angle OAB = (180 - 136) \div 2 = 22°$ (angles in a triangle add up to 180°, and base angles of an isosceles triangle are equal)

    $\angle OAP = 90°$ (angle between tangent and radius is 90°)

    So $b = 90 - 22 = 68°$

    c $OB = OA$ (radii of same circle)

    $\angle OBA = \angle OAB = 14°$ (base angles of an isosceles triangle are equal)

    $c = 180 - 14 - 14 = 152°$ (angles in a triangle add up to 180°)

    $\angle OBP = 90°$ (angle between tangent and radius is 90°)

    So $d = 90 - 14 = 76°$

    d $\angle OBP = 90°$ (angle between tangent and radius is 90°)

    $\angle ABO = 90 - 34 = 56°$

    $e = \angle ABO = 56°$ (base angles of an isosceles triangle are equal)

    $f = 180 - 56 - 56 = 68°$ (angles in a triangle add up to 180°)

    e $g = \angle BAP$ (base angles of an isosceles triangle are equal)

    $g = (180 - 50) \div 2 = 65°$

    $\angle OBP = 90°$ (angle between tangent and radius is 90°)

    So $h = 90 - 65 = 25°$

    f $\angle OTP = 90°$ (angle between tangent and radius is 90°)

    $\angle TOP = 180 - (90 + 32) = 58°$ (angles in a triangle add up to 180°)

    $\angle SOT = 180 - 58 = 122°$ (angles on a straight line add up to 180°)

    $OS = OT$ (radii of same circle)

    So $i = (180 - 122) \div 2 = 29°$ (base angles of an isosceles triangle are equal)

6 $\angle OAC = 90°$ (angle between tangent and radius is 90°)

    $\angle OAB = y°$ (base angles of an isosceles triangle are equal)

    $\angle BAC = \angle OAB + \angle OAC = y° + 90°$

    $\angle ACO = 180 - 90 - y - y = (90 - 2y)°$ (angles in a triangle add up to 180°)

7 $OA = OB$ (radii of same circle); OT is common; $\angle TAO = \angle TBO = 90°$ (angle between tangent and radius is 90°). Therefore triangles OAT and OBT are congruent (RHS).

In congruent triangles, corresponding angles are equal, so $a = b$ and $x = y$.

8 a $a = 63.6°$ **b** $b = 67.4°$ **c** $x = 35.1\,\text{cm}$ **d** $y = 146.6°$

9 16.1 cm

16.3 Angles in circles 1

1 a   **b**

2 $x = 56 + 52 = 108°$ (exterior angle is equal to the sum of the interior opposite angles)

3 a, b Students' own diagrams, e.g.

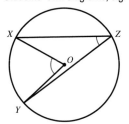

    c Angle at $O = 2 \times$ angle at Z

4 a $a = 140°$ (angle at the centre is twice the angle at the circumference)

    b $b = 48°$ (angle at the centre is twice the angle at the circumference)

    c $c = 250°$ (angles at a point add up to 360°)

    $d = 125°$ (angle at the centre is twice the angle at the circumference)

5 a Students' own diagrams, e.g.

    b 90°   **c** The angle at B is always 90°.

6 a 180°

    b $\angle AOB = 180°$; $\angle ACB = 180 \div 2 = 90°$ because the angle at the centre is twice the angle at the circumference.

7 a $a = 90°$ (angle in a semicircle is 90°)

    $b = 62°$ (angles in a triangle add up to 180°)

    b $c = 16°$ (base angles of an isosceles triangle are equal)

    $d = 180 - 16 - 16 = 148°$ (angles in a triangle add up to 180°)

    $e = 90 - 16 = 74°$ (angle in a semicircle is 90°)

    c $f = 30°$ (angle in a semicircle is 90°, and angles in a triangle add up to 180°); $2f = 60°$

    d $g = 100°$ (angle at the centre is twice the angle at the circumference)

    e $h = 105°$ (angle at the centre is twice the angle at the circumference)

    f $i = 105°$ (angles at a point add up to 360°, and the angle at the centre is twice the angle at the circumference)

8 a The angle at the centre is twice the angle at the circumference, so
$a = 230 \div 2 = 115°$

b Mario calculated the obtuse angle at O and halved it.

9 a $j = 46°$ (angle at the centre is twice the angle at the circumference)

b $a = 138°$ (base angles of an isosceles triangle are equal)
$b = 48°$ (angle in a semicircle is $90°$, and angles in a triangle add up to $180°$)

c $c = 35°$ (angle at the centre is twice the angle at the circumference)
$\angle OBA = 90°$ (angle between tangent and radius is $90°$)
$\angle OBD = 35°$ (base angles of an isosceles triangle are equal)
$d = 90 - 35 = 55°$

10 Sue is correct; angle at the centre is twice the angle at the circumference, and $2 \times 30 = 60°$.

11 $\angle ABC = 90°$ (angle in a semicircle is $90°$)
Using Pythagoras, $AC = \sqrt{80}$ so $r = \frac{1}{2}\sqrt{80}$
Area $= \pi r^2 = \pi \times \frac{1}{4} \times 80 = 20\pi$

12 a $\angle AOC = 130°$ (angle at the centre is twice the angle at the circumference)
Arc length $AC = \frac{130}{360} \times 2\pi \times 24 = 54.5$ mm (1 d.p.)

b $\angle AOB = 50°$ (angles on a straight line add up to $180°$)
Area of sector $AOB = \frac{50}{360} \times \pi \times 24^2 = 80\pi$
Area of triangle $ABC = \frac{1}{2} \times 24 \times 24 \times \sin 50°$
Shaded area $= 30.7 \text{ mm}^2$ (1 d.p.)

13 $\angle ABO = \angle ADO = 90°$ (angle between tangent and radius is $90°$)
$\angle DOB = 125°$ (angles in a quadrilateral add up to $360°$)
$\angle BCD = 62.5°$ (angle at the centre is twice the angle at the circumference)

14 $\angle ABO = 25°$ (base angles of an isosceles triangle are equal)
$\angle BOA = 130°$ (angles in a triangle add up to $180°$)
$\angle AOD = 50°$ (angles on a straight line add up to $180°$)
$\angle ODC = 90°$ (line from centre to midpoint of chord is perpendicular to the chord)
$OA = 5$ cm (radius of circle)
$OD = OA \cos 50° = 5 \cos 50° = 3.2$ cm (1 d.p.)

16.4 Angles in circles 2

1 Minor segment

2 $180° - 2x$

3 a $a = b = c = 50°$ **b** $d = 2x$

c All the angles at the circumference in the same segment are equal, and are half the angle at the centre.

4 a $a = b = 55°$ (angle at the centre is twice the angle at the circumference)

b $c = d = (360 - 170) \div 2 = 95°$ (angles at a point add up to $360°$, angle at the centre is twice the angle at the circumference)

c $e = 42°$ (angles at the circumference subtended by the same arc are equal)
$f = 180 - 90 - 42 = 48°$ (angle in a semicircle is $90°$, and angles in a triangle add up to $180°$)

d $g = 56°$ (angles at the circumference subtended by the same arc are equal)
$h = 34°$ (angle in a semicircle is $90°$, and angles in a triangle add up to $180°$)

5 a i a and b are angles in the same segment subtended by arc ED.

ii c and d are angles in the same segment subtended by arc AC.

iii e and f are vertically opposite angles.

b ABE and CBD are similar triangles.

6 a i $a = 50°$; $b = 260°$; $c = 130°$

ii $a = 100°$; $b = 160°$; $c = 80°$

b $a + c = 180°$

7 a $2x$ **b** $2y$ **c** $2x + 2y = 360°$

d $2(x + y) = 2 \times 180°$, so $x + y = 180°$

e It is true for any angles x and y.

8 a i $a = 85°$ (angles on a straight line add up to $180°$)
$b = 95°$ (opposite angles of a cyclic quadrilateral add up to $180°$)

ii $c = 105°$ and $e = 98°$ (angles on a straight line add up to $180°$)
$d = 75°$ and $f = 82°$ (opposite angles of a cyclic quadrilateral add up to $180°$)

b Exterior angle = opposite interior angle

9 a Students' own drawings.

b Angle x + angle $y = 180°$ because angles on a straight line add up to $180°$.
Angle x + angle $z = 180°$ because opposite angles of a cyclic quadrilateral add up to $180°$.

10 a $a = 72°$ (opposite angles of a cyclic quadrilateral add up to $180°$)
$b = 108°$ (angles on a straight line add up to $180°$)
$c = 93°$ (angles in a quadrilateral add up to $360°$)

b $d = 41°$ and $e = 32°$ (angles at the circumference subtended by the same arc are equal)
$f = g = 107°$ (angles in a triangle add up to $180°$)

c $h = 43°$ (opposite angles of a cyclic quadrilateral add up to $180°$)
$i = 43°$ (angles at the circumference subtended by the same arc are equal)
$j = 137°$ (angles on a straight line add up to $180°$)

d $k = 46°$ and $m = 38°$ (angles subtended by the same arc are equal)
$l = 54°$ (angles in a triangle add up to $180°$)

e $n = 116°$ (opposite angles of a cyclic quadrilateral add up to $180°$)
$p = 26°$ (angle in a semicircle is $90°$)

11 $\angle CBE = 90°$ with students' own reasoning, e.g.
$\angle DBC = 65°$ (base angles of an isosceles triangle are equal)
$\angle BCD = 50°$ (angles in a triangle add up to $180°$)
$\angle DAB = 130°$ (opposite angles of a cyclic quadrilateral add up to $180°$)
$\angle ADB = \angle ABD = 25°$ (base angles of an isosceles triangle are equal)
$\angle CBE = 180° - 65° - 25° = 90°$ (angles on a straight line add up to $180°$, or exterior angle of a cyclic quadrilateral is equal to the opposite interior angle)

12 a Students' own drawings.

b $\angle OAT = 90°$ because the angle between the tangent and the radius is $90°$.
$\angle OAB = 90° - 58° = 32°$
$OA = OB$ because radii of the same circle are equal.
$\angle OAB = \angle OBA$ because the base angles of an isosceles triangle are equal.
$\angle AOB = 180° - 32° - 32° = 116°$ because angles in a triangle add up to $180°$.
$\angle ACB = 116° \div 2 = 58°$ because the angle at the centre is twice the angle at the circumference when both are subtended by the same arc.

13 a 72°   **b** Angle BAT = angle ACB

    c $\angle OAT = 90°$
    $\angle OAB = 90° - x$
    $OA = OB$
    $\angle AOB = 180° - (90° - x) - (90° - x) = 2x$
    $\angle ACB = 2x \div 2 = x$

14 $\angle ACT = 65°$, with students' own reasoning, e.g.
    $\angle ABC = 65°$ (angles in a triangle add up to 180°)
    $\angle ACT = 65°$ (angle between tangent and chord is equal to
    the angle in the alternate segment)

15 $\angle XYZ = 62°$ (angle between tangent and chord is equal to
    the angle in the alternate segment)
    $\angle XOZ = 124°$ (angle at the centre is twice the angle at the
    circumference subtended by the same arc)
    $\angle ZXO = \angle XZO = 28°$ (base angles of an isosceles triangle
    are equal, and angles in a triangle add up to 180°)
    $\angle OXY = 22°$ (angles in triangle XYZ add up to 180°)

16 $OB = OD$ (radii of the same circle)
    $\angle OBD = \frac{1}{2}(180° - x)$ (base angles of an isosceles triangle
    are equal)
    $\angle OBA = 90°$ (angle between tangent and radius is 90°)
    $\angle ABD = 90° - \angle OBD = 90° - \frac{1}{2}(180° - x) = \frac{1}{2}x$

16.5 Applying circle theorems

1 a 5   **b** $5\sqrt{3}$

2 $y = -\frac{1}{2}x + \frac{3}{2}$

3 a $g = 38°$ (angles at the circumference subtended by the
    same arc are equal)
    $h = 98°$ (angles in a triangle add up to 180°)
    $i = 98°$ (vertically opposite angles are equal)
    $j = 44°$ (angles at circumference subtended by the same
    arc are equal)
    b $\angle BCD = 150°$ (opposite angles of a cyclic quadrilateral
    add up to 180°)
    $k = (180 - 150) \div 2 = 15°$ (angles in a triangle add up to
    180°, and base angles of an isosceles triangle are equal)
    c $\angle FEH = 69°$ (base angles of an isosceles triangle
    are equal)
    $i = 69°$ (alternate angles are equal)

4 a $a = 46°$ (angle between tangent and chord is equal to the
    angle in the alternate segment)
    b $b = 35°$ (angle between tangent and chord is equal to the
    angle in the alternate segment)
    $c = 94°$ (angles in a triangle add up to 180°)
    $d = 94°$ (angle between tangent and chord is equal to the
    angle in the alternate segment)
    c $e = 67°$ (angle between tangent and chord is equal to the
    angle in the alternate segment)
    $f = 27°$ (angles in a triangle add up to 180°)
    $g = 86°$ (angle between tangent and chord is equal to the
    angle in the alternate segment)

5 a $AT = BT$ (tangents from an external point are equal
    in length)
    $\angle TAB = (180 - 56) \div 2 = 62°$ (base angles of an isosceles
    triangle are equal)
    $a = 62°$ (angle between tangent and chord is equal to the
    angle in the alternate segment)
    b $\angle BAC = 90°$ (angle in a semicircle is 90°)
    $\angle ABC = 63°$ (angles in a triangle add up to 180°)
    $b = 63°$ (angle between tangent and chord is equal to the
    angle in the alternate segment)
    c $c = 74°$ (alternate angles are equal)
    $d = 74°$ (angle between tangent and chord is equal to the
    angle in the alternate segment)
    $e = 32°$ (angles in a triangle add up to 180°)

6 $\angle OAT = 90°$ (angle between tangent and radius is 90°)
    a $\angle CAO = 180 - 50 - 90 = 40°$ (angles on a straight line
    add up to 180°)
    b $\angle AOB = 360 - 90 - 90 - 48 = 132°$ (angles in a
    quadrilateral add up to 360°)
    c $\angle AOC = 180 - 40 - 40 = 100°$ (angles in a triangle
    add up to 180°, and base angles of an isosceles triangle
    are equal)
    d $\angle COB = 360 - 132 - 100 = 128°$ (angles at a point add
    up to 360°)
    e $\angle CBO = (180 - 128) \div 2 = 26°$ (base angles of an
    isosceles triangle are equal)

7 $\angle ODA = \angle OBA = 90°$ (angle between tangent and
    radius is 90°)
    $\angle DOB = 130°$ (angles in a quadrilateral add up to 360°)
    $\angle DCB = 65°$ (angle at the centre is twice the angle at the
    circumference)
    $\angle OBC = 20°$ (angles on a straight line add up to 180°)
    Reflex $\angle DOB = 230°$ (angles at a point add up to 360°)
    $\angle ODC = 45°$ (angles in a quadrilateral add up to 360°)

8 a $12y = 5x - 169$ **b** $4y = -3x + 75$
    c $3y = 4x + 50$ **d** $15y = -8x - 289$

9 $y = -\frac{2\sqrt{5}}{5}x + \frac{9\sqrt{5}}{5}$ or equivalent

10 a $\angle OAB = 90°$ (angle between tangent and radius is 90°)
    b $3\sqrt{2}$ **c** $x^2 + y^2 = 18$

11 a 4   **b** $4\sqrt{3}$ **c** 6   **d** $2\sqrt{3}$
    e $OC = 8 - 6 = 2 = x$-coordinate of A
    $AC = 2\sqrt{3} = y$-coordinate of A

16 Check up

1

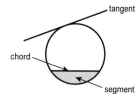

2 a $\angle OBA = 50°$ (angles on a straight line add up to 180°)
    $OA = OB$ (radii of same circle)
    $\angle OAB = \angle OBA$ (base angles of an isosceles triangle
    are equal)
    $a = 180 - 50 - 50 = 80°$ (angles in a triangle add up
    to 180°)
    b $b = 90°$ (angle between radius and tangent is 90°)
    $c = 180 - 56 - 90 = 34°$ (angles in a triangle add up
    to 180°)
    c $AT = BT$ (tangents from an external point are equal
    in length)
    $\angle TBA = \angle TAB$ (base angles of an isosceles triangle
    are equal)
    $d = (180 - 42) \div 2 = 69°$
    $\angle OBT = 90°$ (angle between radius and tangent is 90°)
    $e = 90 - 69 = 21°$

3 $OM = 4\,\text{cm}$

4 a $a = 74°$ (angle subtended by an arc at the centre of a
    circle is twice the angle subtended at any point on the
    circumference)
    $b = 106°$ (opposite angles of a cyclic quadrilateral add up
    to 180°)
    b $c = 90°$ (angle in a semicircle is 90°)
    $d = \angle ACB = 39°$ (angles subtended at the circumference
    by the same arc are equal)

5 a $a = 54°$ (angle between tangent and chord is equal to the
    angle in the alternate segment)
    b $b = 56°$ (angles in a triangle add up to 180°)

6 $\angle ABC = \frac{x}{2}$ (angle at centre is twice the angle at the circumference)
$\angle ADC = 180° - \frac{x}{2}$ (opposite angles of a cyclic quadrilateral add up to 180°)

7 $12y = 5x - 338$

8 $AO = OC = OB$ (radii of same circle)
$\angle ACO = \angle OAC = x$ and $\angle BCO = \angle OBC = y$ (base angles of an isosceles triangle are equal)
$\angle AOD = 2x$ and $\angle BOD = 2y$ (exterior angle of a triangle is equal to the sum of the interior opposite angles)
$\angle ACB = x + y$
$\angle AOB = 2x + 2y = 2(x+y) = 2 \times \angle ACB$

10 a Students' own diagrams, e.g.

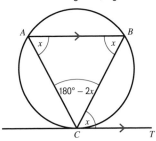

  b Isosceles

16 Strengthen

Chords, radii and tangents

1

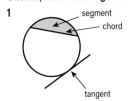

2 a Students' diagrams with OA and OB marked as equal lengths.
  b Isosceles   **c** $x = 20°$; $y = 20°$; $z = 140°$

3 a, b Students' own diagrams.
  c The line only touches the circle at A.
  d Students' own diagrams.   **e** 90°

4 a

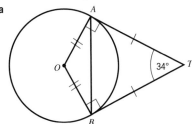

  b Isosceles, 73°, 73°   **c** 90°   **d** 17°

5 a Students' own diagrams.   **b** 6 cm   **c** 2.5 cm

Circle theorems

1 a, b Students' own diagrams with minor arc QR coloured.
  c $\angle QPR$ **d** 65°

2 a 49°   **b** 68°   **c** 284°   **d** 90°

3 They add up to 180°.

4 a Cyclic   **b** $\angle QRS$ **c** 84°

5 A

6 The angles at the circumference are equal.

7 a a and b **b** x and y
  c d and e; f and g

8 a $\angle ACB$ **b** $\angle BDE$ **c** $\angle BHI$ **d** $\angle BMJ$

9 $\angle ACB = 37°$ (angle between tangent and chord is equal to the angle in the alternate segment)
$\angle CAB = 68°$ (angles in a triangle add up to 180°)

Proofs and equation of tangent to a circle at a given point

1 a $\frac{4}{3}$ **b** $-\frac{3}{4}$ **c** $y = -\frac{3}{4}x + c$
  d $c = \frac{25}{4}$ **e** $y = -\frac{3}{4}x + \frac{25}{4}$ or $4y + 3x = 25$

2 a 180°   **b** AC **c** $180 \div 2 = 90°$
  d The angle in a semicircle is a right angle.

16 Extend

1 $a = 60°$ (OBC is an equilateral triangle)
$OA = OB$ so triangle OAB is isosceles (radii of same circle)
$c = \angle OBA = 130 - 60 = 70°$ (base angles of an isosceles triangle are equal)
$b = 180 - 70 - 70 = 40°$ (angles in a triangle add up to 180°)

2 11.0 cm

3 $\angle ABC = 90°$; $\angle ACB = 36°$; $\angle BAC = 54°$

4 $OB = OC$ (radii of same circle)
$\angle OCB = \angle OBC = (180 - 40) \div 2 = 70°$ (base angles of an isosceles triangle are equal, and angles in a triangle add up to 180°)
$OA = OC$ (radii of same circle)
$\angle OAC = \angle OCA = (180 - 110) \div 2 = 35°$ (base angles of an isosceles triangle are equal, and angles in a triangle add up to 180°)
$\angle BCA = \angle OCB - \angle OCA = 35°$
So AC bisects angle OCB.

5 $\angle ODC = 66°$ (opposite angles of a cyclic quadrilateral add up to 180°)
$OC = OD$ (radii of same circle)
$\angle ODC = \angle OCD$ (base angles of an isosceles triangle are equal)
$\angle COD = 180 - 66 - 66 = 48°$ (angles in a triangle add up to 180°)

6 $\angle BCD = 30°$ (opposite angles of a cyclic quadrilateral add up to 180°)
$\angle BOD = 60°$ (angle at the centre is twice angle at the circumference when both are subtended by the same arc)
$OB = OD$ (radii of same circle)
$\angle OBD = \angle ODB = (180 - 60) \div 2 = 60°$ (base angles of an isosceles triangle are equal, and angles in a triangle add up to 180°)
In triangle OBD, all the angles are 60° so it is equilateral.

7 39 cm

8 $\angle CDA = x$ (alternate angles are equal)
$\angle DCA = x$ (angle between tangent and chord is equal to the angle in the alternate segment)
In triangle ACD, $\angle CDA = \angle DCA$, therefore $AC = AD$ and the triangle is isosceles.

9 $AB = 4\tan 30° = \frac{4}{\sqrt{3}}$
$\angle ABC = 120°$ (angles in a quadrilateral add up to 360°)
Area of quadrilateral $ABCD$
$= \frac{1}{2} \times 4 \times 4 \times \sin 60° + \frac{1}{2} \times \frac{4}{\sqrt{3}} \times \frac{4}{\sqrt{3}} \times \sin 120° = \frac{16\sqrt{3}}{3}$
Area of sector $OAC = \frac{1}{6}\pi \times 4^2$
Shaded area = 0.860 cm^2 (3 s.f.)

10 OB is common; $OA = OC$ (radii of same circle); $\angle OAB = \angle OCB = 90°$ (angle between radius and tangent is 90°). Therefore triangles OAB and OCB are congruent (RHS) and $AB = BC$ (corresponding sides).

11 $OA = OB$ (radii of same circle); OM is common; $AM = MB$ (M is midpoint of AB). Therefore triangles OAM and OMB are congruent (SSS). AMB is a straight line so $\angle AMO = \angle OMB = 180 \div 2 = 90°$

12 Students' own proofs that triangles ABO and CDO are congruent using ASA, RHS or SAS

13 Students' own proofs, e.g.
Join OC and let $\angle BAD = x$, then $\angle OCB = x$ (exterior angle of cyclic quadrilateral $AOCD$ equals the opposite interior angle)
$OB = OC$ (radii of same circle)
So triangle OBC is isosceles, hence $\angle OCB = \angle OBC = x$ (base angles of an isosceles triangle are equal)
Hence $\angle BAD = \angle OBC$ (or $\angle ABD$)
Therefore triangle DAB is isosceles, and so $BD = AD$.

14 $(10 - 3y)^2 + y^2 = 10$
$100 - 60y + 9y^2 + y^2 = 10$
$10y^2 - 60y + 90 = 0$
$y^2 - 6y + 9 = 0$
$(y - 3)^2 = 0$
$y = 3$
$x = 10 - 3y = 10 - 9 = 1$
There is only one solution, so only one point of intersection, (1, 3).
Hence $3y + x = 10$ is a tangent to the circle.

15 Students' own proofs, e.g.
$\angle OAC = (90 - 3x)$ (angle between tangent and radius is $90°$)
$\angle OCA = (90 - 3x)$ (base angles in isosceles triangle OAC are equal)
$\angle OCB = x$ (base angles in isosceles triangle OBC are equal)
So $\angle ACB = (90 - 3x) + x = 90 - 2x$
$\angle AOB = 2 \times (90 - 2x) = (180 - 4x)$ (angle at centre is twice angle at circumference)
In quadrilateral $TAOB$, $\angle TAO$ and $\angle TBO$ are both $90°$ (angle between tangent and radius is $90°$)
So $y = 360 - 90 - 90 - (180 - 4x)$ (angles in quadrilateral $AOBT$ add up to $360°$)
Hence $y = 4x$.

16 Students' own proofs, e.g.
$\angle DCT = x$ (base angles in isosceles triangle DTC are equal)
$\angle DBC = \angle DCT = x$ (alternate segment theorem)
$\angle DBC = \angle BCP = x$ (alternate angles in parallel lines BD and PCT are equal)
So $\angle BCP = \angle DTC$
So the lines ADT and BC are parallel (converse of corresponding angles for these lines).

16 Test ready

Sample student answers
Both students worked out the correct value for angle OBT. Student A gave the better answer as they clearly stated the reasons for each part of their calculation.

16 Unit test

1 $\angle OBA = 34°$ (angles on a straight line add up to $180°$)
$OA = OB$ (radii of same circle)
$\angle OAB = 34°$ (base angles of an isosceles triangle are equal)
$\angle AOB = 180 - 34 - 34 = 112°$ (angles in a triangle add up to $180°$)

2 $OA = OB$ (radii of same circle)
$\angle OAB = \angle OBA$ (base angles of an isosceles triangle are equal)
$\angle OAB = (180 - 124) \div 2 = 28°$ (angles in a triangle add up to $180°$)
$\angle OAT = 90°$ (angle between tangent and radius is $90°$)
$\angle BAT = 90 - 28 = 62°$

3 a $90°$ **b** $16\,cm$

4 $\angle BAD = 70°$ (opposite angles of a cyclic quadrilateral add up to $180°$)
Obtuse $\angle BOD = 140°$ (angle at centre is twice angle at circumference)

5 a $\angle BAD = 90°$ (angle in a semicircle is $90°$)
$\angle ABD = 180 - 90 - 19 = 71°$ (angles in a triangle add up to $180°$)

b $\angle ACB = 19°$ (angles subtended at the circumference by the same arc are equal)

6 $\angle OPR = \angle OQR = 90°$ (angle between tangent and radius is $90°$)
$\angle POQ = 360 - 90 - 90 - 42 = 138°$ (angles in a quadrilateral add up to $360°$)
$\angle OPQ = (180 - 138) \div 2 = 21°$ (angles in a triangle add up to $180°$, and base angles of an isosceles triangle are equal)

7 $\angle ABE = \angle ACD$ (angles subtended at the circumference by the same arc are equal)
$\angle BAE = \angle BDC$ (angles subtended at the circumference by the same arc are equal)
$\angle AEB = \angle DEC$ (vertically opposite angles are equal)
Both triangles have the same angles, so they are similar.

8 Students' own proofs, e.g.
$\angle EOB = 180 - 2y$ (angles in a triangle add up to $180°$, and base angles of an isosceles triangle are equal)
$\angle EAB = 180 - x$ (opposite angles of a cyclic quadrilateral add up to $180°$)
$\angle EOB = 2 \times \angle EAB$ (angle at centre is twice angle at circumference)
$180 - 2y = 2(180 - x)$; $90 - y = 180 - x$; $x - y = 90°$

9 Gradient of radius $= -\frac{3}{4}$ so gradient of $PT = \frac{4}{3}$
$3 = \frac{4}{3} \times (-4) + c$; $c = \frac{25}{3}$; $y = \frac{4}{3}x + \frac{25}{3}$ or $3y = 4x + 25$

10 $OP = OA = 12\sin 30° = 6$; $n^2 + 4n^2 = 36$; $n = \sqrt{\frac{36}{5}} = 2.7$ (1 d.p.)

11 $p + r = 180°$, so $r = 180 - p$ (angles on a straight line add up to $180°$)
$p + q = 180°$, so $q = 180 - p$ (opposite angles of a cyclic quadrilateral add up to $180°$)
So $q = r$.

Mixed exercise 5

1 a 312   → **14.1**

b Students' own answers, e.g. Representative/random sample or else Mrs Booth's order could be completely wrong.   → **14.1**

2 a   → **14.2**

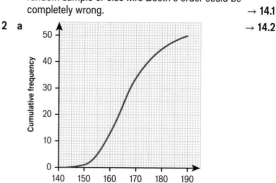

b $13\,cm$ → **14.2**
c $\frac{18}{25}$ → **14.2**

3 AD because angle $AED = 90°$ → **16.3**

4 Angle $QPS = 63°$ as angle $OPR = 90°$, angle $POR = 54°$, angle $POQ = 126°$ and angle $OPQ = 27°$ → **16.2**

5 a No, the median is 3.3 kg, so half of the parcels weigh less than 3.3 kg, not 3 kg. → **14.3**

  b Students' own answers, e.g.
  On average the parcels were heavier on Tuesday than on Monday as the median on Tuesday was higher. The weights of the parcels on Monday were more varied as the IQR is greater than on Tuesday. → **14.6**

6 Sasha should have drawn a dashed line for $y = x$, not a solid line and she hasn't shaded the correct region; she should have shaded the area bounded by the line $y \leqslant 2x + 1$, the line $y < 3$, and the line $y > x$. → **15.2**

7 Students' histograms including correct bar with width from 60 to 75 grams and height 1. → **14.4**

8 $65°, 80°, 100°, 115°$ → **16.4**

9 $108°$ → **14.5**

10 $y = 8 - 2x^2$ → **15.3**

11 Students' own answers, e.g.
  Angle PQS = angle SRP as angles subtended from an arc are equal, PS is a common side in both triangles, angle QSP = angle $RPS = 60°$ as triangle PTS is an equilateral triangle, angle QPS = angle RSP as the angles in a triangle add up to $180°$. Therefore, triangle PQS is congruent to triangle SRP (ASA). → **12.2, 16.4**

12

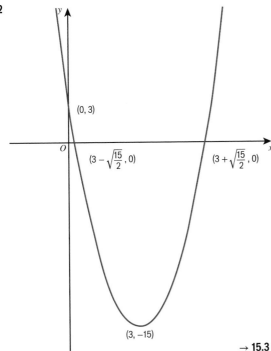

→ **15.3**

13 Students' own answers, e.g.
  Angle $OQX = 90°$ because the angle between the tangent and the radius is $90°$.
  Angle OQP = angle $OPQ = 90 - x$ because the base angles of an isosceles triangle are equal.
  Angle $POQ = 180 - 2(90 - x) = 2x$ because angles in a triangle add up to $180°$.
  Angle $PRQ = 2x \div 2 = x$ because the angle at the centre is twice the angle at the circumference. → **16.4**

14 a Using trigonometry, height $= \frac{1}{2}(8 - x)$

$$\text{Area} = \frac{1}{2} \times (2x - 3) \times \frac{1}{2}(8 - x) > 5$$
$$-2x^2 + 19x - 24 > 20$$
$$-2x^2 + 19x - 44 > 0$$
$$2x^2 - 19x + 44 < 0$$

or area $= \frac{1}{2}ab \sin C > 5$

$$\frac{1}{2}(2x - 3)(8 - x) \sin 30° > 5$$
$$\frac{1}{4}(-2x^2 + 19x - 24) > 5$$
$$-2x^2 + 19x - 24 > 20$$
$$2x^2 - 19x + 44 < 0$$ → **15.4**

  b $4 < x < \frac{11}{2}$ → **15.4**

15 $7y = 3x - 58$ → **16.5**

UNIT 17 More algebra

17.1 Rearranging formulae

1 a $a = \frac{v - u}{t}$ **b** $r = \frac{C}{2\pi}$ **c** $p = qr$

  d $q = \frac{p}{r}$ **e** $h = \frac{2A}{b}$ **f** $s = \frac{\sqrt{t}}{3}$

  g $t = x^2$ **h** $s = \frac{r^2}{3}$

2 a $r(s + 1)$ **b** $y(x + 2)$ **c** $q(p - 1)$

  d $k(a - 4)$

3 a $y = \frac{mz}{5}$ **b** $y = \frac{mz}{5x}$ **c** $y = \frac{2mz}{5x}$

  d $y = \frac{bmz}{5x}$ **e** $y = 2n - x$ **f** $y = an - x$

  g $y = an - 4x$ **h** $y = 4x - an$ **i** $y = \frac{x - an}{2}$

  j $y = b(r - x)$ **k** $y = \frac{b}{3}(r - x)$ **l** $y = \frac{b(x - r)}{k}$

4 a $y = \frac{h}{3 + x}$ **b** $y = \frac{h}{u + x}$ **c** $y = \frac{k}{1 - a}$

  d $y = \frac{k}{b - a}$

5 a $d = \frac{K - n}{3 + n}$ **b** $d = \frac{L + x}{x + n}$ **c** $d = \frac{G - y}{x - y}$

  d $d = \frac{H + ac}{a - b}$

6 a $y = \frac{w - 2}{2x}$ **b** $x = \frac{w - 2}{2y}$

7 a Fred should have factorised $xy + 2x$ first.

  b $x = \frac{H - 7}{y + 2}$

8 a $x = \frac{1}{r}$ **b** $x = \frac{1}{s - 1}$ **c** $x = \frac{1}{t - 7}$

  d $x = \frac{1}{v - b}$ **e** $x = \frac{a}{w - b}$ **f** $x = \frac{1}{y + 1}$

  g $x = \frac{1}{z + b}$ **h** $x = \frac{a}{b + d}$

9 a $x = \frac{1 - e}{e}$ **b** $x = \frac{-f}{f - 1}$ or $\frac{f}{1 - f}$

  c $x = \frac{g}{g - 1}$ **d** $x = \frac{h}{h - 2}$

  e $x = \frac{-j}{j - 2}$ or $\frac{j}{2 - j}$ **f** $x = \frac{-k}{k - 3}$ or $\frac{k}{3 - k}$

  g $x = \frac{l + 1}{l - 1}$ **h** $x = \frac{m + 2}{m - 1}$

  i $x = \frac{n - 2}{n - 1}$ **j** $x = \frac{p + 3}{p - 1}$

  k $x = \frac{q + 1}{q - 2}$ **l** $x = \frac{1 - r}{r - 2}$

10 $t = \frac{x + 2}{x - 3}$

11 $v = \sqrt{\dfrac{2E}{m}}$

12 a $H^2 = x - y$ **b** $x = H^2 + y$

13 a $x = \dfrac{4M^2}{9}$ **b** $x = \dfrac{nL^2}{4}$ **c** $x = \dfrac{T^2 k}{4p^2}$

   d $x = \dfrac{N^2 t}{5}$ **e** $x = \dfrac{zP^2}{y}$ **f** $x = \dfrac{1}{w^2}$

   g $x = \dfrac{16}{y^2}$ **h** $x = \dfrac{49a}{h^2}$

14 a $x = \sqrt{y - n}$ **b** $x = \sqrt{ny}$ **c** $x = \sqrt{ny - 1}$

   d $x = \sqrt{n(y - 1)}$ **e** $x = \dfrac{1}{\sqrt{z}}$ **f** $x = \sqrt{\dfrac{n}{z}}$

   g $x = \sqrt{\dfrac{1+n}{z}}$ **h** $x = \sqrt{\dfrac{n}{z-1}}$ **i** $x = \sqrt{k} - 1$

   j $x = \sqrt{\dfrac{k}{3}} - 1$ **k** $x = 2\sqrt{k} + 1$ **l** $x = \sqrt{2k} + 1$

15 a $x = \sqrt[3]{y}$ **b** $x = \sqrt[3]{\dfrac{z}{4}}$ **c** $x = \dfrac{\sqrt[3]{A}}{4}$

   d $x = \sqrt[3]{2B}$ **e** $x = \sqrt[3]{\dfrac{3C}{2}}$ **f** $x = \dfrac{3}{2}\sqrt[3]{D}$

   g $r = \sqrt[3]{\dfrac{3V}{4\pi}}$ **h** $y = t^3$ **i** $x = \dfrac{y^3}{5}$

   j $k = 2n^3$ **k** $x = yz^3$ **l** $y = \dfrac{x}{z^3}$

17.2 Algebraic fractions

1 a $\dfrac{10}{77}$ **b** $\dfrac{1}{6}$ **c** $\dfrac{54}{35}$ **d** $\dfrac{5}{16}$

2 a $\dfrac{29}{36}$ **b** $\dfrac{85}{99}$ **c** $\dfrac{41}{99}$ **d** $-\dfrac{1}{20}$

3 a $\dfrac{x^2}{6}$ **b** $\dfrac{5x^2}{6}$ **c** $\dfrac{3mx}{10}$ **d** $\dfrac{4}{15y^2}$

4 a $\dfrac{3x}{2y^2}$ **b** $\dfrac{5y^4}{3x^2}$ **c** $\dfrac{3x^4}{2a^2}$

5 a $\dfrac{4}{3}$ **b** $\dfrac{y}{x}$ **c** $\dfrac{1}{x^2}$ **d** $x^4 y^2$

   e $\dfrac{5}{4x^2 y^4}$ **f** $\dfrac{2x}{y^3}$

6 a $\dfrac{x}{3(x+7)}$ **b** $\dfrac{1}{x(x+2)}$ **c** $\dfrac{3x}{2(x+1)}$ **d** $\dfrac{2x}{x-1}$

7 a $\dfrac{10x}{21}$ **b** $\dfrac{x}{4}$ **c** $\dfrac{x}{10}$

   d $\dfrac{4x}{5}$ **e** $\dfrac{13x}{12}$ **f** $\dfrac{19x}{6}$

8 a $10x$ **b** $6x$ **c** $28x$ **d** $12x$

9 a $\dfrac{3}{12x}, \dfrac{4}{12x}$ **b** $\dfrac{7}{12x}$

10 a $\dfrac{11}{18x}$ **b** $\dfrac{1}{20x}$ **c** $\dfrac{13}{18x}$ **d** $\dfrac{17}{15x}$

11 a $\dfrac{5(x-4)}{5\times2} = \dfrac{5x-20}{10}$ **b** $\dfrac{2(x+7)}{2\times5} = \dfrac{2x+14}{10}$

   c $\dfrac{7x-6}{10}$

12 a $\dfrac{5x+8}{6}$ **b** $\dfrac{5x+41}{14}$ **c** $\dfrac{x+67}{36}$ **d** $\dfrac{13x+14}{12}$

13 $\dfrac{9x+24}{10}$

14 a $\dfrac{x+y}{xy} = 1$ **b** $x + y = xy$ **c** $x - xy = -y$

   d $x(1-y) = -y$ **e** $x = \dfrac{-y}{1-y}$ or $\dfrac{y}{y-1}$

15 $u = \dfrac{vf}{v - f}$

17.3 Simplifying algebraic fractions

1 a $6(x+3)$ **b** $x(x+3)$
   c $x^2(x+4)$ **d** $3x(x^2-5)$

2 a $(x-6)(x-3)$ **b** $(x-9)(x+9)$
   c $(2x-5)(x+1)$ **d** $(5x+1)(x+4)$

3 a $\dfrac{1}{y}$ **b** $\dfrac{1}{3}$ **c** $\dfrac{1}{x-7}$

   d $\dfrac{x+2}{x-5}$ **e** $\dfrac{x-3}{x}$ **f** $\dfrac{x}{x-1}$

4 a $x(x-6)$ **b** x

5 a $x+8$ **b** $2x+5$ **c** $\dfrac{1}{x+3}$

   d $3x$ **e** $\dfrac{1}{2x}$ **f** $\dfrac{5}{2x}$

6 a No, because the two terms in the denominator have nothing in common. The denominator cannot be factorised.
   b No, because the denominator cannot be factorised.

7 a $\dfrac{2}{x+5}$ **b** $x+4$ **c** $x+1$ **d** $\dfrac{x-3}{5}$

8 a $\dfrac{x+3}{x-3}$ **b** $\dfrac{x-5}{x+7}$ **c** $\dfrac{x-5}{x+5}$ **d** $\dfrac{2x+1}{x+2}$

9 $\dfrac{2x+3}{x-7}$

10 $\dfrac{2x-1}{3x}$

11 a $\dfrac{2x-3}{3x-2}$ **b** $\dfrac{5x-1}{6x+5}$ **c** $\dfrac{5x-1}{5x+1}$ **d** $\dfrac{x-10}{2x+3}$

12 a $-(x-6)$

   b i -1 **ii** $-\dfrac{6+x}{x+3}$

13 a $-\dfrac{4+x}{x}$ **b** $\dfrac{x-6}{2(x+6)}$ **c** $\dfrac{-3x}{5x+1}$

   d $\dfrac{-2x}{2x-3}$ or $\dfrac{2x}{3-2x}$

14 $\dfrac{(x^2-36)(2x^2+8x)}{x^2+10x+24} = \dfrac{(x-6)(x+6)(2x)(x+4)}{(x+6)(x+4)} = 2x(x-6)$

15 Numerator: $(x+4)(x-3)(x+3)(x-1)(2x)(5x+6)$
Denominator: $(3-x)(3+x)(5x+6)(x+4)(7)(x-1)$
$\dfrac{x-3}{3-x} = -1$; other factors cancel, to leave $-\dfrac{2x}{7}$

17.4 More algebraic fractions

1 a $\dfrac{15x}{4m}$ **b** $\dfrac{75}{4}$

2 a $\dfrac{5}{24x}$ **b** $\dfrac{7x+11}{12}$

3 a $\dfrac{x-4}{x+3}$ **b** $\dfrac{x+2}{x+5}$ **c** $\dfrac{1}{9}$

   d $\dfrac{8}{3}$ **e** $\dfrac{2(x-1)}{x+7}$ **f** $\dfrac{x(x+4)}{x-2}$

4 a i $(x-3)(x+3)$ **ii** $(x+2)(x+3)$

   b $\dfrac{2(x-3)}{x+2}$

5 a $\dfrac{(x-2)(x-3)}{(x+1)(x+4)}$ **b** $\dfrac{7(x+7)}{(x-3)(x-7)}$

6 a $x(x+2)$ **b** $(x+2)(x+3)$
   c $(x+1)(x-1)$ **d** $15x$

7 a $\dfrac{2x+9}{(x+4)(x+5)}$ **b** $\dfrac{7x+1}{(x+1)(x-1)}$

   c $\dfrac{6x+26}{(x-5)(x+3)}$ **d** $\dfrac{7}{(2x-3)(2x+4)}$

   e $\dfrac{x^2+8x+9}{(x+2)(x+1)}$ **f** $\dfrac{5x^2+12x-27}{(x-1)(x+4)}$

8 a $\dfrac{x^2+6x+4}{x^2-4}$ **b** $\dfrac{2x^2+3x-3}{x^2-1}$

   c $\dfrac{2x^2+4x-8}{x^2-16}$ or $\dfrac{2(x^2+2x-4)}{x^2-16}$

   d $\dfrac{9x+15}{x^2-9}$ or $\dfrac{3(3x+5)}{x^2-9}$ **e** $\dfrac{9x+6}{x^2-4}$ or $\dfrac{3(3x+2)}{x^2-4}$

   f $\dfrac{x^2+x-10}{x^2-4}$

9 $a = -4, b = -12$

10 a i $3(x+3)$ **ii** $4(x+3)$

   b $12(x+3)$ **c** $\dfrac{7}{12(x+3)}$

11 a $(x-4)(x+4)$ **b** $\dfrac{x-3}{(x-4)(x+4)}$

12 a $\dfrac{-x}{(3x+5)(x+1)}$ **b** $\dfrac{1-x}{2(x+1)(x+6)}$

c $\dfrac{4x-1}{(x+2)(x+4)(x-7)}$ **d** $\dfrac{-11-3x}{(5-x)(5+x)}$

13 $(x+4) \div \dfrac{x^2-x-20}{x+1} = (x+4) \div \dfrac{(x+4)(x-5)}{x+1}$

$= \dfrac{(x+4)(x+1)}{(x+4)(x-5)} = \dfrac{x+1}{x-5}$

$3 + \dfrac{x+1}{x-5} = \dfrac{3(x-5)+x+1}{x-5} = \dfrac{3x-15+x+1}{x-5} = \dfrac{4x-14}{x-5}$

$a=4,\, b=14,\, c=1,\, d=5$

14 $\dfrac{x^2+3x-2}{10x(x-1)}$

15 $\dfrac{1}{x^2+5x+6} + \dfrac{1}{5x+10} = \dfrac{1}{(x+2)(x+3)} + \dfrac{1}{5(x+2)}$

$= \dfrac{5+(x+3)}{5(x+2)(x+3)} = \dfrac{x+8}{5(x+3)(x+2)}$

$A = 5$

17.5 Proof

1 a Even **b** Odd

2 a B **b** D **c** A, C, E

3 a Identity **b** Equation $(n=3)$

c Identity **d** Equation $\left(n=\dfrac{7}{3}\right)$

4 a LHS $= x^2-3x-3x+9+6x = x^2+9 =$ RHS
b RHS $= x^2+7x+7x+49-6x = x^2+8x+49 =$ LHS
c LHS $= x^2-5x-5x+25-4 = x^2-10x+21$
RHS $= x^2-3x-7x+21 = x^2-10x+21 =$ LHS
d LHS $= 16-(x^2+2x+2x+4) = 16-x^2-4x-4$
$= 12-x^2-4x$
RHS $= 12-6x+2x-x^2 = 12-x^2-4x =$ LHS

5 a $(x-1)(x+1) = x^2+x-x-1 = x^2-1$
b i 9999 (use 100^2-1) **ii** $39\,999$ (use 200^2-1)

6 a $(x+5)(x+2) = x^2+7x+10$
b $x(x+1) = x^2+x$
c $x^2+7x+10-(x^2+x) = 6x+10$

7 $x(3x+4)-5x = 70$
$3x^2+4x-5x-70 = 0$
$3x^2-x-70 = 0$

8 $\dfrac{1}{x^2-x} - \dfrac{1}{x^2+3x} = \dfrac{1}{x(x-1)} - \dfrac{1}{x(x+3)}$

$= \dfrac{(x+3)-(x-1)}{x(x-1)(x+3)} = \dfrac{4}{x(x-1)(x+3)}$

$A = 4$

9 a 2 is a prime.
b Any number less than 1 gives a cube that is less than its square.
c Students' own answers, e.g. $-5-(-2) = -3$,
$-5+(-2) = -7$
d Students' own answers, e.g. $16-4 = 12$

10 a $2n$ is an even number; $2n+1$ and $2n-1$ are the numbers either side of it, which are both odd.
b $2n+2m+1 = 2(n+m)+1$; $2(n+m)$ is a multiple of 2 and so is even, hence $2(n+m)+1$ is odd.

11 a The next even number will be 2 more (because the next number, which is 1 more, will be odd).
b $(2n)(2n+2) = 4n^2+4n = 4(n^2+n)$, which is divisible by 4.

12 $(2n+1)(2m+1) = 4mn+2m+2n+1$
$= 2(2mn+m+n)+1$
As $2(2mn+m+n)$ is a multiple of 2 it is even,
hence $2(2mn+m+n)+1$ is always odd.

13 $2x-2a = x+5$ so $x = 2a+5$
$2a$ is even and 5 is odd; even + odd = odd, so x is odd.

14 a i $\dfrac{1}{30}$ **ii** $\dfrac{1}{12}$ **iii** $\dfrac{1}{56}$

b $\dfrac{1}{90}$

c It will be 1 divided by 99×100.

d i $\dfrac{1}{x(x+1)}$

ii This shows that the difference between two fractions with 1 on the numerators and consecutive numbers on the denominators will be 1 divided by the product of the denominators.

15 $(2n)^2 = 4n^2$, which is a multiple of 4.

16 $(2n-1)(2n+1) = 4n^2-1$, which is 1 less than a multiple of 4.

17 a $n(n^2-1) = n^3-n$
b $n^3-n = (n-1)n(n+1)$
When n is even, $n^3-n = $ odd $\times$ even $\times$ odd $=$ even.
When n is odd, $n^3-n = $ even $\times$ odd $\times$ even $=$ even.

18 $n+(n+1) = 2n+1$
$(n+1)^2-n^2 = n^2+2n+1-n^2 = 2n+1$

19 $n^2-3-(n-3)^2 = n^2-3-(n^2-6n+9) = -3+6n-9$
$= 6n-12 = 6(n-2)$
which is a positive multiple of 6 for all $n>2$.

17.6 Surds

1 a 5 **b** $\sqrt{15}$ **c** $3\sqrt{3}$ **d** $8\sqrt{2}$

2 a $\sqrt{3}$ **b** $\sqrt{30}$ **c** $\sqrt{\dfrac{5}{7}}$

d $\sqrt{25} \times \sqrt{2} = 5\sqrt{2}$ **e** $3\sqrt{2}$ **f** $4\sqrt{3}$

3 a $\dfrac{\sqrt{10}}{10}$ **b** $\dfrac{\sqrt{15}}{5}$ **c** $\sqrt{2}$ **d** $\dfrac{\sqrt{14}}{7}$

4 a $3\sqrt{5}$ **b** $2\sqrt{5}$ **c** $23\sqrt{5}$ **d** $\sqrt{3}$
e $22\sqrt{2}$ **f** $15\sqrt{2}$ **g** $24\sqrt{2}$ **h** $-9\sqrt{7}$

5 a $2\sqrt{3}+2 = 2(\sqrt{3}+1)$
b $3(3+\sqrt{6})$ **c** $3(6-\sqrt{5})$ **d** $5(\sqrt{3}-\sqrt{2})$

6 a $4\sqrt{5}+5$ **b** $11+5\sqrt{7}$ **c** $22+2\sqrt{2}$
d $6-4\sqrt{2}$ **e** $26-8\sqrt{10}$ **f** $52+14\sqrt{3}$

7 $30-10\sqrt{5}$; $a=30,\, b=-10,\, c=5$

8 $a=3$

9 a i $53-6\sqrt{2}$ **ii** $12+4\sqrt{8} = 12+8\sqrt{2}$
b The perimeter of the first shape would be 32 units, which is rational. The perimeter of the second shape would be $8+4\sqrt{8}$ or $8+8\sqrt{2}$, which is irrational.

10 a $\dfrac{3\sqrt{2}+2}{2}$ **b** $2\sqrt{3}-1$

c $\dfrac{19\sqrt{7}-7}{7}$ **d** $\sqrt{5}+1$

11 $a=3,\, b=4$

12 a 4 **b** Rational **c** Rational

13 a $\dfrac{1-\sqrt{2}}{-1}$ or $-1+\sqrt{2}$ **b** $\dfrac{5+\sqrt{3}}{22}$

c $\dfrac{7(4+\sqrt{5})}{11}$ **d** $\dfrac{4(1-\sqrt{6})}{-5}$ or $\dfrac{4(\sqrt{6}-1)}{5}$

e $\dfrac{5+\sqrt{5}}{-4}$ **f** $\dfrac{25+7\sqrt{2}}{31}$

14 a $x = 3 \pm 2\sqrt{2}$ **b** $x = -5 \pm 2\sqrt{3}$

c $x = 8 \pm 2\sqrt{14}$

15 $\dfrac{(2+\sqrt{8})^2}{\sqrt{8}-2} = \dfrac{4+4\sqrt{8}+8}{\sqrt{8}-2} = \dfrac{4\sqrt{8}+12}{\sqrt{8}-2} = \dfrac{4(\sqrt{8}+3)}{\sqrt{8}-2}$

$= \dfrac{4(\sqrt{8}+3)(\sqrt{8}+2)}{(\sqrt{8}-2)(\sqrt{8}+2)} = \dfrac{4(8+5\sqrt{8}+6)}{8-4}$

$= 14+5\sqrt{8} = 14+5\sqrt{4}\,\sqrt{2} = 14+10\sqrt{2} = 2(7+5\sqrt{2})$
$a=2, b=7, c=5$

17.7 Solving algebraic fraction equations

1 a $4x$ b $4x$ c $(x+3)(x+2)$

2 a $\dfrac{x}{20}$ b $\dfrac{2}{x}$ c $\dfrac{10}{x-6}$

d $\dfrac{2x+9}{(x+2)(x+3)}$

3 a $x=-2$ or $x=-4$ b $x=\dfrac{11}{2}$ or $x=1$

c $x=4$ or $x=1$ d $x=1.40$ or $x=-4.06$

4 a $\dfrac{2(2x+1)+5(x+3)}{10}=1$ b $\dfrac{9x+17}{10}=1$

c $9x+17=10$ d $x=-\dfrac{7}{9}$

5 a $x=\dfrac{17}{11}$ b $x=\dfrac{101}{14}$

6 $x=\dfrac{15}{4}$

7 a $x=\dfrac{5}{4}$ b $x=\dfrac{11}{7}$ c $x=-\dfrac{11}{2}$

8 a $x=-\dfrac{5}{3}$ or $x=4$ b $x=\dfrac{3}{2}$ or $x=-2$

c $x=-\dfrac{7}{5}$ or $x=2$ d $x=\dfrac{5}{2}$ or $x=-4$

9 a $\dfrac{x+1}{2x}+\dfrac{x-3}{3x}=\dfrac{3x(x+1)+2x(x-3)}{6x^2}=5$
$3x^2+3x+2x^2-6x=30x^2$
$5x^2-3x=30x^2$
$25x^2+3x=0$
b $x(25x+3)=0$ so $x=0$ or $x=-\dfrac{3}{25}$

10 a $\dfrac{x}{2x-3}+\dfrac{4}{x+1}=\dfrac{x(x+1)+4(2x-3)}{(2x-3)(x+1)}=1$
$x^2+x+8x-12=2x^2-3x+2x-3$
$x^2+9x-12=2x^2-x-3$
$0=x^2-10x+9$
b $(x-1)(x-9)=0$ so $x=1$ or $x=9$

11 a $x=3$ b $x=0$ or $x=8$
c $x=1$ or $x=2$ d $x=-4$ or $x=1$
e $x=3$ or $x=-2$ f $x=\dfrac{11}{2}$ or $x=-1$

12 a $x=0.29$ or $x=-10.29$ b $x=1.21$ or $x=-1.81$
c $x=6.37$ or $x=0.63$ d $x=5.70$ or $x=-0.70$
e $x=4.30$ or $x=0.70$ f $x=5.32$ or $x=-1.32$

17.8 Functions

1 a $y=7$ b $y=22$ c $y=16$ d $y=20$

2 a $H=12t$ b $P=\dfrac{y}{6}$ c $y=(h+3)^2=h^2+6h+9$

3 a 2 b -5 c 20 d $-\dfrac{1}{2}$

4 a 54 b -2 c $\dfrac{1}{4}$ d -250

5 a 5 b 56 c 480
d 2.5 e 600 f -33

6 a $a=3$ b $a=\dfrac{3}{5}$ c $a=-\dfrac{4}{5}$

7 a $a=\pm5$ b $a=\pm2$ c $a=\pm2\sqrt{2}$ d $a=\pm2\sqrt{5}$

8 a $a=0$ and $a=-3$ b $a=1$ and $a=-5$
c $a=-1$ and $a=-2$ d $a=-1$ and $a=-3$

9 a $5x+1$ b $5x-13$ c $10x-8$ d $35x-28$
e $-5x+4$ f $10x-4$ g $20x-4$
h No, because $10x-8 \neq 10x-4$

10 a $f(2)=6-2\times2=2$ so $gf(2)=g(2)=2\times2+7=11$
b i 71 ii -40 iii -58 iv -10

11 a $\dfrac{3}{16}$ b $\dfrac{3}{4}$

12 a i $-4x+13$ ii $37-4x$
iii $4x^2+25$ iv $16x^2-24x+16$
b $gh(x)=g(x^2+7)=10-(x^2+7)=-x^2+3$

13 a $f^{-1}(x)=\dfrac{x-9}{4}$ b $g^{-1}(x)=3(x+4)$
c $h^{-1}(x)=\dfrac{x}{2}-6$ d $k^{-1}(x)=\dfrac{x}{7}+4$

14 a i $\dfrac{x}{4}+1$ ii $\dfrac{x}{4}-1$ iii $\dfrac{7}{4}$ iv $-\dfrac{1}{2}$
b $\dfrac{x}{2}$ c $a=2$

15 $a=5, b=-3$

17 Check up

1 a $y=\dfrac{9-3x}{5x+2}$ b $k=\dfrac{4p^2x}{T^2}$ c $t=\dfrac{3}{x-1}$

2 a -2 b $\dfrac{x-4}{3}$

3 a 26 b -81

4 $a=\pm\dfrac{\sqrt{7}}{2}$

5 LHS $=25-(x^2+2x+1)=-x^2-2x+24$
RHS $=24-6x+4x-x^2=-x^2-2x+24=$ LHS

6 Students' own answers, e.g. $1^3+1^3=2$ or $2^3+4^3=72$

7 a $20\sqrt{2}$ b $23-8\sqrt{7}$

8 a $\dfrac{3\sqrt{5}-\sqrt{10}}{5}$ b $6+3\sqrt{3}$

9 a $\dfrac{x-2}{3}$ b $\dfrac{x-4}{x+1}$

10 a $\dfrac{4x^3}{3y^5}$ b $\dfrac{3(x+10)}{4(x+1)}$
c $\dfrac{4x-11}{(x+4)(x-5)}$ d $\dfrac{16-2x}{(x-6)(x-1)}$

11 $x=-1\pm\sqrt{2}$

13 a $(2n+1)+(2n+3)=4n+4=4(n+1)$
b $2n+(2n+2)+(2n+4)=6n+6=6(n+1)$

17 Strengthen

Formulae and functions

1 a i $y^2=3$ ii $y^2=x$ iii $y^2=\dfrac{x}{k}$ iv $\left(\dfrac{y}{3n}\right)^2=\dfrac{x}{k}$
b $x=k\left(\dfrac{y}{3n}\right)^2$

2 $\div(3y+2)\Big(\ x(3y+2)=1-y \quad x=\dfrac{1-y}{3y+2}\ \Big)\div(3y+2)$

3 a $xy=7+y$ b $xy-y=7$ c $y(x-1)=7$
d $y=\dfrac{7}{x-1}$

4 a $y=1$ b 1
c i 16 ii -24 iii -9 iv 41

5 a $9a-4$ b $a=\dfrac{4}{9}$

6 a 15 b 35 c 50
d 105 e 224 f 0

7 a $f^{-1}(x)=\dfrac{x+5}{3}$
b i $\dfrac{x+9}{2}$ ii $\dfrac{x}{3}+5$ iii $2x-4$ iv $\dfrac{5}{2}x-1$

Proof

1 a $x^2 - 8x + 16$ **b** $x^2 - 8x + 7$
  c $x^2 - 8x + 7$
  d LHS $= x^2 - 8x + 16 - 9 = x^2 - 8x + 7 =$ RHS
2 LHS $= x^2 - x - x + 1 - 16 = x^2 - 2x - 15$
  RHS $= x^2 + 3x - 5x - 15 = x^2 - 2x - 15 =$ LHS
3 a 1, 8, 27, 64, 125
  b Students' own answers, e.g. $4^3 - 2^3 = 64 - 8 = 56$

Surds

1 a 3 **b** $\sqrt{7}$ **c** 4
  d $6\sqrt{5} - 5$ **e** $\sqrt{36}\sqrt{5} + \sqrt{9}\sqrt{5} = 9\sqrt{5}$
2 a $\frac{12\sqrt{3}}{3} = 4\sqrt{3}$
  b $\frac{4 \times \sqrt{11} + \sqrt{11} \times \sqrt{11}}{\sqrt{11} \times \sqrt{11}} = \frac{4\sqrt{11} + 11}{11}$
  c $\frac{8\sqrt{5} - 5}{5}$
3 a $8 + 4\sqrt{7} - 2\sqrt{7} - 7 = 1 + 2\sqrt{7}$
  b $27 - 10\sqrt{2}$ **c** $9 + 3\sqrt{5} - 3\sqrt{5} - 5 = 4$
  d -7 **e** 9
  f i $6 - \sqrt{8}$ **ii** $3 + \sqrt{11}$
4 a $\frac{40 + 8\sqrt{2}}{23}$ **b** $14 - 7\sqrt{3}$ **c** $\frac{42 + 6\sqrt{10}}{39} = \frac{14 + 2\sqrt{10}}{13}$

Algebraic fractions

1 a $\frac{x-8}{x+7}$ **b** $\frac{x+5}{x-2}$
  c $\frac{(x+4)(x-2)}{(x+8)(x-4)}$ **d** $\frac{5(x+1)}{9(x-1)}$
2 a i $\frac{3}{4}$ **ii** $\frac{3}{2}$ **iii** $\frac{1}{x^2}$ **iv** y^4
  b $\frac{9y^4}{8x^2}$
3 a $\frac{4x^2}{9y}$ **b** $\frac{8x}{9y^3}$
4 a i $2(x+5)$ **ii** $(x-5)(x+5)$
  b $\frac{x-5}{2}$
5 a $\frac{8(x+4)}{(x+4)(x+8)} = \frac{8}{x+8}$ **b** $\frac{x+8}{x-9}$
6 a i $3(x+3)$ **ii** $(x+6)(x+3)$
  iii $(x+5)(x+3)$ **iv** $2(x+5)$
  b i $\frac{3(x+3)}{2(x+6)}$ **ii** $\frac{2(x+6)}{3(x+3)}$
7 a $(x+1)(x+2)$
  b i $\frac{3(x+2)}{(x+1)(x+2)}$ **ii** $\frac{5(x+1)}{(x+1)(x+2)}$
  c $\frac{8x+11}{(x+1)(x+2)}$ **d** $x = \frac{5 \pm \sqrt{61}}{2}$
8 $\frac{5-2x}{(x-4)(x-1)}$

17 Extend

1 a Students' own demonstrations, e.g.
  Jack: $1 - 2y = 1 - 2\left(\frac{x-1}{-2}\right) = 1 + (x-1) = x$
  Ruth: $1 - 2y = 1 - 2\left(\frac{1-x}{2}\right) = 1 - (1-x) = x$
  b In Ruth's answer the denominator is positive.
  c $x = \sqrt{\frac{P-3d}{2}}$
2 $R_2 = \frac{RR_1}{R_1 - R}$

3 a $\frac{1}{x} = \frac{1}{v} + \frac{1}{w} - \frac{1}{u}$
  b $\frac{1}{x} = \frac{1}{v} + \frac{1}{w} - \frac{1}{u}$
  $= \frac{uw}{uvw} + \frac{uv}{uvw} - \frac{vw}{uvw} = \frac{1}{uvw}(uw + uv - vw)$
  $x = \frac{uvw}{uw + uv - vw}$
4 a $x = -\frac{5}{2}$ or $x = 4$ **b** $x = \frac{10 \pm 2\sqrt{5}}{5}$
5 $x = 6 \pm \sqrt{31}$
6 a 1.64 **b** 1.44
7 a -21 **b** $15 - 16x$
  c i $\frac{x-3}{-4} = \frac{-x}{4} + \frac{3}{4}$ **ii** $\frac{x-3}{4} = \frac{x}{4} - \frac{3}{4}$
  d $\left(\frac{-x}{4} + \frac{3}{4}\right) + \left(\frac{x}{4} - \frac{3}{4}\right) = \frac{x}{4} - \frac{x}{4} + \frac{3}{4} - \frac{3}{4} = 0$ for all x.
8 a i $x^2 + 13$ **ii** $x^2 + 14x + 55$
  b $x = -3$
9 a i x **ii** x
  b Yes; $fg(x) = x$ and $gf(x) = x$ so the functions are inverses.
  c $fg(x) = \frac{1}{4}(4x+4) - 1 = x + 1 - 1 = x$ and
  $gf(x) = 4\left(\frac{1}{4}x - 1 + 1\right) = x - 4 + 4 = x$
  So the two functions are inverses.
10 Students' own answers, e.g.
  $\frac{49 - x^2}{x^2 - 49} = \frac{(7-x)(7+x)}{(x-7)(x+7)} = \frac{7-x}{x-7} = \frac{-1 \times (x-7)}{x-7} = -1$
11 $(3n+1)^2 - (3n-1)^2 = 9n^2 + 6n + 1 - (9n^2 - 6n + 1) = 12n$,
  which is a multiple of 12 for all positive integer values of n.
12 $\frac{1}{5x^2 - 13x - 6} - \frac{1}{x^2 - 9} = \frac{1}{(5x+2)(x-3)} - \frac{1}{(x-3)(x+3)}$
  $= \frac{x + 3 - (5x+2)}{(5x+2)(x-3)(x+3)} = \frac{-4x+1}{(x-3)(x+3)(5x+2)}$
  $A = -4, B = 1$
13 1.5×10^{11} m
14 a Common ratio $= \frac{1}{\sqrt{x} - 2} = \frac{\sqrt{x} + 2}{1}$
  $1 = (\sqrt{x} + 2)(\sqrt{x} - 2) = x - 4$ so $x = 5$
  b 3rd term $= \sqrt{5} + 2$ and common ratio $= \sqrt{5} + 2$
  4th term $= (\sqrt{5} + 2)(\sqrt{5} + 2) = 5 + 4\sqrt{5} + 4 = 9 + 4\sqrt{5}$
15 $fg(x) = 2(x^2 - 1) + 5 = 2x^2 + 3$
  $gf(x) = (2x+5)^2 - 1 = 4x^2 + 20x + 24$
  $3fg(x) = gf(x)$ so $6x^2 + 9 = 4x^2 + 20x + 24$
  $2x^2 - 20x - 15 = 0$
16 $100a + 10b + c = 99a + 9b + (a + b + c)$
  $= 99a + 9b + 9$
  $= 9(11a + b + 1)$
  $(11a + b + 1)$ is an integer so $100a + 10b + c$ is a multiple of 9.
  Hence 'abc' $= 100a + 10b + c$ is divisible by 9.
17 $5x^2 - 8ax + 6b \equiv 5(x+2b)^2 + 2a - 3$
  $\equiv 5(x^2 + 4bx + 4b^2) + 2a - 3$
  $\equiv 5x^2 + 20bx + (20b^2 + 2a - 3)$
  Comparing x coefficients:
  $-8a = 20b$ so $a = -2.5b$ hence $2a = -5b$
  Comparing constant terms:
  $6b = 20b^2 + 2a - 3$
  $6b = 20b^2 - 5b - 3$
  $0 = 20b^2 - 11b - 3$
  $0 = (5b+1)(4b-3)$
  So $b = -\frac{1}{5}$ or $b = \frac{3}{4}$
  When $b = -\frac{1}{5}$, $a = \frac{1}{2}$ and when $b = \frac{3}{4}$, $a = -\frac{15}{8}$

Sample student answers

1 a The student has not multiplied both sides by $y - 5$ correctly: $x \times (y - 5) = xy - 5x$, not $xy - 5$.

b Putting brackets around '$y - 5$' before multiplying would remind the student to multiply *both* terms in the brackets by x.

2 a The student's mistake is in this step:

$\dfrac{22 - 11\sqrt{5}}{4 + 2\sqrt{5} - 2\sqrt{5} + 5}$; the final $+5$ in the denominator should be -5, because $-\sqrt{5} \times +\sqrt{5} = -5$

b Correct answer is:

$\dfrac{22 - 11\sqrt{5}}{4 + 2\sqrt{5} - 2\sqrt{5} - 5} = \dfrac{22 - 11\sqrt{5}}{-1} = 11\sqrt{5} - 22 = 11(\sqrt{5} - 2)$

17 Unit test

1 $f^{-1}(x) = \dfrac{x}{5} - 4$ or $\dfrac{x - 20}{5}$

2 LHS $= x^2 + 4x + 4x + 16 - 2x - 7 = x^2 + 6x + 9$
RHS $= x^2 + 3x + 3x + 9 = x^2 + 6x + 9 =$ LHS

3 $x = \sqrt{2y} - 3$

4 $\dfrac{1}{3(x - 4)}$

5 $x = \dfrac{3w}{y + 6}$

6 a $10 + \sqrt{2}$ **b** $14 + 6\sqrt{5}$

7 a $\dfrac{3y}{10x^2}$ **b** $\dfrac{37}{20x}$ **c** $\dfrac{x - 1}{x - 7}$ **d** $\dfrac{5}{4}$

8 $2 - \dfrac{x - 1}{x + 2} - \dfrac{x - 4}{x - 2} = \dfrac{2(x^2 - 4) - (x - 1)(x - 2) - (x - 4)(x + 2)}{x^2 - 4}$

$= \dfrac{2x^2 - 8 - (x^2 - 3x + 2) - (x^2 - 2x - 8)}{x^2 - 4} = \dfrac{5x - 2}{x^2 - 4}$

$a = 5, b = 2, c = 4$

9 $a = -4, b = 1$

10 a $x = \dfrac{6}{5}$ or $x = -1$ **b** $x = 4$ or $x = -3$

11 a 12   **b** $a = -\dfrac{1}{2}$ **c** $\dfrac{x - 1}{2}$

12 $n^2 + n = n(n + 1)$
When n is even, $n(n + 1) = $ even $\times$ odd $=$ even.
When n is odd, $n(n + 1) = $ odd $\times$ even $=$ even.
So $n^2 + n$ is never odd.

13 a $\dfrac{1}{1 + \frac{1}{x}} \times \dfrac{x}{x} = \dfrac{x}{x + 1}$ **b** $\dfrac{1}{1 + \frac{1}{9}} = \dfrac{9}{9 + 1} = \dfrac{9}{10}$

c Terms approach 1.

UNIT 18 Vectors and geometric proof

18.1 Vectors and vector notation

1 a 3 right   **b** 5 up

2 a $5\sqrt{3}$ **b** $2\sqrt{13}$

3 $\mathbf{a} = \begin{pmatrix} 4 \\ 2 \end{pmatrix}$; $\mathbf{b} = \begin{pmatrix} -3 \\ 1 \end{pmatrix}$

4

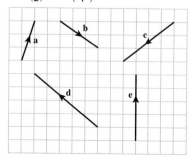

5 a $\begin{pmatrix} 2 \\ 2 \end{pmatrix}$ **b** $\begin{pmatrix} 2 \\ -5 \end{pmatrix}$ **c** $\begin{pmatrix} 4 \\ -3 \end{pmatrix}$

6 $(3, 5)$

7 $(-1, -1)$

8 a **a** and **d**

b No; vectors **b** and **d** are parallel but not in the same direction.

9 6.40

10 a 10   **b** 13   **c** $\sqrt{10}$
d 17   **e** $2\sqrt{13}$

11 a $\begin{pmatrix} -6 \\ -4 \end{pmatrix}$ **b** $2\sqrt{13}$

12 a $AB = 25$
b $AC = 25$, so $AB = AC$ and triangle ABC is isosceles.

13 $(8, 8)$

18.2 Vector arithmetic

1 3 units right and 5 units down

2

3 $\begin{pmatrix} -5 \\ 2 \end{pmatrix}$

4 a

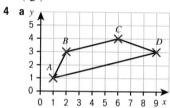

b $\begin{pmatrix} 8 \\ 2 \end{pmatrix}$ **c** Trapezium   **d** $\overrightarrow{AD} = 2\overrightarrow{BC}$

5 a

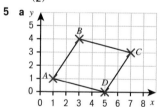

b i $\begin{pmatrix} -4 \\ 1 \end{pmatrix}$ **ii** $\begin{pmatrix} 4 \\ -1 \end{pmatrix}$

c $\overrightarrow{BC} = -\overrightarrow{CB}$

d i $\overrightarrow{AB} = \overrightarrow{DC}$ **ii** $\overrightarrow{AD} = -\overrightarrow{CB}$

6 Parallelogram

7 a $\begin{pmatrix} 5 \\ 2 \end{pmatrix}$ or $\begin{pmatrix} -5 \\ -2 \end{pmatrix}$ or $\begin{pmatrix} 1 \\ -8 \end{pmatrix}$ **b** $(4, 0)$

8

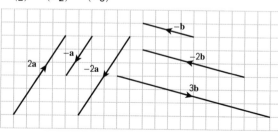

9

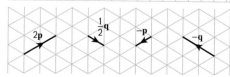

10 a i $\overrightarrow{EF}$, $\overrightarrow{GH}$ and $\overrightarrow{OP}$ **ii** $\overrightarrow{IJ}$, $\overrightarrow{KL}$ and $\overrightarrow{MN}$

b No; $2\mathbf{a}+\mathbf{b}$ is not a multiple of $\mathbf{a}+\mathbf{b}$.

11 a $\begin{pmatrix} -12 \\ 8 \end{pmatrix}$ **b** $\begin{pmatrix} -3 \\ 2 \end{pmatrix}$ **c** $\begin{pmatrix} 6 \\ -4 \end{pmatrix}$ **d** $\begin{pmatrix} 12 \\ -8 \end{pmatrix}$

12 a $\begin{pmatrix} 4 \\ 2 \end{pmatrix}$ **b** $\begin{pmatrix} 6 \\ 3 \end{pmatrix}$ **c** $\begin{pmatrix} -8 \\ -4 \end{pmatrix}$ **d** $\begin{pmatrix} 1 \\ 0.5 \end{pmatrix}$

13 a $\begin{pmatrix} 4 \\ 3 \end{pmatrix}$

b Add the x components and add the y components.

14 a, b

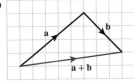

c $\begin{pmatrix} 8 \\ 1 \end{pmatrix}$

d

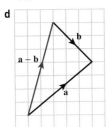

e $\begin{pmatrix} 2 \\ 7 \end{pmatrix}$

15 a $\begin{pmatrix} 9 \\ 2 \end{pmatrix}$

b i $\begin{pmatrix} 1 \\ -1 \end{pmatrix}$ **ii** $\begin{pmatrix} 1 \\ -1 \end{pmatrix}$

c $\mathbf{a}+\mathbf{b}=\mathbf{b}+\mathbf{a}$; adding the x (or y) components of $\mathbf{a}$ and $\mathbf{b}$ gives the same result in either order.

16 a $\begin{pmatrix} -1 \\ -4 \end{pmatrix}$ **b** $\begin{pmatrix} -1 \\ 11 \end{pmatrix}$ **c** $\begin{pmatrix} -1 \\ 8 \end{pmatrix}$

d $\begin{pmatrix} 3 \\ -3 \end{pmatrix}$ **e** $\begin{pmatrix} -2 \\ 10 \end{pmatrix}$ **f** $\begin{pmatrix} -3 \\ 6 \end{pmatrix}$

17 a

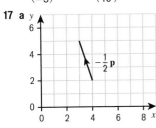

b $\begin{pmatrix} 0 \\ -10 \end{pmatrix}$

18.3 More vector arithmetic

1 a AB and DC; AD and BC

b AB and DC; AD and BC; AM and MC; BM and MD

2 a $\begin{pmatrix} 6 \\ 1 \end{pmatrix}$

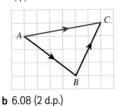

b 6.08 (2 d.p.)

3 a $\begin{pmatrix} -6 \\ 18 \end{pmatrix}$ **b** $\begin{pmatrix} 2 \\ -6 \end{pmatrix}$

4 B and E; $2\mathbf{a}-2\mathbf{b}=2(\mathbf{a}-\mathbf{b})$; $-\mathbf{a}+\mathbf{b}=-(\mathbf{a}-\mathbf{b})$

5 a $\begin{pmatrix} 1 \\ -17 \end{pmatrix}$ **b** $\begin{pmatrix} 16 \\ -6 \end{pmatrix}$ **c** $\begin{pmatrix} -25 \\ 7 \end{pmatrix}$

6 $\begin{pmatrix} -1 \\ 1 \end{pmatrix}$

7 $\begin{pmatrix} 0.5 \\ -0.5 \end{pmatrix}$

8 $\begin{pmatrix} 11 \\ -2 \end{pmatrix}$

9 a $\sqrt{106}$ **b** $6\sqrt{2}$ **c** $4\sqrt{13}$
d $2\sqrt{10}$ **e** $\sqrt{10}$

10 17

11 a $\begin{pmatrix} -1 \\ 4 \end{pmatrix}$

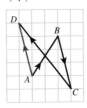

b $\overrightarrow{AC}=\overrightarrow{DB}=\begin{pmatrix} 3 \\ -1 \end{pmatrix}$

12 a $-\mathbf{a}$ **b** $\mathbf{a}+\mathbf{b}$ **c** $\mathbf{a}+\mathbf{b}+\mathbf{c}$

d $\overrightarrow{BC}+\overrightarrow{CD}=\mathbf{b}+\mathbf{c}$

$\overrightarrow{BA}+\overrightarrow{AD}=-\mathbf{a}+(\mathbf{a}+\mathbf{b}+\mathbf{c})=\mathbf{b}+\mathbf{c}$

13 a $\frac{1}{2}\mathbf{a}$

b i $\begin{pmatrix} -6 \\ -4 \end{pmatrix}$ **ii** $\begin{pmatrix} 3 \\ 2 \end{pmatrix}$

14 a $2\mathbf{b}$ **b** $\mathbf{a}+\mathbf{b}$ **c** $\mathbf{a}+2\mathbf{b}$

15 a SR is parallel to PQ and the same length, so $\overrightarrow{SR}=\overrightarrow{PQ}=\mathbf{a}$.

b i $\mathbf{b}$ **ii** $\mathbf{a}+\mathbf{b}$

16 a i $\mathbf{p}+\mathbf{q}$ **ii** $\mathbf{q}-\mathbf{p}$ **b** $\frac{1}{2}(\mathbf{p}+\mathbf{q})$

17 a $\mathbf{b}-\mathbf{a}$ **b** Midpoint of PR

18 a ED is parallel to AB and the same length, so $\overrightarrow{ED}=\overrightarrow{AB}=\mathbf{n}$.

b i $\mathbf{m}$ **ii** $\mathbf{p}$
c i $\mathbf{n}+\mathbf{m}$ **ii** $\mathbf{n}+\mathbf{m}+\mathbf{p}$
d $\mathbf{n}+\mathbf{m}$

19 a $\mathbf{r}$ **b** $-\mathbf{s}$ **c** $\frac{1}{2}\mathbf{r}$ **d** $\mathbf{s}+\frac{1}{2}\mathbf{r}$

20 a $\overrightarrow{AB}$, $\overrightarrow{EF}$ and $\overrightarrow{IJ}$

b i $3\mathbf{p}-3\mathbf{q}$ **ii** $6\mathbf{a}-\frac{13}{2}\mathbf{b}$

21 a $\mathbf{b}-\mathbf{a}$ **b** $\frac{1}{2}(\mathbf{b}-\mathbf{a})$ **c** $\frac{1}{2}(\mathbf{a}+\mathbf{b})$

d

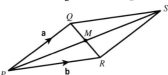

By the parallelogram law, $\overrightarrow{PS}=\mathbf{a}+\mathbf{b}$

M is the midpoint of PS so $\overrightarrow{PM}=\frac{1}{2}(\mathbf{a}+\mathbf{b})$.

18.4 Parallel vectors and collinear points

1 $\begin{pmatrix} -2 \\ 4 \end{pmatrix}$, $\begin{pmatrix} 6 \\ -12 \end{pmatrix}$ and $\begin{pmatrix} 1 \\ -2 \end{pmatrix}$

2 $(2, 5)$

3 a $\begin{pmatrix} -1 \\ 3 \end{pmatrix}$ **b** $\begin{pmatrix} 2 \\ 1 \end{pmatrix}$ **c** $\begin{pmatrix} 0 \\ 0 \end{pmatrix}$

4 $4\mathbf{c} + \frac{1}{2}\mathbf{d}$

5 a $\mathbf{b} - \mathbf{a}$ **b** $\frac{1}{2}\mathbf{a}$ **c** $\frac{1}{2}\mathbf{b}$ **d** $\frac{1}{2}(\mathbf{b} - \mathbf{a})$

6 BC is parallel to PQ and twice the length.

7 a $\mathbf{a} - 2\mathbf{b}$

b $\overrightarrow{GJ}$ is parallel to $\overrightarrow{HI}$ and twice the length.
$\mathbf{a} - 2\mathbf{b} = 2 \times \frac{1}{2}(\mathbf{a} - \mathbf{b})$ so $\overrightarrow{GJ} = 2\overrightarrow{HI}$

8 a $\begin{pmatrix} 1 \\ 5 \end{pmatrix}$ **b** $\begin{pmatrix} -1 \\ -5 \end{pmatrix}$ **c** $\begin{pmatrix} 2 \\ 4 \end{pmatrix}$ **d** $\begin{pmatrix} 1 \\ -1 \end{pmatrix}$

9 $\mathbf{b} - \mathbf{a}$

10 $\mathbf{a} + 3\mathbf{b}$

11 a i $\begin{pmatrix} 1 \\ 4 \end{pmatrix}$ **ii** $\begin{pmatrix} 5 \\ 20 \end{pmatrix}$

b Yes; $\begin{pmatrix} 5 \\ 20 \end{pmatrix} = 5\begin{pmatrix} 1 \\ 4 \end{pmatrix}$

c $\overrightarrow{CD} = 5\overrightarrow{AB}$
This means that $\overrightarrow{CD}$ is 5 times the length of $\overrightarrow{AB}$.

12 a i $\begin{pmatrix} 5 \\ -4 \end{pmatrix}$ **ii** $\begin{pmatrix} 20 \\ -16 \end{pmatrix}$

b $\begin{pmatrix} 20 \\ -16 \end{pmatrix} = 4\begin{pmatrix} 5 \\ -4 \end{pmatrix}$ so RS is parallel to PQ and is 4 times the length.

13 a $\begin{pmatrix} -10 \\ -4 \end{pmatrix}$ **b** $(1, 8)$ **c** $\begin{pmatrix} 2 \\ 3 \end{pmatrix}$

14 a $\begin{pmatrix} 3 \\ 2 \end{pmatrix}$ **b** $\begin{pmatrix} 18 \\ 12 \end{pmatrix}$ **c** $(16, 8)$

15 $\begin{pmatrix} 2 \\ -3 \end{pmatrix}$

16 a $\overrightarrow{AB} = 2\mathbf{b}$ so $\overrightarrow{OC}$ is parallel to $\overrightarrow{AB}$ and the same length.

b $\overrightarrow{BC} = -\mathbf{a}$, so $\overrightarrow{BC}$ is parallel to, and the same length as, $\overrightarrow{OA}$ but the opposite direction.

c Parallelogram

17 a i $\begin{pmatrix} 3 \\ 9 \end{pmatrix}$ **ii** $\begin{pmatrix} 9 \\ 27 \end{pmatrix}$

b Yes; $\overrightarrow{AC} = 3 \times \overrightarrow{AB}$ as $\begin{pmatrix} 9 \\ 27 \end{pmatrix} = 3\begin{pmatrix} 3 \\ 9 \end{pmatrix}$ so the lines are parallel.

c Yes; both lines pass through point A.

d A, B and C are collinear.

18 $\overrightarrow{PQ} = \begin{pmatrix} 3 \\ 3 \end{pmatrix}$ and $\overrightarrow{QR} = \begin{pmatrix} 6 \\ 6 \end{pmatrix} = 2\begin{pmatrix} 3 \\ 3 \end{pmatrix}$ so the lines are parallel.
Both lines pass through point Q so P, Q and R are collinear.

19 a i $2\mathbf{b}$ **ii** $\mathbf{b}$ **iii** $\mathbf{a} + \mathbf{b}$ **iv** $3\mathbf{a} + 3\mathbf{b}$

b $\overrightarrow{OC} = 3 \times \overrightarrow{OX}$ so the lines are parallel. Both lines pass through the point O so O, X and C are collinear.

c $\overrightarrow{XC} = \overrightarrow{XA} + \overrightarrow{AC} = -\mathbf{b} + 2\mathbf{a} + 3\mathbf{b} = 2\mathbf{a} + 2\mathbf{b} = 2 \times \overrightarrow{OX}$ so the lines are parallel. Both lines pass through the point X so O, X and C are collinear.

18.5 Solving geometric problems

1 $\frac{3}{5}$

2 a $\mathbf{p} + \mathbf{q}$ **b** $\mathbf{q}$ **c** $\mathbf{q} - \mathbf{p}$
d $-\mathbf{p}$ **e** $-\mathbf{q}$

---

3 a $\overrightarrow{PR} = 9\mathbf{a} - 6\mathbf{b} = 3(3\mathbf{a} - 2\mathbf{b})$ so PR is parallel to PQ and is 3 times the length.

b P, Q and R are collinear.

4 a $\mathbf{b} - \mathbf{a}$ **b** $\frac{1}{3}(\mathbf{b} - \mathbf{a})$ **c** $\frac{2}{3}(\mathbf{b} + 2\mathbf{a})$

5 a i $\mathbf{b} - \mathbf{a}$ **ii** $\frac{1}{4}\mathbf{b}$ **iii** $\mathbf{a} + \frac{1}{4}\mathbf{b}$ **iv** $\mathbf{b} + \frac{1}{4}\mathbf{a}$

b $\overrightarrow{EF} = \frac{3}{4}(\mathbf{b} - \mathbf{a})$ and $\overrightarrow{AB} = \mathbf{b} - \mathbf{a}$. So $\overrightarrow{EF}$ is a multiple of $\overrightarrow{AB}$; hence the lines are parallel.

6 a $8\mathbf{b}$ **b** $20\mathbf{b}$ **c** $\mathbf{a} + 18\mathbf{b}$

7 a $\frac{1}{2}(2\mathbf{q} + \mathbf{p})$ **b** $n = 4$

8 a i $\mathbf{a} + 6\mathbf{b}$ **ii** $\frac{1}{2}\mathbf{a} + 3\mathbf{b}$

b $\overrightarrow{OD} = 2\overrightarrow{OX}$ so OD and OX are parallel with a point in common, and O, X and D lie on the same straight line. OD is twice the length of OX, so X is the midpoint of OD.

9 a i $\overrightarrow{EF}$ and $\overrightarrow{CB}$ **ii** $\overrightarrow{FA}$ and $\overrightarrow{DC}$

b $\mathbf{b} - \mathbf{a}$ **c** $\overrightarrow{ED}$

10 a $12\mathbf{q} - 3\mathbf{p}$

b $\overrightarrow{PQ} = 6\mathbf{q} - 6\mathbf{p}$ so $\overrightarrow{QN} = 4\mathbf{q} - 4\mathbf{p}$
$\overrightarrow{TN} = \overrightarrow{TQ} + \overrightarrow{QN} = 12\mathbf{q} + 4\mathbf{q} - 4\mathbf{p} = 16\mathbf{q} - 4\mathbf{p}$
$\overrightarrow{TM} = 3(4\mathbf{q} - \mathbf{p})$ and $\overrightarrow{TN} = 4(4\mathbf{q} - \mathbf{p})$ so TM and TN are parallel. Both lines pass through point T, so T, M and N lie on the same straight line.

11 a i $\frac{1}{2}(\mathbf{m} + \mathbf{n})$ **ii** $\frac{3}{4}(\mathbf{m} + \mathbf{n})$ **iii** $\frac{3}{4}\mathbf{n} - \frac{1}{4}\mathbf{m}$

b $3\mathbf{n} - \mathbf{m}$

c $\overrightarrow{MQ} = \frac{3}{4}\mathbf{n} - \frac{1}{4}\mathbf{m} = \frac{1}{4}(3\mathbf{n} - \mathbf{m})$. $\overrightarrow{MQ}$ is a multiple of $\overrightarrow{MR}$ so the lines are parallel. Both lines pass through the point M, so MQR is a straight line.
$\frac{MR}{MQ} = 4$

12 a $2\mathbf{b}$ **b** $2\mathbf{a} + \mathbf{b}$ **c** $4\mathbf{a} + 2\mathbf{b}$

d $\overrightarrow{OS} = \frac{1}{2}(4\mathbf{a} + 2\mathbf{b})$ so S is the midpoint of OT.

e $\overrightarrow{QR} = 3\mathbf{a} - 2\mathbf{b} = \begin{pmatrix} 18 \\ -24 \end{pmatrix}$; $QR = 30$

13 a i $-2\mathbf{p}$ **ii** $-2\mathbf{q}$ **iii** $-2\mathbf{p} + \mathbf{q}$ **iv** $\mathbf{p} - 2\mathbf{q}$

b i $\frac{2}{3}\mathbf{p} + \frac{2}{3}\mathbf{q}$ **ii** $\frac{2}{3}\mathbf{p} + \frac{2}{3}\mathbf{q}$

c $\overrightarrow{OX} = \overrightarrow{OY}$ so X and Y are the same point.

18 Check up

1 a $\begin{pmatrix} 2 \\ 3 \end{pmatrix}$ **b** $\begin{pmatrix} 3 \\ -1 \end{pmatrix}$

2 a $\begin{pmatrix} 3 \\ -5 \end{pmatrix}$ **b** $(3, 7)$

3 $\sqrt{34}$

4

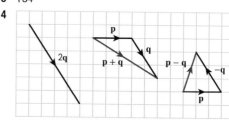

5 $\begin{pmatrix} -2 \\ 3 \end{pmatrix}$

6 a $\begin{pmatrix} 3 \\ 4 \end{pmatrix}$ **b** $\begin{pmatrix} -5 \\ 8 \end{pmatrix}$ **c** $\begin{pmatrix} 12 \\ -6 \end{pmatrix}$

7 $\begin{pmatrix} 2 \\ -1 \end{pmatrix}$

8 a $\mathbf{a}$ **b** $\mathbf{a} + \mathbf{b}$ **c** $\frac{1}{2}(\mathbf{a} + \mathbf{b})$ **d** $\frac{1}{2}(\mathbf{a} - \mathbf{b})$

9 a $3s - 3t$ and $\frac{1}{2}s - \frac{1}{2}t$ **b** $3s + 3t$

10 a $\begin{pmatrix} 4 \\ -3 \end{pmatrix}$ **b** $\begin{pmatrix} -2 \\ 7 \end{pmatrix}$ **c** $\begin{pmatrix} -6 \\ 10 \end{pmatrix}$

11 a i $\begin{pmatrix} 3 \\ 9 \end{pmatrix}$ **ii** $\begin{pmatrix} 9 \\ 27 \end{pmatrix}$

   b $\overrightarrow{AC} = 3\begin{pmatrix} 3 \\ 9 \end{pmatrix}$ so A, B and C are collinear.

12 a i $\frac{1}{2}c$ **ii** $\frac{1}{2}c - \frac{1}{2}a$

   b $\overrightarrow{AC} = c - a = 2\overrightarrow{XY}$; $\overrightarrow{AC}$ is a multiple of $\overrightarrow{XY}$ so the lines are parallel.

13 a $3a - 3b$ **b** $5a - 5b$ **c** $6a - 2b$

15 a 3 possibilities: $(6, -1)$, $(5, -2)$, $(2, -3)$

   b

Q	$\overrightarrow{PQ}$	$\overrightarrow{QR}$	$\overrightarrow{RS}$	$\overrightarrow{SQ}$
$(6, -1)$	$\begin{pmatrix} 4 \\ -3 \end{pmatrix}$	$\begin{pmatrix} 0 \\ 5 \end{pmatrix}$	$\begin{pmatrix} -4 \\ 3 \end{pmatrix}$	$\begin{pmatrix} 4 \\ -8 \end{pmatrix}$
$(5, -2)$	$\begin{pmatrix} 3 \\ -4 \end{pmatrix}$	$\begin{pmatrix} 1 \\ 6 \end{pmatrix}$	$\begin{pmatrix} -3 \\ 4 \end{pmatrix}$	$\begin{pmatrix} 2 \\ -10 \end{pmatrix}$
$(2, -3)$	$\begin{pmatrix} 0 \\ -5 \end{pmatrix}$	$\begin{pmatrix} 4 \\ 7 \end{pmatrix}$	$\begin{pmatrix} 0 \\ 5 \end{pmatrix}$	$\begin{pmatrix} -4 \\ -12 \end{pmatrix}$

18 Strengthen

Vector notation

1 a $\begin{pmatrix} 2 \\ 2 \end{pmatrix}$ **b** $\begin{pmatrix} -2 \\ -2 \end{pmatrix}$ **c** $\begin{pmatrix} -6 \\ 1 \end{pmatrix}$ **d** $\begin{pmatrix} -5 \\ -2 \end{pmatrix}$

2 a $O(0, 0)$

   b i $\begin{pmatrix} 4 \\ 6 \end{pmatrix}$ **ii** $\begin{pmatrix} 8 \\ 2 \end{pmatrix}$

3 a

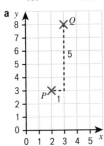

   b $(3, 8)$

4 a

   b $\begin{pmatrix} 3 \\ -1 \end{pmatrix}$

5 a

(right-angled triangle with sides 2, -5)

   b $\sqrt{29}$

   c i 5 **ii** $\sqrt{106}$ **iii** $\sqrt{130}$ **iv** $\sqrt{34}$

Vector arithmetic

1

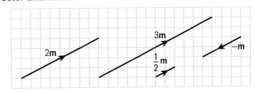

2 a i B   **ii** A

   In B, the arrows for **b** and **c** follow on, end to end.

   b

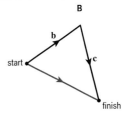

3 a

(diagram of vectors a, b, a+b, −b, 2a, 2a+b, a−b)

   b i $\begin{pmatrix} 4 \\ 6 \end{pmatrix}$ **ii** $\begin{pmatrix} 2 \\ 2 \end{pmatrix}$ **iii** $\begin{pmatrix} 7 \\ 10 \end{pmatrix}$

   c i $\begin{pmatrix} 4 \\ 6 \end{pmatrix}$ **ii** $\begin{pmatrix} 2 \\ 2 \end{pmatrix}$ **iii** $\begin{pmatrix} 7 \\ 10 \end{pmatrix}$

   d They are the same.

4 a i $\begin{pmatrix} 6 \\ 3 \end{pmatrix}$ **ii** $\begin{pmatrix} -1 \\ 2 \end{pmatrix}$ **iii** $\begin{pmatrix} 0 \\ 3 \end{pmatrix}$

   b i $\overrightarrow{AB} + \overrightarrow{BC}$ **ii** $\overrightarrow{AB} + \overrightarrow{BE}$ **iii** $\overrightarrow{BC} + \overrightarrow{CD}$

Geometric problems

1 a

(diagram of vectors a, b, c, d, e)

   b i They are parallel, with **b** being 3 times the length of **a**.

    ii They are parallel, with **c** being the same length but in the opposite direction to **a**.

    iii They are parallel, with **d** being twice the length of **a**.

2 $3a + 3b$, $2a + 2b$ and $\frac{1}{2}a + \frac{1}{2}b$

3 a $\overrightarrow{CD}$ is a multiple of $\overrightarrow{AB}$.

   b $\overrightarrow{CD} = -2\overrightarrow{AB} = -2a$

   c $\overrightarrow{BC} = \overrightarrow{BA} + \overrightarrow{AD} + \overrightarrow{DC} = -a + b + 2a = a + b$

4 a i DC ii AD

b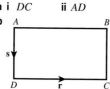

c i $s+r$ ii $s-r$

d i $\overrightarrow{AX}=\frac{1}{2}\overrightarrow{AC}$ ii $\overrightarrow{BX}=\frac{1}{2}\overrightarrow{BD}$

e i $\frac{1}{2}(s+r)$ ii $\frac{1}{2}(s-r)$

5 a, b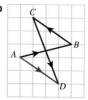

$AD=\begin{pmatrix}3\\-2\end{pmatrix}$

c $\overrightarrow{AC}=\begin{pmatrix}4\\1\end{pmatrix}+\begin{pmatrix}-3\\2\end{pmatrix}=\begin{pmatrix}1\\3\end{pmatrix}$; $\overrightarrow{DB}=\begin{pmatrix}-2\\5\end{pmatrix}+\begin{pmatrix}3\\-2\end{pmatrix}=\begin{pmatrix}1\\3\end{pmatrix}$

$\overrightarrow{AC}$ is parallel to $\overrightarrow{DB}$ and they are equal in length.

6 a i $\begin{pmatrix}6\\3\end{pmatrix}$ ii $\begin{pmatrix}4\\2\end{pmatrix}$

b $\overrightarrow{AB}=3\begin{pmatrix}2\\1\end{pmatrix}$ and $\overrightarrow{BC}=2\begin{pmatrix}2\\1\end{pmatrix}$.

Both are multiples of the same vector and so are parallel. They both pass through the point B so they are collinear.

7 a, b, c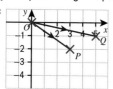

d $\begin{pmatrix}3\\-2\end{pmatrix}$ e $\begin{pmatrix}5\\-1\end{pmatrix}$ f $\begin{pmatrix}2\\1\end{pmatrix}$

8 a $\overrightarrow{AB}=\begin{pmatrix}4\\3\end{pmatrix}$; $\overrightarrow{BC}=\begin{pmatrix}12\\9\end{pmatrix}$

b $\overrightarrow{BC}=\begin{pmatrix}12\\9\end{pmatrix}=3\begin{pmatrix}4\\3\end{pmatrix}$. Both are multiples of the same vector and so are parallel.

c AB and BC are **parallel** and both pass through the point B. So ABC is a **straight** line and A, B, C are collinear.

9 a $\frac{1}{2}a$ b $a-b$ c $\frac{1}{2}(a-b)$

d $\frac{1}{2}(a+b)$ e $\frac{1}{2}b$

f $\overrightarrow{OB}$ and $\overrightarrow{CD}$ are both multiples of **b**, so they are parallel.

10 a $AP=\frac{3}{5}AB$; $BP=\frac{2}{5}BA$

b i $b-a$ ii $\frac{3}{5}(b-a)$ iii $a-b$

iv $\frac{2}{5}(b-a)$ v $\frac{2}{5}a+\frac{3}{5}b$

18 Extend

1 a $j-k$

b i $j-k$ ii $\overrightarrow{JX}=\overrightarrow{KJ}$, and point J is common.

2 a $3b-a$

b $\overrightarrow{AC}=2a$ and $\overrightarrow{MB}=-\frac{1}{2}\overrightarrow{BD}=\frac{1}{2}(a-3b)$

$\overrightarrow{NC}=\overrightarrow{NA}+\overrightarrow{AC}=-2b+2a=2(a-b)$

$\overrightarrow{MC}=\overrightarrow{MB}+\overrightarrow{BC}=\frac{1}{2}(a-3b)+a=\frac{3}{2}a-\frac{3}{2}b=\frac{3}{2}(a-b)$

$\overrightarrow{NC}$ and $\overrightarrow{MC}$ are both multiples of the same vector and so are parallel. They both pass through the point C, so they are collinear, i.e. NMC is a straight line.

3 $\overrightarrow{OR}=\frac{6}{5}(p+q)$; $\overrightarrow{OR}$ is a multiple of $p+q$ so it is parallel to $p+q$.

4 a i $2q-4p$ ii $3(q-p)$ iii $2(q-p)$

b $\overrightarrow{AC}$ and $\overrightarrow{BC}$ are both multiples of $q-p$. Point C is common, so ABC is a straight line.

c 9 cm

5 a $6b-3a$

b $\overrightarrow{AX}=\frac{1}{3}\overrightarrow{AB}=2b-a$

$\overrightarrow{OX}=\overrightarrow{OA}+\overrightarrow{AX}=3a+2b-a=2(b+a)$

$\overrightarrow{OY}=2.5\times\overrightarrow{OX}=5(a+b)$

6 a $(k-2)a+kb$ b $k=6$ c $4a+6b$

7 $\overrightarrow{AB}=\overrightarrow{AO}+\overrightarrow{OB}=-2a+2b=2(b-a)$

so $\overrightarrow{AN}=2k(b-a)$

$\overrightarrow{PM}=\overrightarrow{PO}+\overrightarrow{OM}=-3a+b$

$\overrightarrow{PN}=\overrightarrow{PA}+\overrightarrow{AN}=-a+2k(b-a)=-(2k+1)a+2kb$

MNP is a straight line so $\overrightarrow{PM}$ and $\overrightarrow{PN}$ are parallel and coefficients of **a** and **b** are in the same ratio.

$\frac{1}{2k}=\frac{-3}{-(2k+1)}$; $2k+1=6k$; $k=\frac{1}{4}$

8 $\overrightarrow{AB}=b-a$ and $\overrightarrow{OM}=\frac{1}{2}a+\frac{1}{2}b$

$\overrightarrow{AP}=\overrightarrow{AO}+\frac{3}{4}\overrightarrow{OM}=-a+\frac{3}{4}(\frac{1}{2}a+\frac{1}{2}b)=-\frac{5}{8}a+\frac{3}{8}b$

$\overrightarrow{AN}=\overrightarrow{AO}+\overrightarrow{ON}=-a+kb$

Comparing coefficients of **a** and **b** for $\overrightarrow{AP}$ and $\overrightarrow{AN}$:

$\frac{-\frac{5}{8}}{-1}=\frac{\frac{3}{8}}{k}$; $5k=3$; $k=\frac{3}{5}$

So $ON:NB=\frac{3}{5}:\frac{2}{5}=3:2$

9 $k^2+(3k)^2=(4\sqrt{5})^2$

$10k^2=16\times5=80$ so $k=\pm\sqrt{8}=\pm2\sqrt{2}$

10 a $\begin{pmatrix}2m-1\\1-m\end{pmatrix}$

b $(2m-1)^2+(1-m)^2=10$

$5m^2-6m-8=0$

$(5m+4)(m-2)=0$, so $m=-\frac{4}{5}$ or $m=2$

18 Test ready

Sample student answers

a The answer should read $\overrightarrow{AB}=2n-2m=2(n-m)$ and $\overrightarrow{MN}=n-m$: the student has forgotten the direction of the vectors.

b The answer could be improved by adding a sentence at the end, e.g. This means that AB is parallel to MN and is twice the length.

18 Unit test

1 $\begin{pmatrix}-11\\-18\end{pmatrix}$

2

3 a $2\mathbf{q}$ **b** $\mathbf{p}+\mathbf{q}$ **c** $\mathbf{q}-\mathbf{p}$

4 a $(2,6)$ **b** $\begin{pmatrix}2\\1\end{pmatrix}$

5 a $\begin{pmatrix}-4\\-4\end{pmatrix}$ **b** $(1,5)$ **c** $\begin{pmatrix}1\\2\end{pmatrix}$ **d** $4\sqrt{2}$

6 a i $\mathbf{b}$ **ii** $\mathbf{b}-2\mathbf{a}$

   b $\overrightarrow{CD}=-2\mathbf{a}+\mathbf{b}+\mathbf{a}=\mathbf{b}-\mathbf{a}$

     $\overrightarrow{FX}=\overrightarrow{FE}+\overrightarrow{EX}=\mathbf{b}+\mathbf{b}-2\mathbf{a}=2(\mathbf{b}-\mathbf{a})=2\overrightarrow{CD}$

     $\overrightarrow{FX}$ is a multiple of $\overrightarrow{CD}$ so FX and CD are parallel.

7 $\overrightarrow{PQ}=2\mathbf{r}$ so $\overrightarrow{PX}=\frac{3}{2}\mathbf{r}$

   $\overrightarrow{OX}=\overrightarrow{OP}+\overrightarrow{PX}=\mathbf{p}+\frac{3}{2}\mathbf{r}$

8 a $\overrightarrow{OX}=\overrightarrow{OA}+\overrightarrow{AX}=\overrightarrow{OA}+\frac{2}{3}\overrightarrow{AC}=\mathbf{a}+\frac{2}{3}(\mathbf{b}-\mathbf{a})=\frac{1}{3}\mathbf{a}+\frac{2}{3}\mathbf{b}$

   b $\overrightarrow{OY}=\overrightarrow{OA}+\overrightarrow{AB}+\overrightarrow{BY}=\mathbf{a}+\mathbf{b}+\mathbf{b}=\mathbf{a}+2\mathbf{b}$

     $\overrightarrow{OX}=\frac{1}{3}\mathbf{a}+\frac{2}{3}\mathbf{b}=\frac{1}{3}(\mathbf{a}+2\mathbf{b})$

     $\overrightarrow{OX}$ and $\overrightarrow{OY}$ are both multiples of $\mathbf{a}+2\mathbf{b}$ so OX and OY are parallel. Point O is common, so OXY is a straight line.

9 a $\overrightarrow{MC}=\overrightarrow{MO}+\overrightarrow{OB}+\overrightarrow{BC}=-\frac{1}{2}\mathbf{a}+\mathbf{b}+\frac{3}{2}\mathbf{b}=-\frac{1}{2}\mathbf{a}+\frac{5}{2}\mathbf{b}$

   b $\overrightarrow{NC}=\overrightarrow{NB}+\overrightarrow{BC}=k\overrightarrow{AB}+\overrightarrow{BC}$

     $=k(\mathbf{b}-\mathbf{a})+\frac{3}{2}\mathbf{b}=-k\mathbf{a}+(k+\frac{3}{2})\mathbf{b}$

   c MNC is a straight line so $\overrightarrow{MC}$ and $\overrightarrow{NC}$ are parallel and coefficients of $\mathbf{a}$ and $\mathbf{b}$ are in the same ratio.

     $\dfrac{-\frac{1}{2}}{-k}=\dfrac{\frac{5}{2}}{k+\frac{3}{2}};\ -\frac{1}{2}(k+\frac{3}{2})=-\frac{5}{2}k;\ k+\frac{3}{2}=5k;\ k=\frac{3}{8}$

10

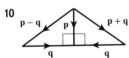

From Pythagoras' theorem, $\sqrt{p^2+q^2}=\sqrt{p^2+(-q)^2}$

UNIT 19 Proportion and graphs

19.1 Direct proportion

1 $y=6x$

2 Yes; the ratio of $x:y$ simplifies to $1:4$ for all pairs of values.

3 a Yes; $C=0.84q$

   b No   **c** No   **d** No

4 a, b

Exchange rates

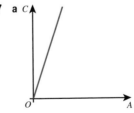

Key
-✕- travelcash.com
✕ currencyexchange

c i $E=1.25S$ (using gradient of line of best fit on graph)

   ii $E=1.1S$ (using gradient of line of best fit on graph)

d travelcash.com

e i Straight line through origin

   ii Equation of the form $y=kx$, where k is a constant

5 a $y=5x$ **b** $y=50$ **c** $x=13$

6 a $y=6.5x$ **b** $y=91$ **c** $x=22$

7 a $x=5$ **b** $x=10.1$ (1 d.p.)   **c** $x=8.125$

8 a $y=\frac{x}{30}$

   b i $y=\frac{1}{6}$ **ii** $y=\frac{9}{10}$

9 $y=150x$

19.2 More direct proportion

1 a 27   **b** 2   **c** 10

2 $A=13.5,\ B=14$

3 $k=2.5$

4 a $F=8a$ **b** $F=160$ **c** $a=14$

5 a The ratio of $P:l$ simplifies to $12:5$ for all pairs of values.

   b $k=2.4$ **c** $P=2.4l$

   d i $P=43.2\,\text{cm}$ **ii** $l=17.5\,\text{cm}$

6 a $d=500t$ **b** $d=2500\,\text{km}$

   c $t=4.5$ hours

   d i The distance doubles.   **ii** The distance halves.

7 a

   b $C=50A$ **c** £4250

8 a $y\propto x^2$ **b** $y=kx^2$ **c** $k=4$

   d $y=100$ **e** $x=2.5$

9 a $y=3.6x^3$ **b** $y=230.4$ **c** $x=5$

10 a $y=25\sqrt{x}$ **b** $y=75$ **c** $x=100$

11 $y=\frac{1}{5}$

12 a $C=0.05s^3$ **b** $C=£6.25$

13 a $T=\frac{R^2}{450}$ **b** $T=50$ minutes

14 $g\propto h^3$

19.3 Inverse proportion

1 2 m

2 $a = \dfrac{5}{b^2}$

3 **a** $A = \dfrac{B}{4}$ **b** $A = 115$

4 **a** A **b** C **c** B

5 **a**
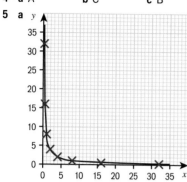

b $k = 8$ **c** $x \times y = 8$

d The product of x and y is the same for each pair of values.

6 **a** $y = \dfrac{10}{x}$ **b** $y = 0.5$ **c** $x = 2.5$

7 **a** $p = \dfrac{3000}{V}$ **b** $p = 2000\,\text{N/m}^2$ **c** $V = 2.5\,\text{m}^3$

d When the pressure doubles, the volume halves.

8 **a** $t = \dfrac{600\,000}{p}$

b No; it takes 4 minutes. (When $p = 2500\,\text{W}$, $t = 240$ seconds.)

9 **a** $s = \dfrac{800}{t}$

b

Speed, s (m/s)	4	10	20	40	80	160
Time, t (seconds)	200	80	40	20	10	5

c
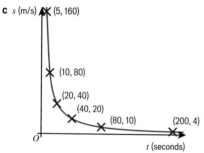

d The time increases more and more as the speed gets closer to 0 m/s.

10 $a = 4$, $b = 10$

11 **a** $y = \dfrac{54}{x^3}$ **b** $y = 0.432$ **c** $x = 2$

12 **a** $y = \dfrac{6}{\sqrt{x}}$ **b** $y = 3$ **c** $x = 1$

13 **a** $h = \dfrac{36}{r^2}$ **b** $r = 0.5$

14 **a** $s = \dfrac{3400}{r^2}$ **b** $s = 192.74$ rev/min

15 $y \propto \dfrac{1}{x^3}$; $28 = \dfrac{k}{a^3}$; $k = 28a^3$ so $y = \dfrac{28a^3}{x^3}$

When $x = 2a$, $y = \dfrac{28a^3}{(2a)^3} = \dfrac{28}{8} = 3.5$

16 $y = \dfrac{24}{x^2}$

19.4 Exponential functions

1 **a** 8 **b** 16 **c** 1
  d $\dfrac{1}{4}$ **e** $\dfrac{1}{16}$

2 **a** 512 **b** 16 384 **c** 524 288

3 **a** $x = 3$ **b** $x = 4$ **c** $x = 4$

4 **a**

x	-4	-3	-2	-1	0	1	2	3	4
y	0.06	0.13	0.25	0.5	1	2	4	8	16

b
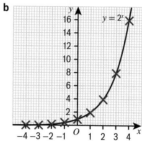

c i $y \approx 11$ **ii** $x \approx 3.3$

d As x decreases, y approaches zero.

5 $y = 1$

6 **a, b**
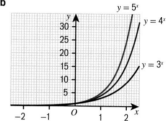

c $(0, 1)$

d $a^0 = 1$ for any non-zero number a.

7
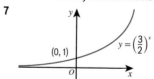

8 **a**

x	-4	-3	-2	-1	0	1	2	3	4
y	16	8	4	2	1	0.5	0.25	0.13	0.06

b
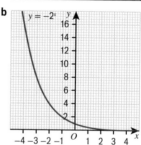

c i $y \approx 0.1$ **ii** $x \approx -3.3$

9 **a** $(0, 1)$

b Yes; when $x = 0$, $y = a^{-x} = a^0 = 1$

10 a

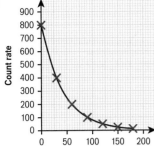

Count rate vs **Time (seconds)**

b Exponential decay   **c** 30 seconds

11 a $10 = ka^1, 250 = ka^3$ **b** $a = 5$ **c** $k = 2$

12 a $k = 0.5, a = 4$ **b** $y = \frac{1}{32}$

13 a $a = 20\,000, b = 0.9$ **b** £14 580

14 a

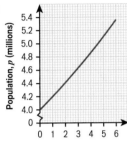

Population, p (millions) vs **Time, t (years)**

b i $p \approx 4.5$ million   **ii** $t \approx 4.6$ years

c The formula is of the form $p = ab^t$, with $a = 4$ and $b = 1.05$

15 a $V = 10\,000 \times 1.02^t$

b

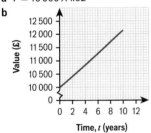

Value (£) vs **Time, t (years)**

c $t \approx 4.8$ years

19.5 Non-linear graphs

1 $18\,\text{cm}^2$

2 B

3 $180\,\text{m}$

4 a 3   **b** 1.5   **c** −1

5 i D   **ii** A   **iii** B   **iv** C

6 a Student B's graph is better; its gradient is closer to that of the curve at point P; it touches the curve at only one point.

b 1

7 a The temperature falls as time increases; the rate of temperature change decreases over time.

b About 0.25 °C/s   **c** About 0.04 °C/s

d The average rate of temperature reduction over the first 300 seconds is greater than the rate of temperature reduction at exactly 300 seconds.

8 a About 7 m/s

b 10 seconds; the graph does not get any steeper after this.

c About 24 seconds

9 a

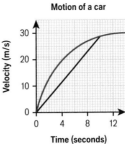

Motion of a car — **Velocity (m/s)** vs **Time (seconds)**

b About $2.9\,\text{m/s}^2$

c About $4.5\,\text{m/s}^2$

d The acceleration decreases over time.

10 a About 40 m

b About 94 m

c About 113 m

11 a About 340 m

b Underestimate; the tops of the trapezia are all underneath the curve so the total area of the three trapezia is less than the area under the curve.

12 a 2.8 m/s   **b** 53 m   **c** $T \approx 3\,\text{s}$

19.6 Translating graphs of functions

1 a $(4, 2)$ **b** $(-3, 2)$

2 a 7   **b** −8   **c** 2

d 1   **e** 0

3 a i 19   **ii** 27

b i $y = 5x + 2 + 2 = 5x + 4$ **ii** $y = 5(x + 2) + 2 = 5x + 12$

4 a A   **b** B

5 a, b

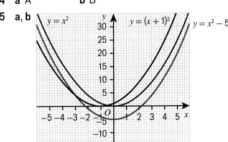

c i $(0, -5)$ **ii** $(-1, 0)$

d i Translation by $\begin{pmatrix} 0 \\ -5 \end{pmatrix}$ **ii** Translation by $\begin{pmatrix} -1 \\ 0 \end{pmatrix}$

6

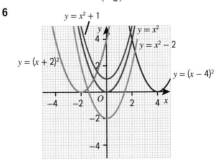

Sketch graphs should be labelled with the following information:

$y = x^2 + 1$: TP $(0, 1)$, y-int $= 1$
$y = x^2 - 2$: TP $(0, -2)$, y-int $= -2$
$y = (x + 2)^2$: TP $(-2, 0)$, y-int $= 4$
$y = (x - 4)^2$: TP $(4, 0)$, y-int not shown

7 a $\begin{pmatrix} 0 \\ 2 \end{pmatrix}$ **b** $\begin{pmatrix} 0 \\ -3 \end{pmatrix}$ **c** $\begin{pmatrix} -1 \\ 0 \end{pmatrix}$

d $\begin{pmatrix} 4 \\ 0 \end{pmatrix}$ **e** $\begin{pmatrix} -5 \\ -2 \end{pmatrix}$

8 $y = f(x - 5)$

9

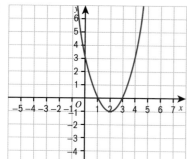

10 a

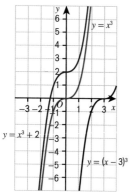

b i $(0, 2)$ **ii** $(3, 0)$

11 a, b

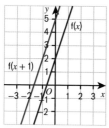

c $y = 3x + 5$

12 a

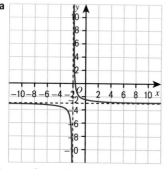

b $x = -2, y = -3$

19.7 Reflecting graphs of functions

1 a $(-1, 2)$ **b** $(1, -2)$ **c** $(-1, 3)$
2 a $-6x - 4$ **b** $-6x + 4$
3 a 9 **b** 1

4 a

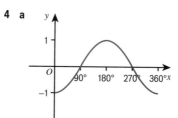

b

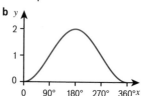

5 a

x	-2	-1	0	1	2
$f(x)$	-10	-6	-2	2	6
$-f(x)$	10	6	2	-2	-6
$f(-x)$	6	2	-2	-6	-10

b

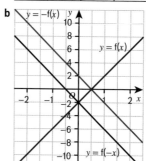

c Reflection in the x-axis
d Reflection in the y-axis

6 a

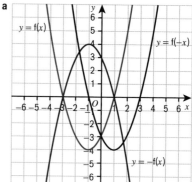

b No; the graphs $y = f(x)$ and $y = -f(x)$ always intersect the x-axis in the same place, and the graphs $y = f(x)$ and $y = f(-x)$ always intersect the y-axis in the same place.

7 a $(2, -4)$ **b** $(-2, 4)$ **c** $(-2, -4)$

8 a

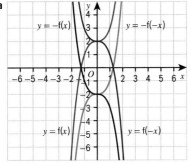

b Rotation of 180°, centre (0, 0)

9 a $y = -g(x)$ **b** $y = -g(x) + \frac{1}{2}$

10 (−2, 4)

19 Check up

1 a $I = 2.5V$ **b** $I = 25$ amps
c $I = \frac{40}{R}$ **d** $I = 10$ amps

2 a $y = 24\sqrt{x}$ **b** $y = 72$ **c** $x = 49$

3 a $c = \frac{352}{d^3}$ **b** $c = 2.816$

4

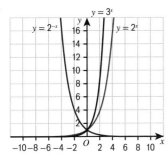

5 a i $a = 3$ **ii** $b = 2$ **b** $y = 48$
6 a $0.75\,\text{m/s}^2$ **b** $0.6\,\text{m/s}^2$ **c** $510\,\text{m}$
7

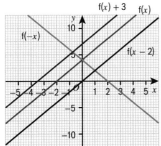

8

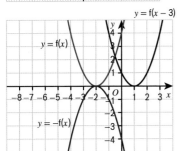

10 a Yes
  b Students' own answers.
  c Students' own answers.

19 Strengthen

Proportion

1 a i Students' own answers, e.g.
$l = 24$, $w = 1$; $l = 12$; $w = 2$; $l = 8$, $w = 3$; $l = 6$, $w = 4$
  ii Inversely proportional
  iii

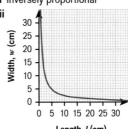

b i Students' own answers, e.g.
$l = 1$, $A = 5$; $l = 2$; $A = 10$; $l = 3$, $A = 15$; $l = 4$, $A = 20$
  ii Directly proportional
  iii

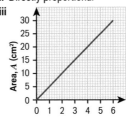

2 a i $A \propto B$ **ii** $A = kB$
  b i $C \propto \frac{1}{D}$ **ii** $C = \frac{k}{D}$
  c i $M \propto N^2$ **ii** $M = kN^2$
  d i $D \propto \frac{1}{G^3}$ **ii** $F = \frac{k}{G^3}$
  e i $H \propto \frac{1}{\sqrt{T}}$ **ii** $H = \frac{k}{\sqrt{T}}$
  f i $R \propto S^3$ **ii** $R = kS^3$

3 a $20 = 2k$ **b** $k = 10$ **c** $F = 10a$
  d $F = 40$ **e** $a = 6$

4 a $10 = \frac{k}{2}$ **b** $k = 20$ **c** $a = \frac{20}{b}$
  d i $a = 4$ **ii** $b = 4$

5 a $k = 5$ **b** $d = 5t^2$ **c** $d = 245$ **d** $t = 3$

Exponential and other non-linear graphs

1 a

t	0	1	2	3	4
n	1	2	4	8	16

b

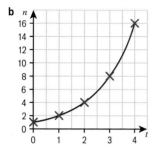

c B $(n = 2^t)$ **d** Increase   **e** Exponential growth

f

t	0	1	2	3	4
n	8000	4000	2000	1000	500

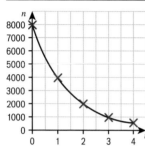

E $(n = 8000 \times 2^{-t})$
Decrease
Exponential decay

2 a $4 = ab^0$ **b** $a = 4$ **c** $8 = 4b$ **d** $b = 2$
e $y = 4 \times 2^x$ **f** $y = 32$

3 a i 20 m/s   **ii** 40 m/s   **iii** 60 m/s
b 0.4 m/s^2

4 a 7.5 km   **b** 2.5 km

Transformations of graphs of functions

1 a

x	−4	−3	−2	−1	0	1	2	3	4
$f(x)$	16	9	4	1	0	1	4	9	16

b, d

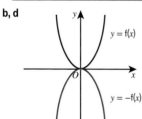

c

x	−4	−3	−2	−1	0	1	2	3	4
$-f(x)$	−16	−9	−4	−1	0	−1	−4	−9	−16

e $f(x)$ is reflected in the x-axis.

2 a

x	−3	−2	−1	0	1	2	3
$f(x)$	−2	−1	0	1	2	3	4

b, d

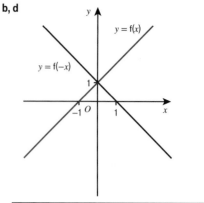

c

x	−3	−2	−1	0	1	2	3
$f(-x)$	4	3	2	1	0	−1	−2

e Reflection in the y-axis.

3 a

x	−3	−2	−1	0	1	2	3
$f(x)$	−10	−8	−6	−4	−2	0	2

b, e, h

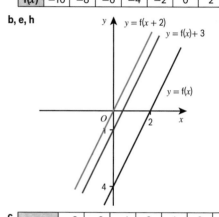

c

x	−3	−2	−1	0	1	2	3
$f(x) + 3$	−7	−5	−3	−1	1	3	5

d 3 greater
f $y = f(x)$ is translated by $\begin{pmatrix} 0 \\ 3 \end{pmatrix}$

g

x	−3	−2	−1	0	1	2	3
$f(x + 2)$	−6	−4	−2	0	2	4	6

i $y = f(x)$ is translated by $\begin{pmatrix} -2 \\ 0 \end{pmatrix}$

19 Extend

1 a At A the ball is travelling upwards and decelerating.
At B the ball has reached its maximum height.
At C the ball is accelerating towards the ground.
b The speed at A is the same as the speed at C.
c The velocities at A and C have the same magnitude, but one is positive and one is negative.

2 i B   **ii** D   **iii** A   **iv** C
3 a A   **b** £2642.86
4 a $D = \dfrac{6390}{r^2}$ **b** $D = 10.2$ cm (1 d.p.)

c $r = 10.0$ cm (1 d.p.)   **d** $\dfrac{d}{4}$ cm

5 a

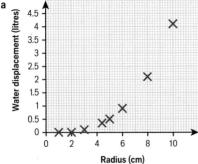

b $W \propto r^3$ **c** $W = 0.004r^3$ **d** 16.4 litres

6 a The profits increased over the period. The increase was greatest between months 1 and 2, and then it was slower but fairly steady for the next 4 months.

b Total profit over the 6-month period

7 a $(-4, 0), (2, 0)$ **b** -9

c

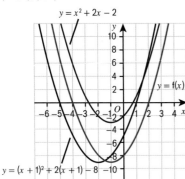

8 a

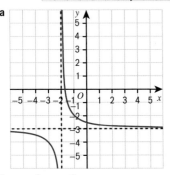

b $x = -2, y = -3$

9

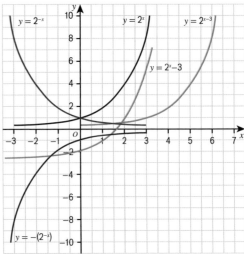

a i $(0, 1)$ **ii** $y = 0$
b i $(0, -1)$ **ii** $y = 0$
c i $(0, -2)$ **ii** $y = -3$
d i $(0, \frac{1}{8})$ **ii** $y = 0$

10

t	0	1	2	3	4	5
v	4	6	10	16	24	34

Areas of trapezia are 5, 8, 13, 20 and 29
Total area = distance travelled = 75 m

19 Test ready

Sample student answers

a The other intersection with the x-axis, and the minimum point

b No

c

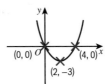

19 Unit test

1 a $T = 960\,\text{s}$ **b** $T = 560\,\text{s}$

2 a A   **b** D

3 a $y = \frac{34.56}{x^3}$ **b** $x = 4$

4 $y = \frac{4}{9}\sqrt{6}$

5 a $A(-3, -9), B(-6, 0), C(0, 0)$
  b $A(3, 9), B(6, 0), C(0, 0)$
  c $A(-1, 9), B(-4, 0), C(2, 0)$

6 a Answers between 17 and 25 m/s^2
  b Answers between 3650 and 3800 m

7

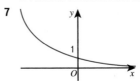

8

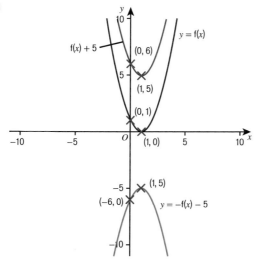

9 a i $a = 4$ **ii** $b = 3$ **b** $y = 324$

10 a Students' own answers, e.g.
$A\ y = 4^x$ $B\ y = -4^x$
$C\ y = 4^{-x}$ $D\ y = -4^{-x}$
  b Students' own answers of the form
  $y = f(x),\ y = -f(x),\ y = f(-x)$ and $y = -f(-x)$

11 Students' own answers.

Mixed exercise 6

1 12, 19.2, 11.25 → **19.1**

2 i B **ii** G **iii** H **iv** E → **19.4**

3 a 10.2 (to 3 s.f.) → **18.1**

  b $3\begin{pmatrix} -3 \\ 3 \end{pmatrix} - \begin{pmatrix} -4 \\ 10 \end{pmatrix} = \begin{pmatrix} -9+4 \\ 9-10 \end{pmatrix} = \begin{pmatrix} -5 \\ -1 \end{pmatrix} = \frac{1}{2}\mathbf{c}$,

  so $3\mathbf{a} - \mathbf{b}$ is parallel to $\mathbf{c}$. → **18.2**

  c 14.8 (to 3 s.f.) → **18.1, 18.3**

4 £77.40 → **19.3**

5 $\frac{20-x}{15}$ → **17.2**

6 No, $\overrightarrow{QR} = 2\mathbf{a} + 3\mathbf{b}$ and $\overrightarrow{QS} = 4\mathbf{a} + 8\mathbf{b}$. $\overrightarrow{QS}$ is not a multiple of $\overrightarrow{QR}$, so QRS is not a straight line. → **18.4**

7 $\frac{13x+15}{2x(x+3)}$ or $\frac{13x+15}{2x^2+6x}$ → **17.4**

8 $\frac{a}{x+y} = \frac{1}{x-y}$
  $ax - ay = x + y$
  $ax - x = ay + y$
  $x(a-1) = y(a+1)$
  $x = \frac{y(a+1)}{a-1}$ → **17.2**

9 $\frac{x-4}{2x-1}$ → **17.3**

10 The graph of $y = 2^x$ crosses the y-axis at 1, not 2, and the curve doesn't touch the x-axis but just gets close to it. → **19.4**

11 $(2x+3)(3x+1)(x-4)$
  $= (2x+3)(3x^2 - 11x - 4)$
  $= 6x^3 - 22x^2 - 8x + 9x^2 - 33x - 12$
  $= 6x^3 - 13x^2 - 41x - 12$ → **17.5**

12 $5x + 18$ → **17.8**

13 $(2n-1)^2 + (2n+1)^2 = 4n^2 - 4n + 1 + 4n^2 + 4n + 1$
  $= 8n^2 + 2 = 2(4n^2 + 1)$
  which is a multiple of 2 because n is an integer, so it is always an even number. → **17.5**

14 1.2 → **19.2**

15 → **17.8**

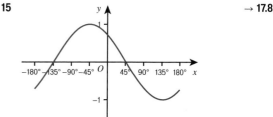

16 $p = -\frac{9\sqrt{11}}{2}$ → **19.3**

17 $\overrightarrow{EF} = 3\mathbf{a} + 2\mathbf{b} = \overrightarrow{CB}$
  $\overrightarrow{BF} = -\mathbf{a} + 3\mathbf{b} = \overrightarrow{CE}$
  Therefore, $BCEF$ is a parallelogram. → **18.4**

18 a $f^{-1}(71) = \sqrt{\frac{71+4}{3}} = 5$ → **17.8**

  b $x = -1$ or $x = \frac{2}{3}$ → **17.8**

19 a $\overrightarrow{OM} = 2\mathbf{p} + 3\mathbf{q}$ → **18.4**

  b $\overrightarrow{PZ} = \frac{2}{3}(-4\mathbf{p} + 3\mathbf{q})$
  $\overrightarrow{PN} = -4\mathbf{p} + 3\mathbf{q}$
  $\overrightarrow{PN}$ is a multiple of $\overrightarrow{PZ}$ so $\overrightarrow{PZ}$ is parallel to $\overrightarrow{PN}$ and they both pass through point P; therefore PZN is a straight line. → **18.4, 18.5**

20 $M(1, 3), N(4, -2)$
  Gradient $MN = -\frac{5}{3}$, gradient $AC = -\frac{5}{3}$
  The gradient of $MN =$ gradient AC so MN is parallel to AC. → **18.4**

21 Approximately 3 m/s → **19.5**

22 $3n^3\left(\frac{2}{n} + 4\right) + 6n(5n - 2n^2)$
  $= 6n^2 + 12n^3 + 30n^2 - 12n^3$
  $= 36n^2 = (6n)^2$ → **17.5**

23 a → **19.6**

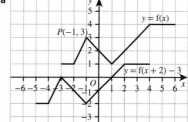

  b (1, 3) → **19.7**

24 a $y = x^2 - 6x + 5$ → **19.6**

  b $y = -x^2 + 4$ → **19.7**

25 $\frac{5x-35}{x^2-3x-28} \div \frac{x-4}{x^3-16x}$

  $= \frac{5(x-7)}{(x-7)(x+4)} \div \frac{x-4}{x(x-4)(x+4)}$

  $= \frac{5(x-7)}{(x-7)(x+4)} \times \frac{x(x-4)(x+4)}{x-4}$

  $= 5x $ → **17.4, 17.5**

26 $x = -1.76$ or $x = 22.76$ $\qquad \rightarrow$ **17.7**

27 $\dfrac{1}{1+\sqrt{2}} \times \dfrac{1-\sqrt{2}}{1-\sqrt{2}} = \dfrac{1-\sqrt{2}}{-1} = -1+\sqrt{2}$

$\dfrac{3}{4+\sqrt{2}} \times \dfrac{4-\sqrt{2}}{4-\sqrt{2}} = \dfrac{12-3\sqrt{2}}{14}$

$-1+\sqrt{2} - \dfrac{12-3\sqrt{2}}{14} = \dfrac{-26+17\sqrt{2}}{14}$ $\qquad \rightarrow$ **17.6**

28 $\overrightarrow{AB} = -\mathbf{a}+\mathbf{b}$, $\overrightarrow{OC} = -\mathbf{a}+\mathbf{b}$, $\overrightarrow{CP} = \dfrac{5}{3}\mathbf{a}-\mathbf{b}$

$\overrightarrow{CQ} = \mathbf{a} - \dfrac{3}{5}\mathbf{b}$

$CP : CQ = \dfrac{5}{3}\mathbf{a}-\mathbf{b} : \mathbf{a}-\dfrac{3}{5}\mathbf{b}$

$5(5\mathbf{a}-3\mathbf{b}) : 3(5\mathbf{a}-3\mathbf{b})$

$\qquad 5 : 3$ $\qquad \rightarrow$ **18.5**

Index